Imaging Animal Industry

Iowa and the Midwest Experience

Series editor, William B. Friedricks,
Iowa History Center at Simpson College

Imaging Animal Industry

American Meatpacking in Photography and Visual Culture

Emily Kathryn Morgan

UNIVERSITY OF IOWA PRESS, IOWA CITY

University of Iowa Press, Iowa City 52242

uipress.uiowa.edu
Printed in the United States of America

Printed on acid-free paper

Library of Congress Cataloging-in-Publication Data
Names: Morgan, Emily Kathryn, author.
Title: Imaging Animal Industry: American Meatpacking in Photography and Visual Culture / Emily Kathryn Morgan.
Description: Iowa City: University of Iowa Press [2024] |
Series: Iowa and the Midwest Experience | Includes bibliographical references.
Identifiers: LCCN 2023049680 (print) | LCCN 2023049681 (ebook) | ISBN 9781609389635 (paperback) | ISBN 9781609389642 (ebook)
Subjects: LCSH: Packing-houses—United States—History. | Photography, Industrial—United States—History. | Slaughtering and slaughter-houses—United States—Public opinion. | Meat industry and trade—United States—History—Sources.
Classification: LCC TS1971.U6 M67 2024 (print) | LCC TS1971.U6 (ebook) | DDC 664/.90290973—dc23/eng/20240304
LC record available at https://lccn.loc.gov/2023049680
LC ebook record available at https://lccn.loc.gov/2023049681

Contents

Acknowledgments

I WOULD NOT HAVE been able to undertake this project without the expertise and guidance of numerous archivists and librarians. In Iowa, the staff of the State Historical Society of Iowa have provided valuable assistance and sound advice. I thank Mary Bennett and Charles Scott, of the Iowa City Research Center, and Becki Plunkett, of the Des Moines Research Center, for their assistance in locating and reproducing materials for me and answering my questions. Thanks also to Janet Weaver and the staff of the Iowa Women's Archive at the University of Iowa and to the staff of the Special Collections and Archives at the University of Iowa Libraries, particularly Denise Anderson and Jillian Sparks. At the Brucemore Historic Site in Cedar Rapids, Jessica Peel-Austin guided me toward materials in the Farmstead Foods Archive and offered patient assistance over the course of my project. Tom Munson of the Sioux City Public Museum gave me access to a trove of useful photographs. I deeply appreciate the assistance I have received over the years from the archivists and librarians of the Special Collections and University Archives at the Iowa State University Library. I have benefited tremendously from the advice and guidance of Amy Bishop, Olivia Garrison, Becky Jordan, and Laura Sullivan. I am grateful to film archivist Rosie Rowe for pointing me toward, and granting me access to, meatpacking-related films in the library's collection. At the Wisconsin Historical Society, archivists and staff helped me to make sense of the organization's extraordinary collections related to the meat industry. Lee Grady, Kyle Krause, Andy Kraushaar, Simone Munson, and Lisa Saywell offered guidance and assistance with several labor collections. Jonathan Nelson alerted me to materials related to the history of Oscar Mayer & Co. in Madison, Wisconsin, graciously granting me access to items still being processed for accession. In Chicago, the staff of Special Collections at the Chicago Public Library, including Michelle McCoy, Debra Orellana, and Glenn Humphreys, assisted me with many of the materials that made their way into this book. I am particularly grateful to archivist Morag Walsh for helping me to understand some of the nuances of copyright law and pointing me toward resources that have proved useful over the course of this project. I also received excellent assistance and guidance from the staff in the Department of Special Collections at the University of Chicago Library; the staff of the Special Collections and University Archives at the University of Illinois Chicago; and the archivists and librarians of the Abakanowicz Research Center at the Chicago History Museum. At the John W. Hartman Center for Sales, Advertising & Marketing History at Duke University in North Carolina, I benefited from Joshua Larkin Rowley's deep knowledge of the center's collections. Hope Ketcham-Geeting and Neal Shipe also assisted with my many requests and queries.

The first funding I received for this project was a Research Grant for Authors from the State Historical Society of Iowa. This award not only funded my early research but provided me with the encouragement to pursue the topic further, leading to this book. In particular I wish to thank Dr. Marvin Bergman, editor of the *Annals of Iowa*, for his time, encouragement, advice, and support during the early stages of my research and for his patience in shepherding one of my articles through to publication. My travel to Madison, Wisconsin, was funded by a Grant to Scholars from the Friends of the Library at the University of Wisconsin–Madison. I am grateful to Libby Theune for her assistance with the award. An Alvin Achenbaum Travel Grant from the David M. Rubenstein Rare Book and Manuscript Library at Duke University permitted me to travel to Durham, North Carolina, to conduct research at the John W. Hartman Center for Sales, Advertising & Marketing History. Throughout this project, my research has benefited tremendously from the support I have received from Iowa State University. I must thank Sandra Norvell of the university's Institute for Design Research and Outreach. I have benefited tremendously over the years from her critical eye, pragmatism, and good advice. A Research Grant from the university's Center for Excellence in the Arts and Humanities funded my travel to Chicago to conduct research in four archives there. Two Foreign Travel Grants from the university's Faculty Senate permitted me to present this research at international conferences, valuable experiences that have strengthened the book immeasurably. Publication of this book has been made possible by a Publication Subvention Grant from the Office of the Vice President for Research at Iowa State University.

I am so grateful for the incredible support of my colleagues and friends at Iowa State University. A talk I gave as part of the Margaret Sloss Center for Women and Gender Equity's Feminist Friday series generated fruitful, intense, informative conversation; many thanks to Ruxandra Marcu-Rowe for inviting me and to Andra Castle for facilitating the talk. Thanks to Carlton Basmajian of the Center for Excellence in Arts and Humanities for inviting me to present my research to a diverse, cross-campus audience. The faculty and staff of the Department of Animal Science and the Iowa State University Meat Laboratory generously opened their classrooms and facilities to me. My thanks especially to Michael Holtzbauer, Steven Lonergan, Jeff Mitchell, Sherrlyn Olsen, and Jodi Sterle. I deeply appreciate the support, encouragement, and enthusiasm of my colleagues in the College of Design and the Department of Art and Visual Culture. Barbara Caldwell, Cameron Campbell, John Cunnally, Jennifer Drinkwater, April Eisman, Carol Faber, Raluca Iancu, Brenda Jones, Jena Jones, Sarah Kyle, Ingrid Lilligren, Seda McKilligan, Kimberly Moss, Joe Muench, Teresa Paschke, Ulrike Passe, Andrea Quam, Chuck Richards, Luis Rico-Gutierrez, Anna Segner, Austin Stewart, Olivia Valentine, and Barbara Walton have all provided support, advice, and encouragement over the course of this project, and I feel lucky to be part of such an enthusiastic, encouraging, and supportive community. Thanks to Kimberly Zarecor for her mentorship, friendship, and support of me and my research over the years. Thanks, too, to my students, who inspire and energize me, even when it seems like they aren't listening.

Conversations with scholars and researchers in a wide variety of fields have helped me to work through knotty problems, pointed me toward useful resources, and enriched my understanding of this research tremendously. My thanks to Heather Belnap, Paul Godfrey, Michael Gray, Jennifer Greenhill, Peter Hamilton, Roger Horowitz, Luis Nadeau, Jeffrey Richmond-Moll, Gary Saretzky, Peter Scholliers, Vanessa Schwartz, and Jason Weems. Discussions with Nancy Gebhart never failed to energize me, and her questions and support have helped me to think about this project in new ways. Christina Lindeman has been a superb and consistent writing buddy, listening to my groans, giving invaluable suggestions, and helping me to stay grounded in art history. My collaborations with Camille Bellet have challenged and surprised me, and work on our cooperative project has fed and nurtured my enthusiasm for this one. I am so grateful for the mentorship and support of Britt Salvesen across projects, years, and space. Her invitation to speak at the Visual Studies Research Institute in 2019 provided a much-needed opportunity to hone this research at a critical juncture. Sarah Moore has offered vital guidance, encouragement, and support, and I appreciate her consistent, good-humored willingness to help me and answer my many questions. Keith McElroy has been a model of grace and generosity. James Swensen's extremely helpful comments and suggestions on a draft of the manuscript for this book improved it immensely, and I deeply appreciate his time and his thoughtful, honest assessments. I would also like to thank the anonymous reviewer of the manuscript. Portions of chapter 2 are adapted from my article "*Waterloo Packer*: Selling Slaughter," which appeared in *History of Photography* in 2018. I thank the editor, Luke Gartlan, and the two anonymous reviewers who commented on drafts of the article. At the University of Iowa Press, I appreciate the support and confidence of William Friedricks, who first approached me about this book, and James McCoy, who has shepherded the manuscript through the publication process.

I would not have been able to complete this book without the love and support of my family. My parents, Peter and Sally Morgan, and my brother, Evan Morgan, have offered unconditional encouragement and guidance and have been consistent and excellent listening ears when I have needed it. I appreciate the support and interest of my parents-in-law, Stan and Jane Gassmann, who pointed me toward some genuinely helpful research resources on the Iowa meat industry. My husband, Aaron Gassmann, has been kind and patient throughout, and I would be lost without his encouragement and good advice. He and my children, Vivian and Charles, have lived with this project (and its effects on our dinner table) for a long time. This book is for them.

Imaging Animal Industry

THE ALLBRIGHT-NELL CO.

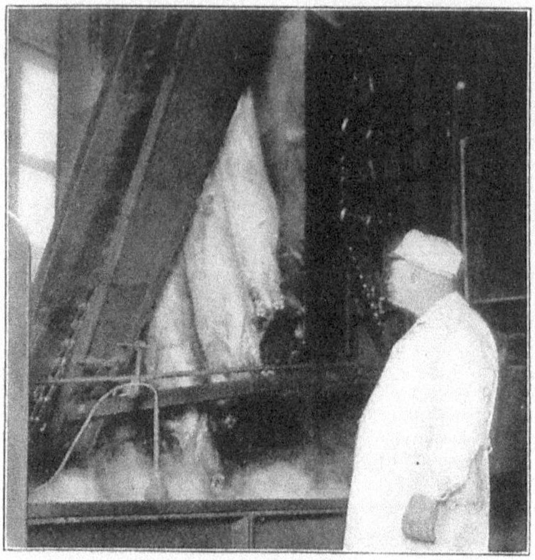

Hog Carcasses Being Conveyed from Scalding Tub into ANCO Dehairing Machine

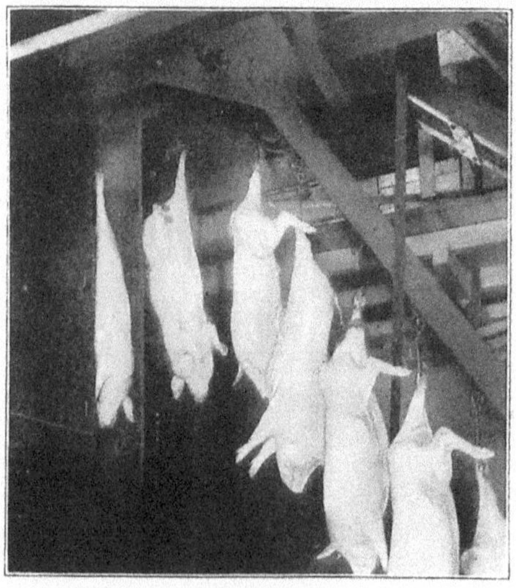

Hog Carcasses Being Delivered from ANCO Dehairing Machine

9

Figure 1. ANCO, *ANCO Machinery*, 9. Trade Catalog Collection, Special Collections, Chicago Public Library.

Wonders of the Flesh

IN A 1923 PHOTOGRAPH, a man watches pigs rise into the air. Resolutely matter-of-fact, the photograph hardly engages in a flight of fancy. In its way, however, it does represent the fulfillment of a fantasy: trumpeting human inventiveness, the photograph celebrates human ability to overcome animal materiality, to transform animal bodies from flesh into commodity (fig. 1). Published by the Allbright-Nell Company (ANCO), a maker of large-scale machinery for the American meat industry, the photograph is one of a suite of images produced to show ANCO's products in action in meatpacking facilities across the US.[1] Arranged sequentially in ANCO's 1923 catalog, two per page, they depict industrial-scale animal slaughter as a series of logical, orderly steps, in which machines and humans work together to transform animal bodies into food for human consumption. Captioned "Hog carcasses being conveyed from scalding tub into ANCO dehairing machine," this photo shows a group of hog carcasses, three in number but standing in for countless more, rising out of a scalding bath designed to loosen their bristles from their skin. The dehairing machine toward which they rise will remove the bristles entirely, rendering their skin hairless and their bodies increasingly object-like, the last step before gutting and cleaning.

The man in the image, a packinghouse worker, watches the elevation of the three carcasses. Standing in one-quarter profile with his back nearly to the viewer, he functions as a rückenfigur, a figure often appearing in paintings and other images as a symbolic stand-in for the viewer.[2] Confronting both the animal bodies on the chain and the advanced machine that will remove their bristles, he exhibits an appreciative, slightly bemused equanimity. In this way he models an appropriate reaction not only to the scene but indeed to the entire process of industrial animal slaughter: admiration for the human ingenuity that delivers such useful, inventive devices; dispassionate evaluation of both the raw animal material and the slaughter processes that render it usable and consumable. The second photograph on the page, also showing elevated hog carcasses but without a human figure, invites viewers to try for themselves, to assume a similar attitude of reverent observation.

Visual culture played an important but underrecognized role in the establishment and growth of industrial-scale meat production in the nineteenth and twentieth centuries. Pictures and illustrated texts functioned as useful means of documenting and recording the activities of industrial-scale slaughter; they permitted exchange of ideas and information; and they were instrumental—and viewed by the industry *as* instrumental—in propagating and encouraging Americans' familiarity with, even embrace of, large-scale meat production practices and products.

Americans did not need encouragement to value the eating of meat. Meat has long been viewed, in the US and elsewhere, as a core desire, both luxury and necessity—a seasonal product, the availability of which waxed and waned with the cycles of agricultural production, abundance, surplus, and scarcity, making it sometimes readily available, other times tantalizingly beyond reach for those without means. Slaughtered, processed, and preserved in the fall, a pig could feed a rural family well into the months of winter, the butchery itself an occasion for feasting. Urban families visited the butcher for their chops, bacon, and, on special occasions, beefsteaks and roasts. But Americans had to be sold on the idea of meat as a product of unquestionable, absolute everydayness, central to the table at breakfast, lunch, and dinner, year-round.[3] More important, they had to be sold on the value and wholesomeness of meat produced at industrial scale, in vast, unprecedented quantities. Visual culture proved essential to this endeavor.

From the later nineteenth century, the American meat industry afforded an abundance, even an excess, of product. In order to prosper economically, to maintain and expand this scale, the meat industry had to prosper culturally as well. Indeed, it had to play an active role in entrenching itself, in fostering development of a culture that regarded meat eating as a fundamental right. From the late nineteenth century into the middle of the twentieth, photographs and other forms of visual culture became increasingly indispensable to this campaign, playing a central role in the endeavor to sell factory-produced meat not only as a product but as an idea. Photographs helped Americans to visualize and to conceive the massive and burgeoning business of large-scale meat production. More important, they helped the industry to exert control over how the American public perceived the very idea of factory-produced meat.

Though always an interest of the meat industry, positive public image became imperative in the wake of controversy or scandal. Offered a particularly effective or sensational description of the industry's abuses and oversights—such as what Upton Sinclair presented in his sensationalistic 1906 novel *The Jungle*—Americans evinced an alarming tendency to allow their imaginations (or their consciences) to run away with them.[4] But by commissioning and distributing visual images, the meat industry found it could generate an idealized vision of large-scale meat production through an ideally controllable means. Though not the only image-making technology deployed by the meat industry, photographs proved particularly useful. Photos appeared simultaneously candid and manipulable, showing *a* truth and making it into *the* truth, a quality that could be enhanced if the photographs were carefully sequenced and/or deployed in conjunction with text.[5] With pictures and illustrated texts, it was possible to present the industry as a paragon of clean, efficient modernity, capable of producing a valuable product at massive scale in order to feed not only the nation but, via export, the world.

Photographs are made by the action of light reflecting off a surface—metal, cloth, skin, flesh, bone, meat, whatever—through a lens and onto the light-sensitive material inside a camera. The medium appears to have a single-minded, mechanistic insistence on showing precisely what is before it, whatever falls in front of the camera's lens, without misrepresentation.[6] For an industry such as meatpacking, so thoroughly saturated in gore, the camera might thus seem a less than ideal tool for creating promotional and celebratory imagery. Yet there are myriad ways to manipulate both a photograph itself and the message it can be made to carry. Photography is as capable of dissimulation, as capable of serving the needs, whims, ideals, and prejudices of its makers, as any other representational medium.[7] The meat industry gravitated toward photography for much the same reason that many other individuals and industries did at and after the turn of the twentieth century: it offered the speediest, most modern, and most economical means of creating visual representations, for whatever purpose the industry might need them. As a bonus, the industry—again, much like many other parties—benefited from the fact that photography not only permits but facilitates a particularly effective kind of image control.[8]

Every photograph results from a host of choices and biases, both conscious and unconscious, on the part of the photographer, printer, users, and other parties: selective framing and point of view; cropping to diminish or control the amount of information present in an image; printing for subtle tonality versus contrast.[9] Photographers and their employers make countless decisions about framing, inclusions, and omissions; but this decision-making process is often invisible in the final image. Photographs deployed in series, or in conjunction with text, gain even more messaging power.[10] Sequencing and selective editing permit presentation of image-text narratives that seem quite open about the slaughter process, even as they elide a great deal. Retouching, such as by drawing on a negative or by altering an image before it is converted to a plate for reproduction in a periodical or book, may also do a great deal to mitigate the potential shock factor, even in seemingly unflinching depictions of slaughter operations. The very nature of the photographic process alters perception of the subject matter: even in the most detailed and unsparing black-and-white photograph, blood is not red.

Photography possesses an even greater capacity to promote, propagandize, and comment than other visual media precisely *because* it is perceived as an absolutely trustworthy medium. Photographic images and photographically illustrated texts of packinghouse operations carry implicit authority and believability. Rendered in an authoritative medium, the promotional message itself gains authority. This is not to say that every deployment of photography, whether by the meatpacking industry or by any other entity, represents a deliberate, cynical endeavor to outwit the audience. The makers and users of many powerfully persuasive photographs and photo-texts are themselves often wholly convinced of the medium's inherent truthfulness as well. Messaging carries that much more power when the creators themselves believe.

The meat industry followed and participated in a burgeoning circulation of photographic imagery by and about American industry in the early to mid-twentieth century.[11] Photographs, as machine-made images, seemed an especially appropriate way for American industries to represent themselves. Audiences and photographers alike have tended to regard photographs as transparent, trustworthy images, capable of telling a truth better than other types of pictures. This seeming transparency lends itself well to corporate America's desire to enshrine its own goals as incontrovertible truths and, by extension, to fetishize its products as secular icons.[12] With the assistance of photographic and other imagery, meat has been fetishized like any other industrial product, sold not just as sustenance for the body but as an idea and an ideal.

The vision presented in images of the meat industry has always been only a glimpse, however: even with the presence and assistance of the lens, massiveness and complexity combine to facilitate invisibility. The impossibility of visualizing such a diffuse industry, of imagining not only meat production but the huge and intricate networks—of animal breeding, husbandry, and transport; of labor recruitment, training, and management; of product transport and distribution; of public and community relations—within which the industry exists and on which it subsists, have proven beneficial. Throughout its history, the meat industry, like many large industries, has been simultaneously ubiquitous and cryptic, its scale so large that one cannot stand back far enough to see it properly.[13] For the partial view we do get, we must often thank the camera and other tools of visual culture; but the hands in which such tools have rested have often been those of the industry itself. Controlled narratives are clean narratives, attractively single-minded, lacking the messy multivocality of a more democratic storytelling process; but in them we see also what Allan Sekula calls "the tyranny of a single anecdote."[14]

In some ways, this is true of any heavy industry that has used photography. Auto manufacture, for instance, is similarly diffuse and has similarly benefited from selective imaging. Charles Sheeler's famed 1927 photographs (and paintings) of the Ford Motor Company's new River Rouge Plant deploy photography's sharp focus and detailed rendering to present a heavily edited vision of the plant, a massive and heroic industrial landscape. Raw materials come into the River Rouge Plant via dedicated rail spurs and private canals, and finished automobiles leave the plant, the factory responsible for shaping rough nature into finished culture.[15]

The meat industry, however, contends with the added complication that its base material is not only raw but living; that it enters—must enter—the plant under its own volition. Disassembly, not assembly, characterizes the activity of the production line in a meatpacking plant: whole animals become parts for consumption.[16] A story like the one Sheeler told of the

River Rouge Plant proves more difficult to tell of the meat industry: cars are easier to fetishize than meatloaf. In meatpacking, the raw material is more picturesque than the finished product, the animal nicer to look at and more interesting to contemplate than the flesh alone. Visual culture's challenge has been to valorize and heroicize meat making, just as it has done with machine making, to highlight the aesthetics of slaughter.

The success of image-making technologies, photography foremost among them, in meeting this challenge aided the meat industry's success. Photos and other images assisted the industry's endeavor to establish itself as a provisioner for national prosperity, not just contributing to but literally sustaining the rise of progress.[17] A nation, like an army, marches forward on its stomach. Photographs helped the meat industry to convince Americans that the forward march of progress depends on cheap, abundant meat, provided only by an industry operating at massive scale.

Industrial-Scale Meat Production in the US

Meat production on an industrial scale is a midwestern invention. Settled in the mid- and later nineteenth century primarily for the very purpose of large-scale grain production, the fertile Midwest proved so effective at generating grain surplus that it quickly became clear the region produced too much—more than consumers, even international consumers, could eat, and far too much for grain production alone to be profitable for every midwestern farmer. Seeking to add value, farmers found that by feeding grain to animals, they could effectively convert a low-value agricultural product—grain—into a higher-value one: meat. As historian William Cronon notes,

> The wagons and cars full of grain that jammed Chicago-bound roads and railroads were not the only means by which hinterland farmers could send their crops to market. Corn in particular became more profitable in its travels by undergoing an alchemist's transmutation into denser, more valuable substance: kegs filled with whisky or animals that could carry themselves to market.[18]

From the very start of large-scale settlement of the Midwest, massive surplus and wide-scale distribution were always the goals. Pork proved the most feasible meat for industrial-scale production in the earliest days—roughly the 1850s onward—since it was more easily preserved than other meats by smoking or pickling.[19] The earliest midwestern centers for large-scale pork production were along waterways. Cincinnati, Ohio, gained the moniker "Porkopolis" for its large-scale pork production. After it was smoked or pickled in barrels, most of the city's pork floated to larger distribution centers downriver.[20]

Once the railroad began to penetrate the midwestern and western prairies, joining previously far-flung settlements into complex networks, meat production centers arose in other cities with limited access to water transport. This period, in which the rise of the railroads both drove and was driven by widespread movement of American populations in the post–Civil War era, saw the emergence of new centers for the trade first of agricultural surplus, then of livestock, and then, increasingly, of meat. Chicago provided the supreme example.[21] Growing by leaps and bounds in the first half of the nineteenth century, Chicago became a point of interface and trade between the more settled (by Euro-Americans, anyway) lands of the eastern US and the prairie wildernesses to its west. By virtue of its rail connections, it constituted an ideal hub for the trade of animals and meat. Chicago's Union Stockyards opened in 1865, overseen by a board constituted of meatpackers and railroad men.[22] Originally designed around a trade in live animals, the character of the stockyards and the surrounding area, which came to be known as Packingtown, shifted over the 1870s and 1880s to accommodate the increasing importance of the meat trade. More and more meatpackers took their operations to Chicago, building on the lands adjacent to the yards.[23] "By the mid-1880s, twenty-nine major packinghouses and several small firms [were] located near the stockyards."[24] In Chicago, the necessary components for the mass production of meat—systems for transporting animals and meat; systems for managing and slaughtering animals; and systems for packaging and preserving meat—grew up together. Siegfried Giedion writes of how, in response to and in support of this rise, "the great plains west of the Mississippi . . . were transformed, in a little over a decade, into a gigantic reservoir of cattle."[25] In the Union Stockyards, cattle, pigs, sheep, and horses coming from this western "reservoir" moved directly off railcars and into the massive, ever-growing grid of holding pens.[26]

One crucial technological development supporting the rise of the Midwest as a meat-making center, not just an animal-raising one, was the refrigerated train car, as Giedion notes. As soon as it became possible for animals to be slaughtered in the Midwest and meat shipped from there—as soon as it became possible to centralize the entire chain of meat production, from animal husbandry to slaughter to cutting to the making of by-products—meat production went from being a trade to an industry, companies burgeoning to huge size virtually overnight and a heavily rationalized division of labor becoming vitally necessary to the maintenance of production at massive scale.[27]

Although both traditional slaughter/butchery processes and mechanized, compartmentalized slaughter/meatpacking processes begin with a whole live animal and end with a variety of foodstuffs and by-products, the two sets of practices hardly resemble one another in their specifics. The craft form gave rise to the mechanized form; but through translation and evolution the mode of practice, the scale of production, and the very nature and perception of the end product have been radically altered.

In craft-scale slaughter and butchery, large animals such as cattle and pigs would typically be slaughtered and processed one by one. A single worker or a small team of workers would perform all the tasks necessary to slaughter, dress (eviscerate), clean, and split the carcass, in preparation for chilling and further portioning. Depending on the time period, the size of the animal, and the mode of slaughter, the animal might be stunned before slaughter. (Cattle have always been stunned prior to bleeding, simply because they are so large and potentially dangerous; but stunning of pigs was not mandated in the US until 1958, when humane slaughter laws came into effect for conventional slaughter.[28]) Subsequently, the animal is hung head-down and bled, usually by piercing an artery in the neck. A beef or sheep carcass may then be laid flat, or it may be left hanging while the team, working around the animal, carefully skins it. A pig carcass retains its skin, but immediately after bleeding the carcass is scalded in a hot water bath to facilitate removal of the hair.

Once the team has removed the hide or hair, they hang the carcass head-down again. They remove the head and open the carcass precisely and carefully, taking the internal organs out intact to ensure the meat is not contaminated. The animal, once an organic whole, is separated into offal and meat, dirty and clean, polluted and pure, through a process at once rationalistic and ritualistic.[29] The carcass is split and washed, and the halves are sent to a cooler to chill, still hanging, in preparation for further cutting or processing. Throughout the process, members of the small group performing the slaughter orbit around the carcass, coordinating with one another to perform the myriad tasks, some of them done simultaneously, that are involved in converting the animal from creature to meat. In modern facilities, inspectors are present throughout the process to witness the slaughter and processing and to inspect the carcass, viscera, and head.[30]

Industrial-scale meat processing both does and profoundly does not resemble the process of craft-scale slaughter and butchery. The starting point and the end result of the slaughter process—from live animal to hanging, split carcass, ready to chill—remain the same; and indeed, the steps necessary to arrive at the end result remain roughly the same, too: stun the animal, bleed the animal, skin or dehair the carcass, dress the carcass in proper sequence to avoid contamination, then split, wash, and chill the carcass. But the assembly line demands compartmentalization of each task and division of some tasks into multiple, separate series of discrete motions, performed in a repetitive sequence that has been studied and refined for maximum efficiency. Rather than working as a group on a single animal, the workers stand in, and embody, a linear progression—a disassembly line. Removal of an organ or organ system like the digestive system becomes the work of several people, who operate in sequence rather than as a group, cuts made carefully, one after another, each worker making only their assigned cuts.[31] In its passage from living creature to hanging, split carcass, the animal might pass through the hands of a few dozen people (with many more to come, after the carcass is chilled); each worker, in turn, performs the same few operations—or even the same single operation—on well over one thousand animals each day.[32] Such compartmentalized mass production has long characterized American meat production: the processes of animal disassembly famously informed how Henry Ford conceptualized the assembly lines in his automobile plants.

European nations were experiencing growth and change in their own meat trades right alongside the growth in the American market in the later nineteenth and early twentieth centuries. European solutions to the problem of how to manage large-scale production of healthy meat tended to be different from American approaches, however. Where the US's evolving system depended on private and corporate commerce—a situation that fostered the rise of the massive meat factories that characterized the American system, each centered around its compartmentalized disassembly line—European nations tended to adopt solutions including centralized, publicly run facilities.[33] The previously existing systems in many European nations, some of them in use for hundreds of years, had tended to involve small slaughterhouses serving localized populations, a single village or a single city neighborhood, producing only enough to serve a market determined by proximity more than anything else.[34] During the later eighteenth and throughout the nineteenth centuries, European governments began to consider how to regulate such a diffuse, casually overseen industry, given the potential for a poorly run slaughterhouse to become a locus for deadly contagion. In Europe,

> the reform of slaughterhouses was about the creation of an efficient, safe, and preferably invisible supply network that would provide growing populations with meat while protecting them from the threats related to animal slaughter. It involved both the relocation of slaughterhouses and the reorganization of the practice of butchering. More precisely, it entailed the centralization of abattoirs and the municipal and state supervision of butchering.[35]

France undertook regularization of its meat trades with decrees for the establishment of public slaughterhouses—called *abattoirs*, a neologism derived from the word for "felling"—issued by Napoleon I.[36] Such facilities "were regarded not as sources of income, but as centers where the animals could be slaughtered under supervision."[37] In the earlier nineteenth century, Paris had five such facilities; during the massive public works project overseen by Baron Haussmann under Napoleon III at mid-century, these were replaced by the single huge central facility of La Villette.[38] The quantity of meat produced at La Villette was massive by comparison with any other European slaughterhouse of the time; but the method of production stood in stark contrast to that practiced in American slaughterhouses. "Mass slaughtering" was unknown; animals were felled one at a time, each in a separate booth. Into the later nineteenth century, the slaughter and butchering was done by hand, not by machine. "In this curious symbiosis of handicraft with centralization lies the peculiarity of this establishment as well as of many others in Europe."[39]

By contrast with the European model, the history of the American meat industry is a history of private industry, with little governmental involvement in the business of slaughter and, at least in the early years, little governmental oversight as well. The packinghouses that arose in Cincinnati during its heyday as "Porkopolis" were privately owned, and so were those that grew up around the Union Stockyards in Chicago. Even the Union

Stockyards themselves were a corporate endeavor: it was not Chicago but the rail industry that conceived and established the Union Yards, before turning them over to a corporation created for their management.[40] As governmental oversight of the industry increased in the late nineteenth and early twentieth centuries through establishment of agencies and implementation of enforced inspection protocols, inspectors worked with, and ultimately within, industry-built and industry-run facilities. Where European facilities like La Villette were effectively government-controlled platforms designed to permit and facilitate inspection, in the US the situation was reversed: inspection came (and still comes) to industry.[41]

Public attention turned to American industrial-scale meat production with urgency and force after the 1906 publication of *The Jungle*, Sinclair's novel detailing labor abuses and unsanitary conditions in the meat industry.[42] A muckraking journalist with socialist sympathies, Sinclair undertook to examine the meat industry and to expose working conditions there, with the aim of increasing public support for workers' rights. Drawing on his own investigations and on other published accounts, Sinclair detailed shocking conditions and offered horrifying stories about filth, vermin, contagious illness among workers, contaminated and diseased meat, and ghastly workplace accidents.[43] Though the work was fiction, the publishing company recognized—and played up—its proximity to fact: publishers traveled to Chicago and toured Packingtown to fact-check the book, and bacteriologists and physicians were drafted to write articles supporting Sinclair's assertions.[44] Sinclair's hoped-for reforms of working conditions largely did not materialize: packinghouses remained dangerous, difficult, low-paying workplaces.[45] But the public outcry did lead directly to the passage of both the Meat Inspection Act and the Pure Food and Drug Act (which created the Food and Drug Administration), signed on the same day in 1907 by President Theodore Roosevelt.[46] The experience oriented the meat industry to the threat of negative public attention. Images, in subsequent years, constituted one important means of combating and deflecting such attention: through careful and selective use of photography and other visual imagery, the meat industry could control its public image and, by extension, its future.

Visualizing COVID-19 in the Meat Industry

Although this book focuses primarily on images created up to the mid-twentieth century, the meat industry became suddenly highly visible again during the global COVID-19 pandemic. In the US, the spread of COVID-19 forced a public and, for many Americans, eye-opening reckoning with the conditions of industrial-scale meat production. Early in the pandemic, workers in most industries were sent home to shelter in place indefinitely; but meatpacking became, by order of President Donald Trump (with the input of the American meat industry), essential work and meatpacking workers essential workers, required to continue reporting daily in order to ensure consistent, unbroken availability of meat in global food supplies.[47] Part of an endeavor to curb the growth of food insecurity in the US and globally during the pandemic, the move nonetheless had devastating impacts. Laboring in close quarters on meatpacking disassembly lines, many workers became ill with COVID. Official death tolls among meatpacking workers number in the hundreds, but the impact of meat-industry-related infections was far greater, as workers carried COVID back to their families and communities, perpetuating and extending the pandemic. Even as it continued to feed the populace, to reassure global consumers that the food supply would not be interrupted by the pandemic, the meatpacking industry fed the virus too.

The industry is everywhere and nowhere. Located in small- and medium-sized towns throughout the Midwest and West, slaughterhouses, processing facilities, and canning plants occupy anonymous, attention-deflecting structures. Connected to markets by networks of rails and roads, these facilities depend on the free movement of raw materials (i.e., animals), product, and even workers who can flow across the landscape unfettered and unremarked. During the COVID-19 pandemic, the very ubiquity of the industry came to seem both a promise and a threat, the spread of the virus enabled by the same networks that ferried cheap, ready supplies of meat. The American public became intensely aware of the industry, its working conditions, its diffusion throughout the midwestern and western landscapes, and the makeup of its workforce in unprecedented ways. The meat industry workforce of today, largely composed of non-white workers, mirrors other populations disproportionately affected by COVID.[48] Public attention to the meat industry during the pandemic often highlighted these disparities, noting that the population consuming all this meat and the population dying to produce it were not necessarily the same.[49]

Visual culture played little part in these stories. Most articles about conditions inside meatpacking plants were illustrated with an image of the outside of the plant, or a stock image of an anonymous meatpacking plant interior.[50] A May 2020 article from *Business Insider* carried a purpose-made photograph of the exterior of a Tyson pork processing facility with two workers exiting, clothed in personal protective gear. But even such insipid images as this one, showing nothing of interior conditions and only hinting at the company's virus-containment measures, proved a rarity.[51] Some more extensive pieces contextualized the spread of the virus within the meat industry by ruminating on the history and structure of American meatpacking and incorporating images of industrially bred animals; but even here, the images themselves were often stock photographs.[52] One Reuters article about the spread of COVID at a Canadian facility carried an image showing a worker disinfecting the interior, but close reading of the caption under the image revealed it to be a repurposed photograph from 2008, when the plant had closed down for disinfection after an outbreak of listeriosis.[53] Across coverage of the meat industry during the pandemic, outlets seemed at a loss over how to show to readers the impact of COVID's rise and spread.

The phenomenon was not limited to the meat industry. In May 2020, Sarah Elizabeth Lewis remarked on the general dearth of photographs showing people sickened by or dying from COVID. "Visualization is a powerful tool," she notes. "It can help us more deeply understand the severity of the situation as we work to curb the virus. But the visuals we need most in this time are difficult to come by."[54] In the absence of revelatory images, she continues, "the virus is harder to combat."[55] Without images depicting the effects of COVID on meatpacking workers and their communities, public reports on the pandemic among this population could seem remote, unreal, even irrelevant to many Americans' lived experiences.[56] Although it is beyond the scope of this book to address at length how the COVID-19 pandemic within the American meat industry was visualized, I will note that on this front, the meat industry itself offered little to go on. Even under the intense light of COVID coverage, images of packinghouse interiors proved scarce, and visual understanding of conditions within the industry remained remarkably opaque. The conditions of controlled visibility that I describe in this book, the careful and coordinated deployment of visual imagery by the meat industry that occurred over the course of the early to mid-twentieth century, are now bygones.

Hunting, Slaughter, and Meat in Art History: Precedents and Models

Photographs of the meat industry in some ways extend long traditions in the West of imaging, and imagining, the slaughter and consumption of meat animals. But all killing of animals has not been depicted equally. Hunting is among the most ancient and universal subjects for visual culture. It tends to be glorified and often heavily symbolic: the province of deities, kings, and noblemen, for whom hunting epitomizes and extends their dominion. When they slaughter their quarry, they do so decisively and mercifully, enacting with the single animal body the entirety of human prerogative over all animal bodies (and perhaps even some human bodies, too). Slaughter associated with butchery, with the everyday making of meat, on the other hand, tends to be unromantic, humble, even debased. Butchery tends to be shown matter-of-factly, performed by anonymous and unglorified figures —when it is shown at all.

Most animal slaughter does not qualify as hunting, as Matt Cartmill points out: "We define hunting . . . as the deliberate, direct, violent killing of unrestrained wild animals; and we define wild animals in this context as those that attack or shun human beings. The hunt is thus by definition an armed confrontation between humanness and wildness, between culture and nature."[57] Hunting for meat and for sport appears in prehistoric cave paintings and ancient Assyrian relief sculptures. Symbolic treatment of the hunt—the hunt as a metaphor rather than as the literal representation of humans pursuing an animal—seems to begin nearly as early as depictions of the hunt itself.

Over five thousand years ago, a Sumerian scribe used a cylinder seal to press a hunting scene into a slab of clay before then covering the slab with writing. Partially subsumed under the cuneiform, the hunting scene shows a priest-king using dogs to hunt or herd boars, enacting as he does so his role as the good shepherd who protects both his animal flocks and his human subjects.[58] In an Egyptian drawing made nearly two thousand years later, a New Kingdom pharaoh slaughters a lion representing, as the accompanying inscription notes, all Egypt's enemies—"every foreign land."[59]

Ancient Greek artists painted vases with hunting scenes rich in symbolism and layered religious significance: Artemis both pursuing and protecting the deer of the forest; Hercules pursuing mythic animals for his labors.[60] By the medieval period in Europe, texts and images often treated the hunt as an elaborate, multilayered metaphor: for the pursuit of love, both secular and sacred; for the pursuit and exercise of power.[61] In Baroque paintings, genteel society hunt boars, stags, and foxes, the images treating the chase and the kill with all varieties of symbolism: hunted animals connote the damned, the desperate, the pitiable, and the controlled.[62]

All these instances, from the ancient to the more recent, share in common the tendency to use the hunt as a commentary on humanity and human qualities rather than on the animal. In enacting the hunt, humans enact and fulfill their social roles, concretizing and realizing their obligations to, or their power over, one another. In the cylinder seal, the Sumerian priest-king appears as subduer and controller of the animal world, justifying his position as protector, and thus ruler, of his people. In the Egyptian image, a lion dies, but in the collective imaginary, the pharaoh slays all the enemies of Egypt. Artemis and Hercules negotiate over the Ceryneian Hind, enacting—and shaping, in the process—paradigms for the performance of femininity and masculinity.

The very right to hunt symbolically, rather than for mere satisfaction of base desires and belly hungers, becomes a critical marker of social status. Hunting as love, hunting as pursuit, hunting as metaphor are the province of aristocracy in European medieval and Renaissance art, while peasants hunt only to provision themselves (if they hunt at all). Royal hunters in the Limbourg Brothers' *Très Riches Heures* hunt wild animals through the forest, but peasants in the book venture into the forests only to find, check on, and tend their pigs.[63] Woodsmen in Winslow Homer's watercolors, chasing deer into the water to trap them, enact a brutal, ungentlemanly hunt, pursuing the deer as a crucial means of sustenance, not as the noble quarry more befitting a sporting gentleman.[64] Indeed—with the possible exception of Homer, who in his late watercolors increasingly focused on the experience of the hunted animal qua animal—hunting in art overwhelmingly speaks to the human experience, to human concerns.[65] The animal is almost never only an animal, and the chase almost never only a chase.

Depictions of workaday slaughter, butchery, and consumption of meat constitute a separate, but related, tradition, nearly as ancient as that of the hunt but far less exalted. The ancient Egyptians cataloged animal slaughter in Old Kingdom tomb reliefs as early as 3000 BCE, and by the Middle Kingdom, they could model a dedicated slaughter space fairly completely: the tomb of Meketre, at Thebes, included a model slaughterhouse in a purpose-built box (fig. 2). A group of miniature men labor inside the shop-in-a-box, several of them working together to slaughter two cattle, who lie bound on the floor, and a goose, closer to the door. Behind them, two more men tend fires to be used for processing the animals, while at the far back is a low enclosure for sheltering live animals or storing supplies, with a ladder leading to a loft space above the enclosure where cuts of meat hang to cure.[66] The space appears efficiently and logically designed for the purposes of animal slaughter and the sale of product, speaking to the long-standing everydayness of the practices. The bound animals themselves, too, are everyday creatures, not noble quarry but humble beasts, fulfilling the purpose for which they were bred.

Humility reigns in these images of butchery, in contrast to the glories of the hunt. A Greek black-figure vase from the late Archaic period depicts a man and a boy working together to butcher an animal. The man is no Herakles: naked except for his loincloth, he butchers a hooved animal with the assistance not of a god but of the nude boy. Cuts of meat and accoutrements of everyday slaughter surround them, reinforcing the regularity and routineness of this activity.[67] In German printmaker Hans Schäufelein's late medieval depiction of a farmer slaughtering a hog, the everyday matter-of-factness of the activity becomes clear in how the farmer hauls the animal out of the barn, gripping the hind legs and tail with a look of keen concentration on his face (fig. 3). Surrounded by a bucket, an axe, and other tools of slaughter, the farmer's wife waits behind him, holding her ladle and pot with equanimity, ready to break the animal down into meat, fat, and other useful substances.[68] It is this mundaneness that most clearly distinguishes depictions of butchery, butcher shops, and slaughterhouses from depictions of the hunt. While genteel Europeans hunt boars and stags in paintings, burghers and peasants slaughter and butcher oxen, workaday counterparts to the glorious animals of the hunt.

Over 3,700 years after the ancient Egyptian slaughterhouse model was created, the entry for "butcher" in Denis Diderot and Jean le Rond d'Alembert's *Encyclopedia* demonstrates that, with regard to the routine slaughter of animals, not much had changed. The entry shows a dedicated

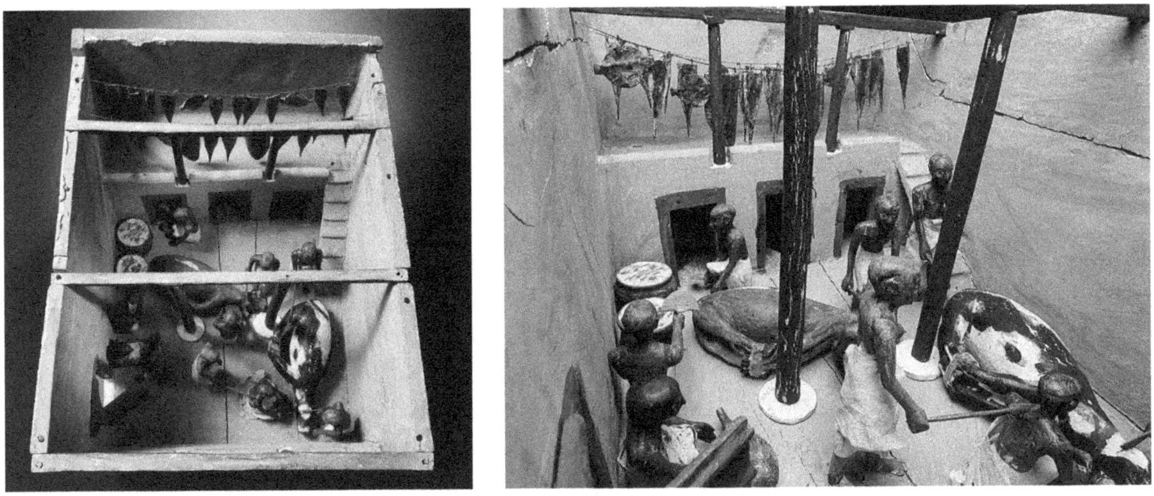

Figure 2.
Model of a Slaughter House, ca. 1981–1975 BCE, Egyptian, from the Tomb of Meketre at Thebes, Middle Kingdom. Metropolitan Museum of Art, Rogers Fund and Edward S. Harkness Gift, 1920 (accession number 20.3.10).

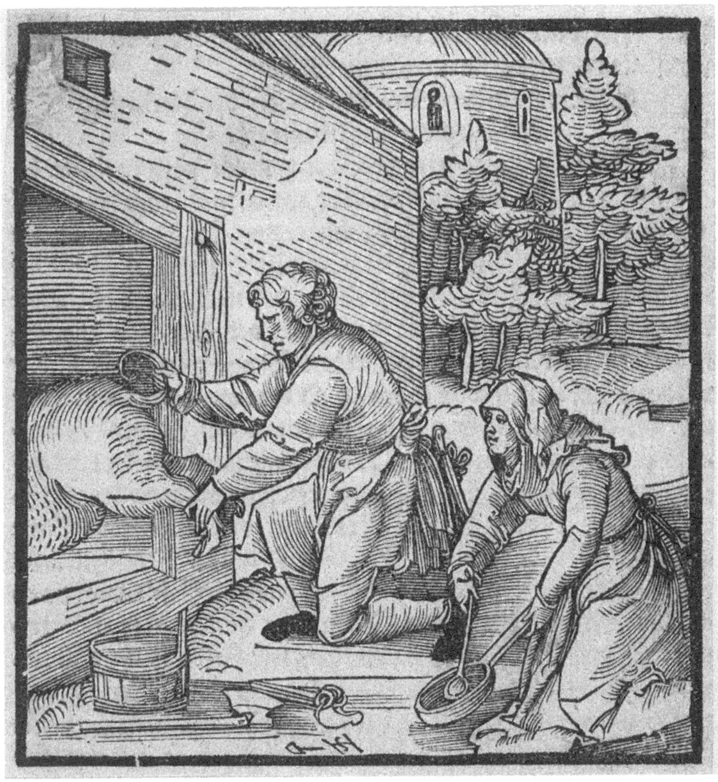

Figure 3.
Hans Schäufelein, *A Farmer Slaughtering a Hog* [. . .], 1517, woodcut, from *Hymmelwagen auff dem, wer wol lebt* [. . .], by Hans von Leonrodt. Metropolitan Museum of Art, Purchase, Anne and Carl Stern Gift, 1959 (accession number 59.507.157[12]).

slaughtering space in which one pair of men restrains and stuns an ox, while another pair of men wait to assist with hoisting, bleeding, and dressing the animal, and a fifth man dresses a sheep's carcass (fig. 4). In the background at right, two cattle carcasses hang, dressed and stretched, while in the background at left, two live cattle stand waiting. The accompanying text states outright the everydayness of meat:

> Butcher's meat is the most ordinary food after bread, and therefore one of those which most often affects health. The police cannot therefore watch too carefully over this object: but they will take all the precautions that it entails, to be satisfied that the cattle destined for the butchery are healthy; that they are killed and have not died of disease, or suffocated; that the flesh is cleaned properly, and that the meat is cut in due time.[69]

Although the text argues for proper regulation of the butchering trade due to its importance to the French national diet, the language attests in the process to the same workaday, unremarkable nature of animal slaughter that the ancient Egyptian model also embraced.

Butchers in paintings hang, cut, and peddle meat, dripping in their stained aprons; bent rustics roast joints over flames and wave chops on the ends of knives. Images of butchery and images of meat are often related, and sometimes inseparable. In certain traditions, too, meat imagery carries its own robust symbolism. Northern European art of the sixteenth century inaugurates meat and butchery paintings as their own genre, beginning with Pieter Aertsen's massive 1551 painting *A Meat Stall with the Holy Family Giving Alms*.[70] Aertsen's work is, ostensibly, a history painting: the Holy Family appears in the background, giving alms to the poor as they flee toward Egypt. But the painting, "long . . . recognized as the first still-life of viands and victuals in the history of art," emphatically embraces the painting of everyday objects.[71] Dominated by a partially skinned cow's head in the foreground, the composition incorporates the disarticulated bodies of cattle and swine, carcasses of fowl and fish, and ropes of sausage, among a host of other edibles. Cruciform symbols abound, in keeping with the Christian theme: crossed fish, crossed trees, even a pair of pretzels hanging from a nail. In Aertsen's painting, as in a host of still lifes to follow, "food for the body is a visual metaphor of food for the soul."[72] But it is, at the same time, hardly *only* a metaphor.[73] "Aertsen portrayed the properties of the animal flesh itself in exquisite sensory detail: the translucence of tripe extruding

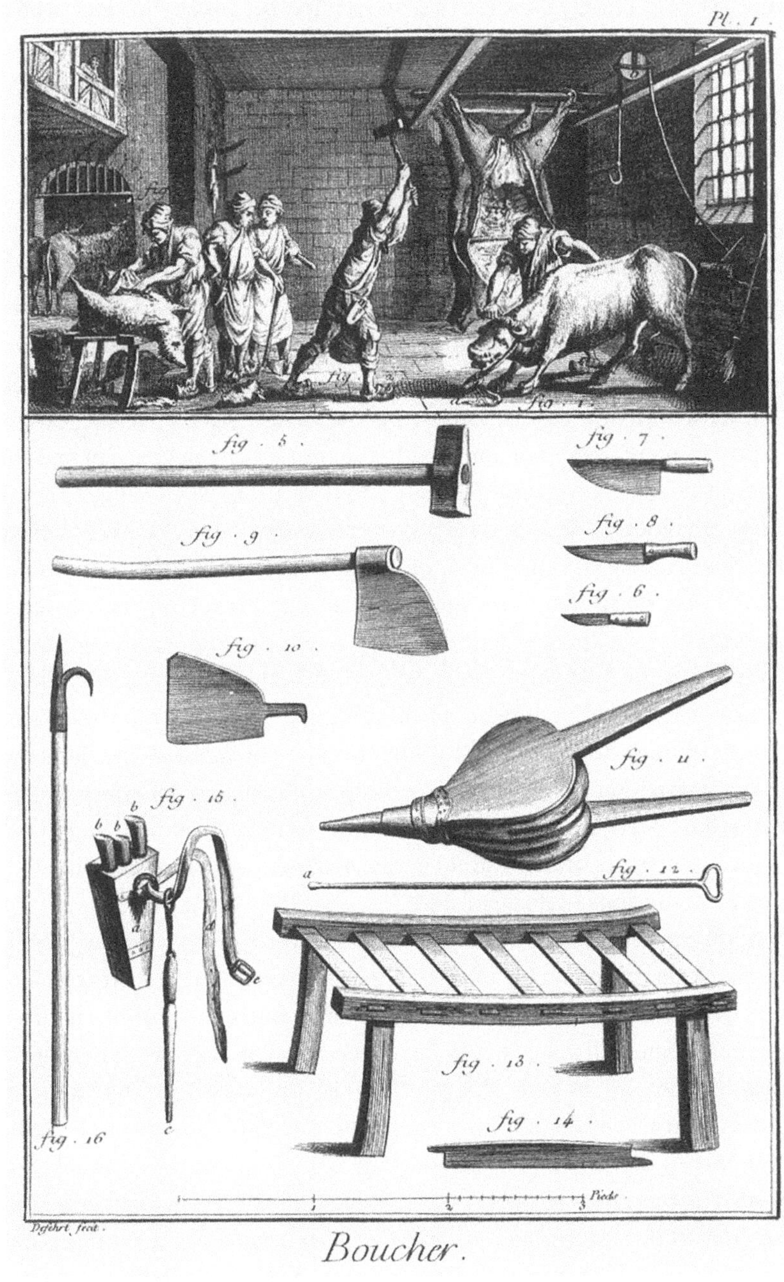

Figure 4. "Boucher," in Diderot and d'Alembert, *Encyclopédie*, vol. 2, plate 1.

slime; the soft, bloody muscle of a severed joint; the silken shimmer of cooling, just congealing animal fat, contrasted with the chalkiness of rendered suet."[74] The metaphoric becomes even richer by virtue of how it is overlaid onto and conflated with the literal.

Rembrandt's *Slaughtered Ox*, too, is meat *and* metaphor—an ox stretched on a rack to figure the crucified Christ but also simply to await processing by the absent butcher, off finishing his dinner or stropping his knife.[75] Dutch and Flemish still-life paintings incorporate cuts of meat that both are and are not what they seem to be: artfully composed depictions of delicacies and desirables, lending a sheen to the lives of those who owned and displayed them and asserting the supreme relevance of the physical, material world; and cogent statements on the temporality of success and the proximity of decay, rot, and death, the supreme irrelevance of material goods in the face of the eternal. Raw and cooked, roasts on silver platters shine with blood or rendered fat, delectable and decaying; viewers salivate over the promise of satiation, even as they recall that their own bodies, too, are only flesh.[76] Both resolutely physical and resolutely antimaterial, paintings of meat stand as extraordinary contradictions, testament to the transcendent beauty of the everyday and, simultaneously, to the hideous truths hidden in what appears to be sublime.[77] Even within the stolidly factual confines of Diderot and d'Alembert's *Encyclopedia*, the depiction of animal slaughter carries an existential symmetry: two hanging carcasses on one side, dead; two standing cattle on the other side, living; and the restrained ox crossing the threshold between them, beneath a sledgehammer arrested at the apex of its deadly arc.

All these representations from the Western tradition, whether of hunting, slaughter, or meat consumption, share in common their treatment of the activities as *small-scale* pursuits. One man or one small group of men and dogs pursue and kill one fox. Butchers keep a few animals, slaughtering and dressing them one by one. Not until the rise of industrial-scale slaughter, with its continuous disassembly line, does it become possible to imagine, much less to depict, the endless dispatching of animal after animal, each handled by hundreds of workers on its passage from creature to meat, each human worker performing the same small set of operations on thousands of animals. In addressing the visual culture of the meat industry, then, it is necessary to revisit an ancient subject, considering how modernity and industrialization necessitate that it be envisioned anew.[78]

Relatively few studies have undertaken the task of studying how industrial-scale meat production has been imagined and imaged by visual culture. Michael Garval's "Visions of Pork Production, Past and Future, on French Belle Époque Pig Postcards" looks at popular images of pigs in French visual culture at the turn of the twentieth century. "Scrutinizing pig postcards," Garval notes, "[reveals] a good deal not only about the surprising scope and impact of visual media during this period, but also about changes in pork production that were characteristic not only of the French but also of the broader western food system's incipient modernization."[79] During this period, slaughter was increasingly industrialized, and Garval finds that the creators and buyers of these postcards used them to express social anxieties over this profound shift, the postcards serving as "strategies for creating an effective distance from slaughter's disturbing realities."[80]

Jason Weems's "Scale, a Slaughterhouse View" focuses on photographs of the Chicago meat industry, examining panoramas and other attempts to show the huge scale of operations in Packingtown. Some of these photographs depict the Union Stockyards, while others attempt to show aspects of the trade in animals and meat, including stages in the slaughter process, in greater detail. Overall, Weems writes, "The program that emerged for visualizing The Yards marked an attempt to simultaneously comprehend, celebrate, and disarm the massive-scale slaughtering system, and to rationalize its diminishment of living bodies (and beings) as inviolable wholes," an attempt Weems contends was unsuccessful: "The failure of the imagery to [achieve this rationalization] embodied, in drained blood and rendered flesh, the incommensurabilities of scale that existed between living bodies and industrial systems."[81] Weems describes how, at the turn of the century, tourists flocked to the Union Yards and Packingtown, and guidebooks abounded. Even in encountering the Union Yards in person and in image, even in seeing the landscape both from above and at ground level, however, it ultimately proved difficult to reconcile "one's own sense of corporeality and spirituality" with the massiveness of the packinghouse endeavor.[82]

At the end of the day, however, many questions regarding how human beings represent animals and animality founder on this very issue, the extent to which it is possible to separate animality from humanity. Does our recognition of our own animality augment or diminish our identification with nonhuman animals? Do our base desires to eat animals, to serve our own corporeal needs, interfere with our ability to empathize with them, whether in pictures or in reality? This issue becomes complicated, as authors including John Berger and Steve Baker have pointed out, by the frequency with which representations of animals are actually about humans, humanity, human concerns, human metaphor.[83] In death, too—indeed, in death perhaps more often than in any other state—animals symbolize something beyond themselves. Their lives and their deaths are rarely shown as their own; they are, instead, ours, writ large and repeatedly. Animals have rarely been allowed simply to *be*, in and of themselves.[84]

Even wildlife photography that aims to show animals ostensibly in a wholly natural, undisturbed state "is a compensatory fantasy haunted by a desire for an unmediated relation with animals."[85] Wildlife photography uses human technologies to imagine, and to image, an absence of human technologies. Wildlife becomes more symbolically laden the more difficult it is to catch in images. Here, the human desire takes precedence over the animal existence.[86]

Methods, Materials, and Motivations

In the present day, the camera has a reputation as a tool not for assisting the meat industry but for critiquing it, via the surreptitious gathering of still and video images and the subsequent, often confrontational, deployment of such imagery by animal rights groups.[87] Visual images of meat production today are often used to reveal poor conditions inside meatpacking facilities and large-scale animal-handling facilities; to convince people of the negative effects of meat eating; and to serve, overall, as a means of exposure, revelation, and critique.[88] In response to such activity, many US states today have laws explicitly banning cameras in meat production or animal agriculture facilities. Frequently written by or with the guidance of industry organizations and lobbying groups, such "ag-gag" laws, as critics refer to them, have come into existence under, and in direct response to, assumptions that the normative use of cameras in such facilities will be for critique and exposure, that the presence of a camera not only cannot help the industry but will almost certainly injure it.[89] The 2004 Animal and Ecological Terrorism Act, the model legislation on which a number of state laws have been based, lists among its prohibited acts not only the destruction of animal-handling properties or facilities but also "entering an animal or research facility to take pictures by photograph, video camera, or other means with the intent to commit criminal activities or defame the facility or its owner," effectively equating picture-taking with criminal vandalism.[90]

When I first began looking at historic images of the meatpacking industry, I assumed that they had been produced for similar reasons—that any photograph of the meat industry would normatively serve to expose and critique its practices. It quickly became clear that this was not the case: many historic images had been made with the acquiescence or outright sponsorship of the meat industry. I began endeavoring to place these images within their proper contexts: to think about why they might have been made, by whom, and for whom; and to consider how and where they circulated, what public(s) encountered them. I found that images of meat production in the early to mid-twentieth century are not rare. The meat industry not only tolerated but embraced the camera as a useful tool, and the general circulation of direct or even gruesome pictures occasioned little public notice or revulsion. Historically, photography's *controlled presence*, not its total absence, has provided crucial support for the meat industry's endeavors to cultivate and maintain positive public perceptions of its activities.

In the past, the meat industry deployed photographs liberally to serve its own needs: to expand its appeal, to increase market share, to cement its status as an inevitable, indispensable component of the American corporate and culinary tradition—indeed, of the American identity. Advertising constitutes a tremendously important way the industry has used images, but it is by no means the only way. Images have served as tools for education, training, and research, aiding and abetting the circulation of ideas within the industry. Photographs have helped the industry tell a story about itself to itself—a crucial part of how the industry constructs its self-image, which then feeds its public image.

In looking at images, this book builds a visual history of the American meat industry from the end of the nineteenth century into the mid-twentieth century. It complements and builds upon existing studies of the meat industry.[91] Few studies up to this point, however, have addressed the roles that photographs and other visual images have played in supporting and expanding industry activities.

Throughout the book I deploy images as case studies, drawing on archives and publications, including books and magazines. I focus primarily on materials created between the later nineteenth century and the mid-twentieth century, roughly 1880 to 1960, a time period during which American meat production grew rapidly into an industrial-scale practice and the meat industry quickly developed a high degree of sophistication in its use and circulation of images to promote, defend, and propagate its products and its interests. Various industry stakeholders created or circulated these materials. Meatpacking companies both large and small have produced and deployed images in large numbers as part of their overall endeavor to shape both the actual fabric of the industry and its public perception. So have industry organizations and interest groups, who are among the most prolific producers and users of visual imagery. Other stakeholders include the myriad entities critical to meat industry operation, such as manufacturers of machinery and equipment for slaughter and meat production. Labor unions, too, have produced and used photography freely.

I began exploring this material in part because of my curiosity about uses of photography for critique, rather than promotion, of meat production. Yet this book ultimately engages relatively little with images created by individuals and organizations interested in animal rights. In looking at the role of photography at a time when the American meat industry was rapidly expanding, consolidating, and industrializing, I found that animal rights groups were simply not a significant source of friction for the industry during the period.[92] The growth of opposition to industrial-scale meat production on the basis of issues related to animal rights and animal liberation did not occur until later in the twentieth century, during a period when the meat industry's public visibility—though not, of course, its actual presence or its (tremendous) economic significance to American industrial production—was already diminishing.[93] Uses of photography for animal rights and animal liberation activism constitute a fascinating topic of study in their own right, and in recent decades, animal rights groups have been among the most enthusiastic users of photography for critique and exposure of the meat industry. But such imagery plays little role in this book, as it played little role in how the meat industry chose to conceive, envision, and promote itself in the first decades of the twentieth century.[94]

This book does not, in itself, argue either for or against meat eating. It is not attempting to promote particular patterns of behavior. In looking at the visual culture of American industrial-scale meat production, this book engages in close reading and analysis of archival and circulating imagery and texts, a process involving, at its center, critical evaluation of both visual and textual messaging. This book attempts to identify and to bring to the surface, through analysis and close reading, some of the core values and motivations the industry has expressed in producing and circulating visual materials. In promoting its products and its interests, the meat industry first has to tell a story about itself to itself, defining its own values internally in order to sell itself and its products more effectively. In so doing, it both generates and recirculates stereotypes of human behavior, too: gendered assumptions and gender hierarchies, racial distinctions, class distinctions. The industry has, historically, promoted a vision of its labor force as anonymized and instrumentalized. And it depends, at its heart, on representations that promote a vision of animality as fundamentally lesser than humanity, existing to support human needs and human interests. This book is interested in showing how photography and visual culture have contributed to propagating these values.

Today there is perhaps no animal *less* symbolically freighted than the industrially bred, industrially raised, industrially slaughtered meat animal. Far from encapsulating the human experience, such animals have in American culture a nonpresence. They are multitudinous but invisible, nonbeings living their millions of brief, micromanaged but simultaneously obscure nonlives. They live in nonplaces, vast spaces almost wholly unvisited, even unthought-of, by most consumers.[95] To think of these animals symbolically would require us to think of them *at all*—an endeavor mostly to be avoided, as far as animal and meat industries are concerned.

This book hopes, then, to begin to think about the unthinkable and to look at the unwatchable: to consider visual representations of meat and slaughter within the context of an industry that not only denies animal symbolism but often rejects the very beingness of animals themselves, replacing it with something closer to thingness. Depicted primarily or solely as raw material, animals lose their wholeness, their animality. In the packinghouse, animal bodies are dismantled, literally disintegrated. The disintegration of their animality in visual representations mirrors and prefigures this physical process. Photos of livestock, packinghouses, and industrial-scale meat production processes not only represent but facilitate the dissolution of animal meaning into digestible meaninglessness, of bodies into flesh into meat.

CHAPTER ONE

Inside Views

Machinery, Perspective, and the Controlled Narrative

THROUGH AT LEAST the middle of the twentieth century, meatpacking companies and industry organizations tended to regard the camera as a trusted and crucial partner in representing their products, their operations, and their endeavors. Visions of meatpacking sponsored by the industry itself—by meatpacking companies, slaughterhouse machine and tool manufacturers, and industry organizations, among others—often present meat making as a state-of-the-art process, in which huge machines and impressive tools facilitate the wondrously speedy transformation of organic matter. This is particularly evident in material created for internal circulation: photograph albums, in-house periodicals, even trade catalogs and other media designed to sell the activities of the meat industry within the meat industry itself. Analyzing publications designed for internal circulation permits understanding of the meat industry's general self-image—how it envisioned and represented itself to itself. Photographs and photographically illustrated texts designed by insiders for insiders give a sense of how packing companies viewed themselves, their activities—including the core activity of animal slaughter and processing—and their overall business model.

This chapter examines individual photos, photo books, and other photo-texts that were produced by meatpacking companies, trade organizations, and auxiliary manufacturers of supplies, equipment, and by-products, primarily for circulation within the industry. The subject matter includes both unpublished and published materials: unique images and documents like photo albums; annual reports or similar documents released to narrow audiences of shareholders; and catalogs and brochures for trade audiences. Looking at these materials, it becomes possible to see what individual organizations and the meat industry as a whole regarded as their key values, the key components of meat industry identity. Meatpacking companies and industry organizations like stockyards have tended to value, among other things, their capacity to produce at scale, efficiently and with little waste; and, by extension, their ability to provide foodstuffs in quantity, to feed a population in the US and beyond that increasingly clamored for inexpensive, readily available meat. Meat industry organizations have also placed high value on modernity and modernization, often commissioning albums or viewbooks to commemorate new construction, new technologies, and new facilities. Where workers appear in these images, they function as necessary activators of the spaces, the machinery, and the processes of slaughter. The labor of slaughter appears dignified but mechanistic, and workers become engines of the activity of slaughter. Worker and machine cooperate to process animal bodies, which are treated as raw material, their animality suppressed or wholly denied.

As Sekula notes, while photography may be, and often has been, turned with ease toward the ends of capitalist exchange, the medium itself is not inherently capitalist.[1] But because it came of age in the early to mid-nineteenth century at the same time as industrial-scale enterprise, the medium "stood ready to play a central role in the development of a culture centered on the mass marketing of mass-produced commodities."[2] Considerations of mass marketing generally focus on publicly directed messaging, but the internal messaging and self-imaging practiced by large corporations also matter tremendously in consolidating and quantifying industry power. Advertising often operates just as much by encouraging consumer recognition of a brand's or company's perceived identity and values as it does by selling the actual attributes or benefits of a product.[3] A similar process of corporate identity formation occurs internally as well, within an organization and also among organizations within an industry. This corporate identity formation, much like the formation of an individual identity, depends in part on the organization's ability first to project a desired self-image, including organizational aspirations and values; and then to see this desired image reflected back to itself, in what becomes a cycle of image creation, adjustment, confirmation, and enactment.[4]

Beyond the meat industry, many corporations order, sponsor, or collect photographic images for practical purposes: documentation, record keeping, training, commemoration, public relations, community and labor relations.[5] In collecting, sequencing, and circulating images, companies abstract, aestheticize, and narrativize their own operations. Rosalind Shipley writes that the photographic documentation generated by a company could be designed to "affirm the company's mythology" but that in the process it might also "create its own narrative, spinning an intricate web of truth and fiction."[6] Controlled circulation of images allows a company to control its own image, to guide how it is viewed internally, by its own workers, and externally, by other companies and by members of the public.[7] By virtue of its repeated—and repetitious—manifestation in images, the rightness of company endeavor becomes not only inevitable but normative: ubiquitous, inescapable, and incontrovertible. Analysis of corporate albums and booklets, then, aids in unpacking the processes by which corporate activities—including industrial-scale animal slaughter—become concretized and normalized in the American collective imagination.

Circulation of photo albums in corporate settings served in part to cement organizational identity and to reinforce organizational values. David Gray writes that the Seth Seiders Syndicate, an early purveyor of corporate motivational posters and other materials, issued to its sales managers a printed album containing pictures of the company's main offices in Chicago and New York. Sales managers then used the album to train new salesmen. Already attuned to the power of images as key components of its motivational materials, the Seth Seiders Syndicate also depended on pictures to train its own workforce. The pictures

> modeled visually the attitudes and behaviors that the Syndicate desired, placed a human and familial face on a large and potentially impersonal organization, offered visual corroboration that the organization was a fully synchronized community of employees, and proffered to employees, especially the newly hired, a rationale for adopting as their own the "Big Vision" of their employer.[8]

The company used the album as "an important form of aesthetic training through which those who sold the Syndicate's products—namely, the salesmen—learned to embody their employer's very identity."[9]

Close examination of such materials also reveals the unintentional messaging, the bedrock assumptions and power dynamics on which organizations construct their identities. The meat industry makes its very livelihood out of the exercise of absolute power over the bodies and lives of nonhuman animals. Studying images of meatpacking, then, illuminates not only industrial and corporate identity formation but also how images may be used to envision, represent, and thus perpetuate power relationships. This includes the companies' ultimate power possession, the power to arbitrate animal lives, to take them at will and en masse. David Nye writes of the General Electric photograph collection that "the corporation's creation and control of such materials is a metaphor for its cultural hegemony."[10] With regard to animal life, the power held by meatpacking companies is considerably more than hegemonic, and images serve as more than metaphors: they serve as a means by which power is legitimated, multiplied, and thus perpetuated.

In coining the word *biopower*, Michel Foucault labeled a type of power that developed during and out of the Enlightenment. Previously, Foucault writes, the sovereign had exercised the right to take away an individual's life via the death penalty, but as the absolute power of the sovereign diminished and a more diffuse state power developed, the function of power shifted. Where once it had dictated when life could be taken away, this new power —biopower—now served to administer life.

> The disciplines of the body and the regulations of the population constituted the two poles around which the organization of power over life was deployed. . . . This bio-power was, without question, an indispensable element in the development of capitalism; the latter would not have been possible without *the controlled insertion of bodies into the machinery of production* and the adjustment of the phenomena of population to economic processes.[11]

In *The Disciplinary Frame*, John Tagg makes clear how photography, particularly in conjunction with the legal system, became an instrument for the administration of biopower over the course of the nineteenth century. As legal codes and policing forces and policies were regularized and professionalized during that period, imaging—capturing—the bodies of prisoners with photography became an integral part of the process. Where once only rulers and the powerful had the privilege to be depicted in images, Tagg notes, this new

> regime of sense . . . guaranteed a new order of truth and turned on a new structure of documentation whose institutionalized effect was to reverse the political axis of representation, making it no longer a sign of power and prestige to be recorded, but a sign of subjection. It is . . . only within the closure of this discursive frame . . . that we can understand how photog-

raphy . . . could come to function as an instrument of surveillance, record, evidence, and truth.[12]

Foucault and Tagg address primarily the effects of power over human lives. Applying these models to the human-animal relationship, particularly to the human relationship with meat animals, requires a more nuanced understanding.[13] If *biopower* is power over life, then with regard to present-day practices of animal breeding, feeding, and behavioral monitoring, it would be reasonable to refer to something approaching *total biopower*. Control over and regulation of meat animal lives not just at their end point but from their very conception has become, for all intents and purposes, absolute. "Large-scale killing requires equally large-scale breeding," Dinesh Wadiwel notes.[14] Slaughter constitutes the point, in the sense of a single moment, when the exercise of biopower achieves its apex; slaughter constitutes the point, in the sense of the ultimate purpose, toward which entire systems of animal management and control have directed their energies.

Indeed, packing companies exercise over the animals that constitute their raw material a total power exceeding even that of Foucault's pre-Enlightenment sovereign: whereas he meted out death as a punishment for wrongdoing, meatpacking companies depend on it as the instrument of and reason for their very existence. But for power to be exercised continually and successfully over animal bodies in the packinghouse, without interruption, it cannot appear like power at all. It must look like the natural order, and, more important, it must look *like order*, not like chaos.

In these systems of animal control, photography has performed, and continues to perform, a highly valuable service.[15] For circulating a sense of order, of business as usual, visual images and particularly photographs have proven as vitally necessary to industrial-scale meat production as to Tagg's legal "regime of sense," extending and reinforcing human dominion over animal bodies. Nicole Shukin's *Animal Capital: Rendering Life in Biopolitical Times* also addresses the power dynamics inherent in representations of animals, whether in writing, images, advertising, or metaphor. Shukin describes how representations perform a crucial function in perpetuating human-animal power dynamics. Humans, she writes, have arrogated the power to represent animals as a component of their broader systems of control. Indeed, representation *is* control: mimetic representations of animals —in images, texts, or other forms—do not just show power relationships;

rather, "mimesis *constitutes* the real workings of power, at least partially."[16] Drawing on the dual meaning of the term *render*, she demonstrates how the right to "render" animals, in the sense of drawing or representation, exists parallel to and in support of the activity of "rendering" animals, in the sense of wringing value from the raw materials sourced in their bodies.[17] Widespread circulation of animal representation, then, serves to reproduce, reinforce, and circulate human-animal power relationships as well.

It is useful here to borrow from W. J. T. Mitchell, who uses the phrase "cloning terror" to describe how representations of terrorist activities both multiply and perpetuate not only the feeling of terror but terrorism itself. Mitchell describes how photographs of terror and violence serve to terrorize on initial viewing and then to reterrorize on repeated viewings, thus actively generating and sustaining a feeling of terror. Images of botched responses to terror, like the Abu Ghraib prison photographs, likewise serve as effective recruiting tools for extremist groups, thus multiplying the number of people involved in subsequent terrorist activity.[18] In both situations, reality spawns images, and images in turn spawn reality.[19] "The imagination is a constitutive core of . . . reality," Mitchell notes.[20]

In a similar manner, by representing the human-animal power dynamic in photographs and photo-texts as a natural fact, rather than a cultural artifact, the American meat industry uses images to clone *power*, perpetuating and reinforcing the dynamic. Photography engages in a kind of cosmic sense-making, its powers of documentation put to the task of visualizing not just moments but the vital processes by which humans order and construct their worlds—the human mastery, dominance, and ingenuity required both to make and to unmake, each of which appears vitally important to human progress.

Bodies into Machinery

When Foucault writes, in *The History of Sexuality*, of "the controlled insertion of bodies into the machinery of production," he is speaking metaphorically about intersections of the human body with the disciplinary and reproductive mechanisms of capitalism. But with industrialized animal slaughter, the animal body is literally inserted into the machinery of the meat factory. Large-scale animal slaughter and meat production depend on the development of a host of specialized machines and tools, each designed both to accommodate and to disarticulate the animal body. These run a gamut from traditional and general-purpose tools, such as knives, to huge machines designed to perform a specific role within the compartmentalized industrial disassembly line. Unsurprisingly, photographs of machinery and tools have thus constituted a significant subset of the pictures made inside packinghouses, the machine-made image tracking the progress of mechanized animal disassembly.

Mechanization of slaughter, Giedion writes, proceeded slowly, with the processes of slaughter becoming compartmentalized but otherwise remaining largely unchanged for many decades from how they had been in the early manual stage.[21] Machines might be invented to address individual steps in the process of slaughtering and dressing large animals—machines for removing pigs' hair, for instance, or for raising the animals off the floor and into a head-down position appropriate for bleeding—but many of these came very early in the industrial period, the 1860s and 1870s. Inventions plateaued as it became clear that the slaughter of the organic animal could not be fully mechanized.[22] This remained true for many years even after Giedion's observations, slaughter of large animals becoming primarily mechanized only in fits and starts, and in some places continuing to resist full mechanization.

Most factories depend on at least some organic raw material to make their products. Rubber cultivation, extraction, and processing enabled the rise of the automobile industry. The manufacture of bread depends on living yeast cultures.[23] The raw material in a packinghouse, however, is not only organic but moving under its own power—possessed of a physical will at the least, if not a spirit. Indeed, law requires that the raw material in a packinghouse should walk right up to the point of slaughter; sick and dying animals cannot be slaughtered alongside healthy animals.[24] Rubber may come from a tree; yeast may reproduce and grow according to its own rhythms. But neither one moves into a factory on its own legs, or—more to the point—possesses the capacity to *resist* assimilation into the production line.

A great deal of energy thus must be exerted in packinghouses to cause the raw material to enter the factory willingly and actively, and then to stop moving at precisely the right time—to possess sufficient will to enter the facility, but not to resist what happens to it once there. Routinizing the introduction of the live animal into the production line, the start of

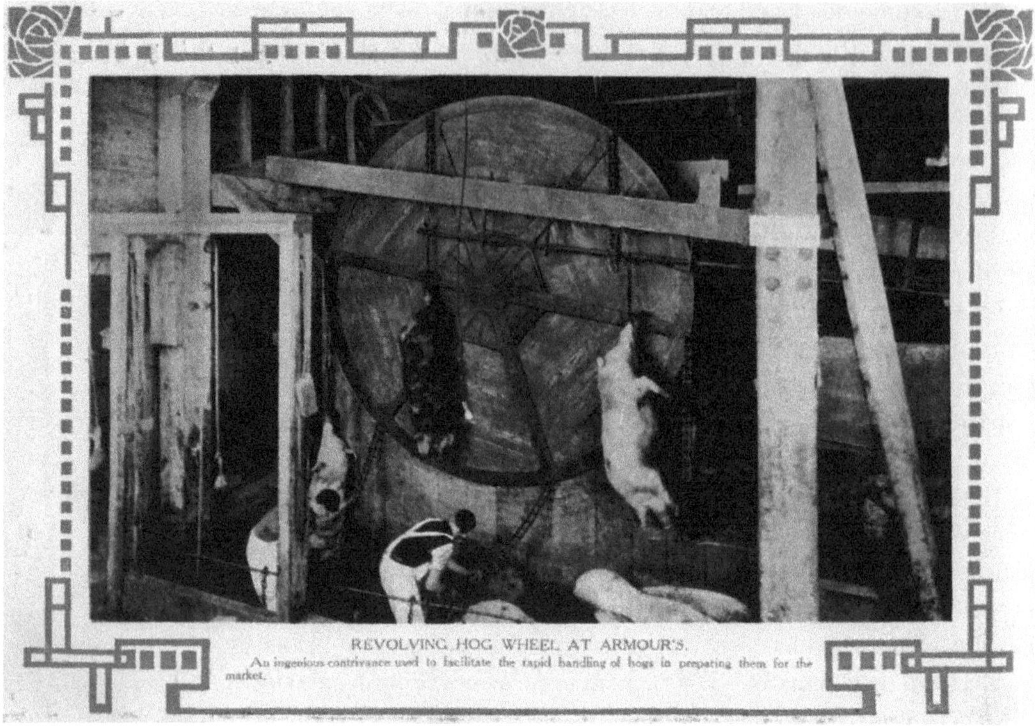

Figure 5. "Revolving Hog Wheel at Armour's," in *Souvenir of Chicago in Colors* (Chicago: V. O. Hammon, 1908), 22. Chicago City-Wide Collection, box 41, folder 20, Special Collections, Chicago Public Library.

the process of production, is a critical stage, since all subsequent acts of cutting and processing depend on it.[25] The animal must be introduced to the production line in two stages: first, getting the animal to walk into the packinghouse; and second, mechanizing the process by which the animal is elevated into a head-down position, appropriate for bleeding.

The individual idiosyncrasies of a given species must be taken into account during elevation. Pigs have historically presented the greatest challenge in this respect: they are larger and harder to manage than sheep, but before passage of the Humane Methods of Slaughter Act of 1958, they were not stunned prior to bleeding and thus, unlike cattle, remained sentient throughout the process of elevation and up to the point of death.[26] A great deal of ingenuity was directed at how to elevate the animals into a head-down position while minimizing danger to the workers responsible for doing the job. One contraption from the late nineteenth century, patent drawings of which were included by Giedion in his book, incorporated a trained pig, called a "Judas" animal, into its design.[27] The Judas pig led a small group of others into an enclosed chamber, where it walked to the front to receive a food reward while, behind it, the other animals found themselves being raised into a head-down position as the floor beneath them dropped away.[28]

Another, far more widely adopted machine dispensed entirely with the Judas animal, allowing the machine—staffed and served by human functionaries—to absorb all the work of elevating the pigs. These machines, called Hurford wheels or simply "hog wheels," began to appear in packinghouses around the turn of the twentieth century (fig. 5).[29] Such wheels presented quite a sight, subjugating the animal and allowing its body to be assimilated more readily into the disassembly line. Visitors came in droves to Chicago's Packingtown, encouraged by guidebooks and enticed in part by accounts of the existential sublimity of slaughter at such massive scale.[30] For many, the hog wheel epitomized the metaphysics of mechanized death, a juggernaut relentlessly driving the conversion of animals from creatures to carcasses. A description of the wheel anchors Sinclair's account in *The Jungle* of his protagonist's initial encounter with the packinghouse during a guided tour:

> They went into a room from which there is no returning for hogs. It was a long, narrow room, with a gallery along it for visitors. At the head there was a great iron wheel, about twenty feet in circumference, with rings here and there along its edge. Upon both sides of this wheel there was a narrow space, into which came the hogs at the end of their journey. . . . The wheel had stopped while men were cleaning up. In a minute or two, however, it began slowly to revolve, and then the men upon each side of it sprang to work. They had chains which they fastened about the leg of the nearest hog, and the other end of the chain they hooked into one of the rings upon the wheel. So, as the wheel turned, a hog was suddenly jerked off his feet and borne aloft. At the same instant the ear was assailed by a most terrifying shriek . . . followed by another, louder and yet more agonizing—for once started upon that journey, the hog never came back; at the top of the wheel he was shunted off upon a trolley, and went sailing down the room. And meantime another was swung up, and then another, and another, until there was a double line of them, each dangling by a foot and kicking in frenzy—and squealing. . . . Meantime, heedless of all these things, the men upon the floor were going about their work. Neither squeals of hogs nor tears of visitors made any

difference to them; one by one they hooked up the hogs, and one by one with a swift stroke they slit their throats. . . . It was all so very businesslike that one watched it fascinated. It was porkmaking by machinery, porkmaking by applied mathematics. . . . One could not stand and watch very long without becoming philosophical, without beginning to deal in symbols and similes, and to hear the hog squeal of the universe.[31]

The wheels solved the persistent problem of how to elevate agitated hogs quickly to the head-down position appropriate for bleeding, taking the animal from living being to drained, denuded object in just a handful of minutes. They varied in size, with smaller wheels present at smaller facilities, while large facilities were served by machines that towered over animals and workers alike. Turning continuously, the wheel would catch up pigs by a hind leg that a worker had placed in a shackle, then lift them off their feet. At the apex of the rotation the shackle would transfer from the wheel onto a downward-angled rail, where gravity would draw the animal, still in a head-down position, toward the "sticker," the worker responsible for puncturing the throat.[32] This arrangement epitomized the human-animal-machine triangulation central to all industrial packinghouse operations: workers fed animals to a machine of heroic scale, which in turn delivered them to a human—the sticker—who (ideally) operated like a machine. Once the hogs had been dehaired, they were as close as possible to uniform, the process of wheel–sticker–scalding bath rendering them literally raw material for all subsequent packinghouse processes. After this sequence, packinghouse operations were drained of metaphysics just as the animals themselves had been drained of blood: hairless and bloodless, the creatures were pork, not pigs. The hog wheel and the scalding immersion enacted a reverse baptism, a reverse transubstantiation: from beings, they became flesh.

Of all the machines in turn-of-the-century packinghouses, hog wheels show up again and again in images of meatpacking, serving a kind of metonymic function as stand-ins for the size, scale, and constant movement of the industry as a whole. One image, created circa 1910 at a Cedar Rapids, Iowa, facility called Sinclair & Co., compresses the whole of the critical wheel–sticker–scalding bath sequence into a single remarkable image (fig. 6).[33] In the background, a live pig hangs on the wheel, about to be delivered onto the rail. The sticker stands near the dead center of the image, but far from the camera, at the back of the room, pausing just a moment for

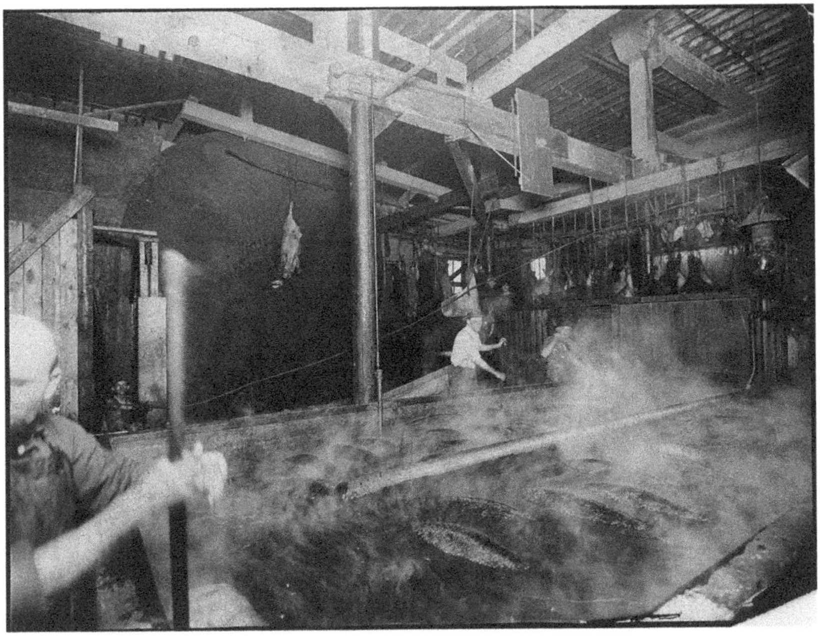

Figure 6.
Hog slaughter area, Sinclair & Co., Cedar Rapids, IA, ca. 1910. Farmstead Foods Collection, Brucemore Historic Site, Cedar Rapids, IA. Courtesy of Brucemore.

the photograph. To the right of the sticker, behind a low wall, pigs hang to bleed in a row, only their hind legs visible. The pigs' path subsequently takes a turn, just outside the frame at the right of the photograph, and then the bodies pass before the lens again, now quite close, submerged in the scalding bath and visible only as oval shapes breaking the water's dark, steaming surface. Cut off by the leftmost margin of the image, a worker looms in the foreground with a long-handled stick, preparing to hoist the next pig out of the bath and pass the carcass along for dehairing.

The image testifies to the inexorable logic of the process, the constant transit of animals from life to death in a space at once enormous and compressed, choked with men and machines. This image provides a revealing look at the mechanistic underpinnings of mass animal slaughter. Still, a great deal is elided: sound and smell, among other things. Photographs like this one operate to reinforce the logic of the packinghouse in part because they are mute, because they still movement, eliminate stench and noise, minimize chaos. Even in this photograph of the hog wheel, bleeding area, and bath, which compresses so much of the hog's initial path through the packinghouse into a single picture, the point at which the animal actually

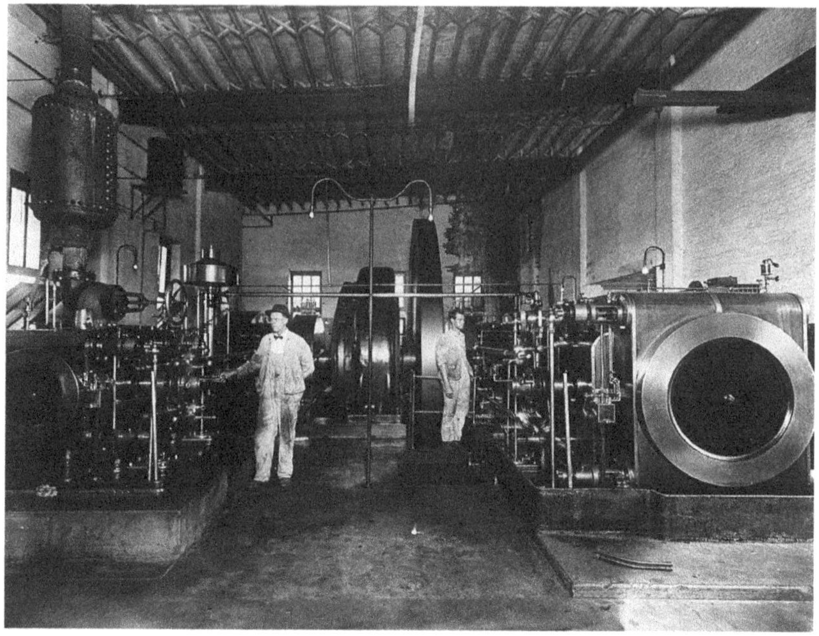

Figure 7. Engine room, Sinclair & Co., Cedar Rapids, IA, ca. 1910–1920. Farmstead Foods Collection, Brucemore Historic Site, Cedar Rapids, IA. Courtesy of Brucemore.

passes from life to death is obscured—placed in the dark back corner of the image and momentarily suspended as the workers pause to wait for the photograph to be made. The photo itself necessitated, and thus depicts, a momentary reprieve, even as it confirms the relentlessness of the process of slaughter and disassembly.

Another, quite different type of machine also shows up repeatedly in visual records from the American meat industry. The dynamos and drivers required to run meatpacking plants, though not themselves part of the disassembly line, occasioned a great deal of fascination as literal engines of success. Prior to the early 1930s, individual packinghouse machines had no internal power sources of their own. Every machine had to be connected to every other and, via networks of drive belts, to huge central engines that ran entire factories. Although such engines show up in photographs from other locales and other industries—the massive Corliss engine that drove most of the machines at the 1876 Philadelphia Centennial Exposition features prominently in images and accounts of that event—the meat industry was among the first to deploy centralized drivers in its factories in a systematic manner.[34] In another photograph from Sinclair & Co., workers stand among the machines, dwarfed by comparison (fig. 7). But if the individual human seems puny by comparison with the machine, the massiveness of human ingenuity looms large. The gigantic machines testify to the size of the operation, stand-ins for the success of industrial-scale meatpacking as an ambition and an endeavor. Through these images, Sinclair & Co. demonstrates how monumental and how unstoppable its endeavors have become, driven by huge machines and the huge industrial will they represent.

Corporate Daybooks and Albums: Sequential Visions

Corporations circulate photographic imagery both formally and informally: within phototextual publications such as reports and industry periodicals as well as via circulation of snapshots and photo albums. Study of the former type of material proves somewhat easier than study of the latter, simply because books and pamphlets often survive in multiple copies whereas snapshots and photo albums are usually unique and unduplicated. Still, surviving examples seem to indicate that creation of house albums was relatively common. House diaries or corporate daybooks might include solely business records, or they might incorporate news clippings or other materials in the manner of a scrapbook.[35] The Chicago History Museum retains a series of daybooks from Armour and Company's meatpacking plant in Kansas City, written right at the turn of the century. Labeled *Harry Schultz Journals*, the books begin in 1898 and continue into the early years of the twentieth century. They include brief, handwritten annotations for each day of business, noting the weather, number and types of animals processed, visitors to the plant's office, and major news events. Numerous and varied handwriting styles, probably those of clerks and assistants, indicate collective, corporate authorship of the daybooks. Although no photographs are inserted among their pages, the Schultz daybooks are interleaved with news clippings; the book from 1898, for instance, contains numerous clippings detailing events in the Spanish-American War.[36] Incorporating photographs into such daybooks would have been a logical extension.

Other surviving albums, as well as individual, unbound snapshots, indicate that by the end of the nineteenth century, packinghouses regularly employed or commissioned photographers to produce a wide variety of visual records. They collected and compiled more casual snapshots into albums, sometimes alongside or at least in the same binding with professionally made images. Exterior images and staff pictures were most common, but photographers also depicted interiors and disassembly lines as early as such interior photography was feasible, roughly by the last decade of the nineteenth century. Cameras became smaller and more portable before the turn of the century, permitting increased access to difficult environs like factory interiors; and photographic materials, including emulsions and papers, became increasingly uniform and straightforward to use. Emulsions also increased in light sensitivity, and thus in exposure speed, making it increasingly likely by the later 1920s that a skilled photographer could make sharp, clear action pictures even in a relatively dim interior.[37] Early photos of packinghouse interiors reflect these technological limitations: they had to be done with the full knowledge of everyone in the photo and with at least implicit permission of packinghouse management. Workers had to pause in their labors for the duration of the exposure, and the endless disassembly line, including the moving chain on which the animals hung, had to stop, even if for only a moment.

The archive of the Morrell Meat Packing Company, housed in Special Collections and Archives at the University of Iowa, incorporates records from this multinational corporation, founded and headquartered in the UK but with several American branch locations.[38] One photo from a Morrell facility in either Iowa or South Dakota, taken around 1900, shows a group of workers pausing from the labor of processing cattle (fig. 8).[39] At the back of the room, close to the left side of the image, a carcass hangs with its skin still attached, next to a wheelbarrow presumably containing internal organs. In an area not visible to the camera, the carcass will be placed on a pallet, or "bed," on the ground to be partially skinned and further dressed, then suspended again and moved along a rail for more skinning and cleaning, toward the man who stands farthest to the right. Just behind him, the snaking overhead rail extends through a pair of tall, swinging doors, beyond which lies the cooler.

The image speaks to the ongoing absorption of new processes and new technologies into older modes of slaughter. Men stand with knives and handsaws, practicing butchery as something like a craft.[40] But the sheer number of carcasses, and the increasingly compartmentalized division of labor, bespeaks larger-scale production. Both the overhead suspension rail and the light bulbs hanging from the ceiling are relatively new developments. If invention here seems to augur increasing ease, this is false or at least misleading prophecy; as the processing lines mechanize, the pace of

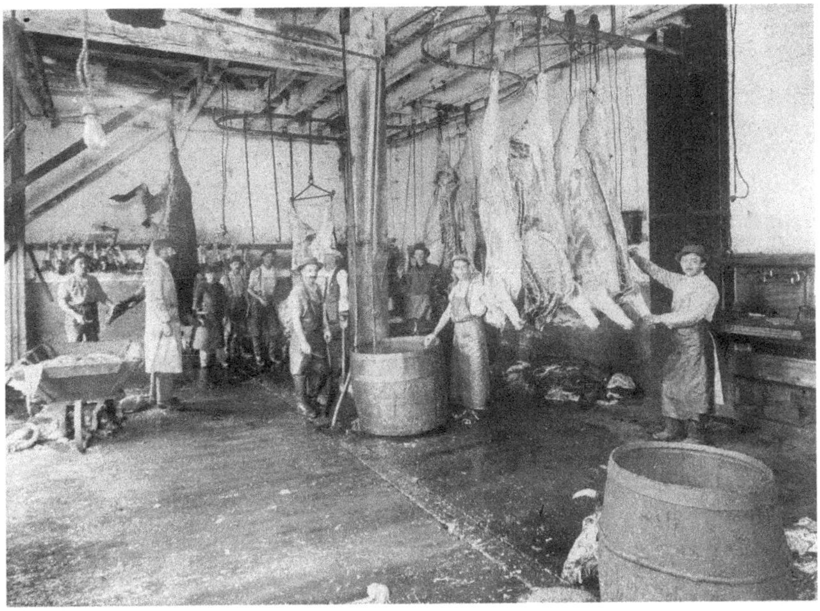

Figure 8.
Workers cleaning carcasses, ca. 1900–1910. Morrell Meat Packing Company Collection, collection MSC 0100, Special Collections and Archives, University of Iowa Libraries.

the line increases. Invention demands more, not less, from the body of the individual worker.

The image also testifies to the ongoing stresses placed on photography, another new technology, in representing this ongoing modernization. Although it improved tremendously over popular print methods, there was still only so much the medium could do in such a difficult environment. Rendering of faces is incomplete, to somewhat otherworldly effect: during the long exposure, many of the men moved their eyes, which thus rendered as unsettling blank spots. Correcting for the failure of the photographic process, someone came back in with a pen and put little dots on the men's faces. In some ways, the gesture reads as the opposite of the apotropaic erasure of eyes and faces that one sees in ancient manuscripts, defaced over time by fearful readers: here the eyes have been reinserted where the photograph effectively took them out, reassigned their power to mark the subjects as possessors of a gaze, not just as subjects of one.[41] This attitude of adjustment toward the photographic image, toward the idea that the photograph showed a truth but simultaneously needed fixing, is quite common at the turn of the century, and serves as a reminder that the turn toward

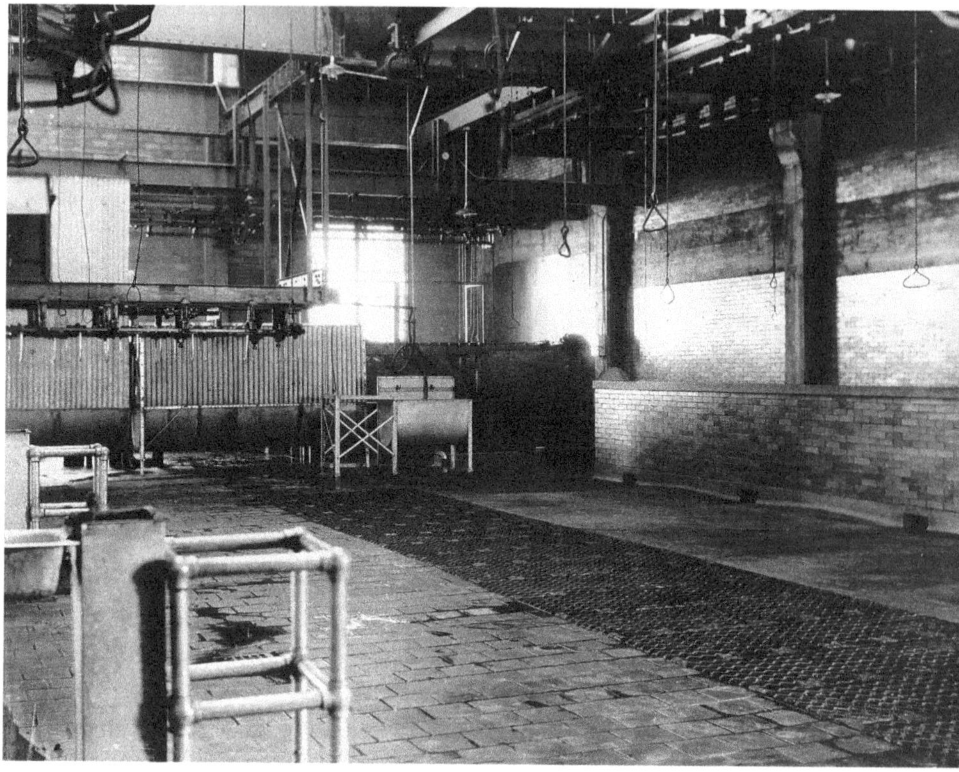

mechanization, whether of slaughter or of image making, was a slow turn, a process of evolution.

Also among the records in the Morrell archive are pages from a photograph album, now unbound and stored flat.[42] Most of the images in the album depict a single Morrell facility at Sioux Falls, South Dakota, in 1925. Some photos depict a new beef killing and processing facility, others an existing pork processing facility with some new rooms or additions. In virtually every image, the aim seems to have been to show the space itself. Although workers, animals, and machinery populate some of the spaces, where they appear, they seem incidental to the facility itself, captured in wide shots.

The sequence begins with a panorama, comprising two photographs joined at the short edge (fig. 9). The inscription identifies this as a "Section of killing floor in new beef house." Proud of the ultramodern, hygienic

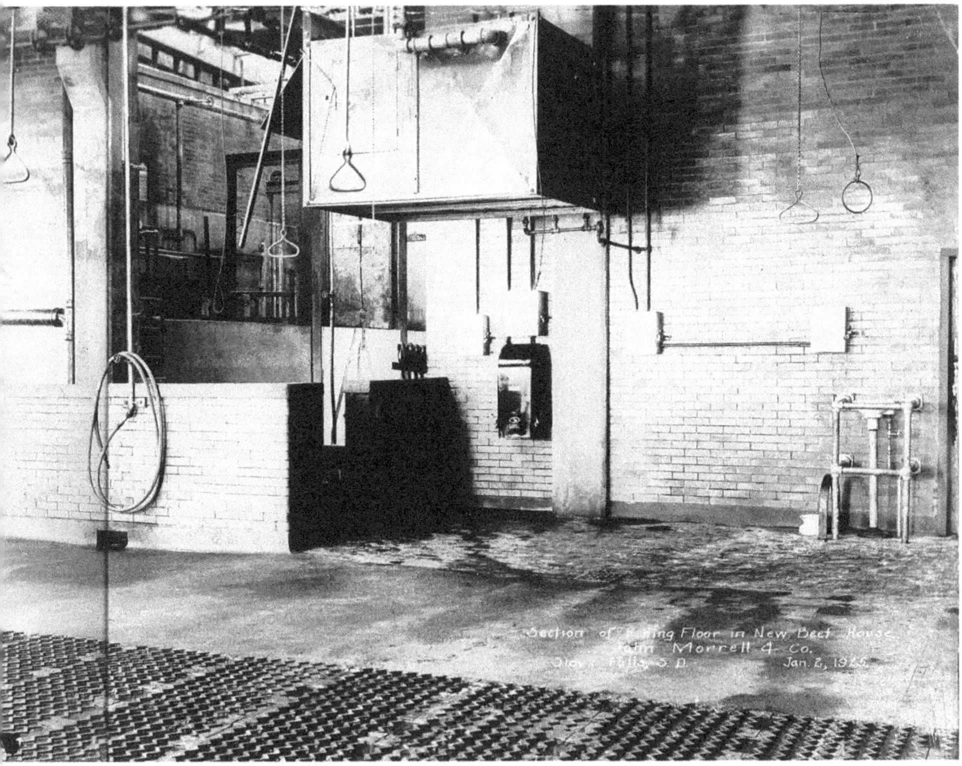

Figure 9.
"Section of Killing Floor in New Beef House, John Morrell & Co., Sioux Falls, SD, Jan. 2, 1925." Morrell Meat Packing Company Collection, collection MSC 0100, Special Collections and Archives, University of Iowa Libraries.

space, the company commissioned these photographs showing it empty and scrubbed clean, before the day's work had begun. Depopulated, the space appears virtually uninflected, potentially useful for a wide variety of purposes. Only details give away the intended function: rings dangle here and there from the ceiling, gesturing at the need to hang beef carcasses for slaughter, dressing, and hide removal; floors slope toward large drains, evidence of the need for regular, thorough cleansing in a space designed and designated to accommodate polluting activity. All the work occurring in this space was skilled labor, albeit dirty and physically demanding.[43] In this initial image, however, the hangers dangle empty and the hoses are coiled on their hooks. Light streams in a window at the left side of the image, its low angle indicating that the image must have been made in the early morning. (It could not have been sunset, because on January 2—the date inscribed on the image—the room would still be full of workers at dusk,

Figure 10. "North Half of Killing Floor—New Beef House, John Morrell & Co., Sioux Falls, SD, Jan. 2, 1925." Morrell Meat Packing Company Collection, collection MSC 0100, Special Collections and Archives, University of Iowa Libraries.

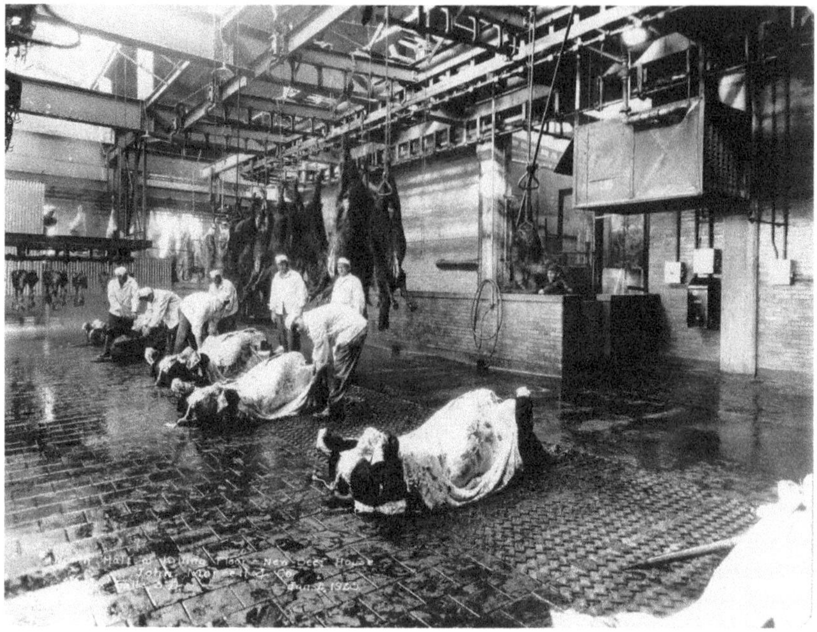

the workday continuing well after the early nightfall of a northern winter.) Here, before the day's work has started, the room appears clean and freshly washed, the puddled water gleaming on the brick floor.

A second image shows the same room, now no longer empty (fig. 10).[44] The picture depicts just half the space, but in greater detail. Two workers behind a short wall at center right pause from the work of slaughtering and dressing a carcass to lean on the wall and acknowledge the camera. On the nearer side of the wall, a row of nearly intact carcasses hangs head-down. More lie on the floor, belly-up, attended by several workers who have begun removing the hides. These men, too, pause for the picture. Some straighten to look at the lens, and others pantomime their jobs, posing motionless in a bent posture. The floor shines slickly, now thick with smeared blood. After skinning, the carcasses are rehung and then passed along to the next segment of the disassembly line, just visible in the far background of the image.

In producing this pair of images depicting the same space, first empty and then activated, Morrell & Co. indicated its pride in the modernity of its new building. The first image highlights the machinery, the efficient

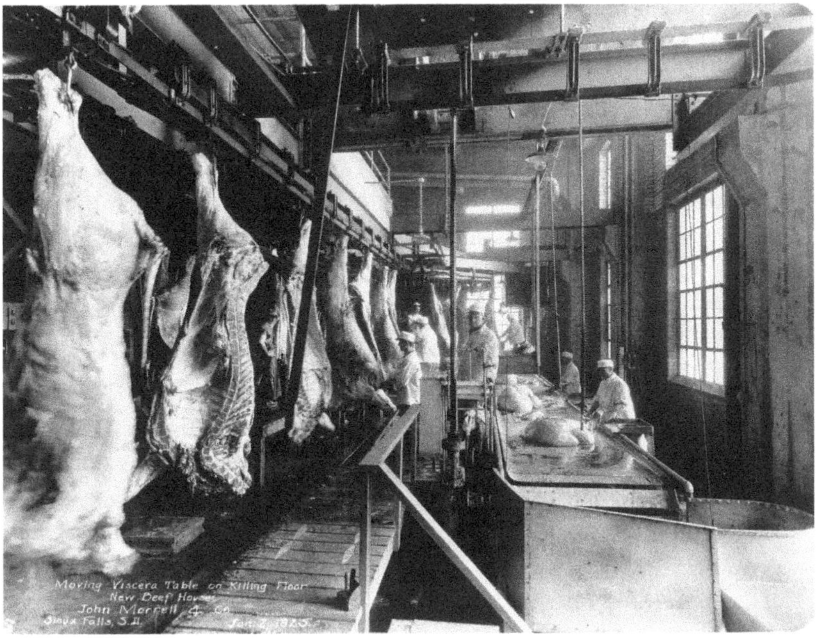

Figure 11.
"Moving Viscera Table on Killing Floor, New Beef House, John Morrell & Co., Sioux Falls, SD, Jan. 2, 1925." Morrell Meat Packing Company Collection, collection MSC 0100, Special Collections and Archives, University of Iowa Libraries.

layout, the openness and versatility of the central area. The clean, bright, empty room stands ready to receive the day's labors. Ample light fills ample space. But to show the full capability of the space, the album has to depict it in action, as well, making clear how work occurred within the space, how modern machinery in modern spaces facilitated efficient slaughter and processing of cattle into beef. The pair of images together encapsulates and perpetuates Morrell's ideal: sanitary slaughter on an enormous scale, conducted by skilled and contented workers. In this way, the paired images implicitly refute persistent perceptions of slaughterhouse spaces, such as those circulated in Sinclair's novel *The Jungle*, as dark, filthy, and disease-ridden.

Subsequent images in the album move deeper into the packinghouse. The next photograph takes the viewer into an area that appeared in the background of the first pictures—the next station on the moving overhead rail (fig. 11). A row of beef carcasses hangs at left, stripped of skin. The photo depicts evisceration, in the deep background, and then the splitting of the dressed carcasses, done by workers near the center of the image and closer to the camera. After removal, the viscera are deposited onto the "moving viscera table" identified in the caption. The table serves a dual function and takes on a dual significance in the album. First, it represents a branching point in carcass processing, the moment when the internal organs are separated from the flesh and begin their own journey apart from, but parallel to, the carcass from which they were removed. The animal's flesh, cleansed of the polluting viscera, becomes meat; and the viscera, separated from the meat, become—or move closer to becoming—by-products. Second, the table holds the viscera so federal meat inspectors can examine them for irregularities indicative of disease. A diseased animal would be pulled from the disassembly line, condemned, and destroyed.[45] With this image, then, Morrell & Co. demonstrates not only the modern mechanization of its facilities but also its careful compliance with federal inspection laws. The camera's viewpoint supports this implicit messaging: the photographer shows the table not by itself but in its spatial and functional relationship to the hanging carcasses. The photograph demonstrates how the company's modern, efficient production line accommodates the inspectors and the regime of order, cleanliness, and consumer protection that they represent.

The subsequent image, labeled "Sectional view of coolers—new beef house," takes viewers into the carcass-chilling room, where sides of beef hang to chill before they are shipped out or further broken down into smaller cuts (fig. 12). At the far end of the space, two workers move a carcass into the space, bringing it in from the kill line depicted in the prior images. This detail demonstrates how the cooling room is connected to the other spaces that appear in earlier photographs, showing that the cooling room is one more stop—albeit a lengthier one—in the carcass's progress from living animal to edible product. A pleasing, nearly symmetrical image, the photo offers a magisterial view down the cooling room. The repetition and sheer volume in perspectival packinghouse views like this one convey order and abundance. Weems notes that such images "serve to bring industrial scale under the beholder's mastery," marking "the point at which the animal body is fully and irrevocably rendered from animate being to industrial product."[46] The following image in the Morrell album reinforces this sense: it shows the hide cellar, containing stacks of hides built from the skins of thousands of animals. The piles testify to the capacity of this large packinghouse, and by extension the meat industry overall, to dispatch and process animals in staggering numbers, day in and day out.[47]

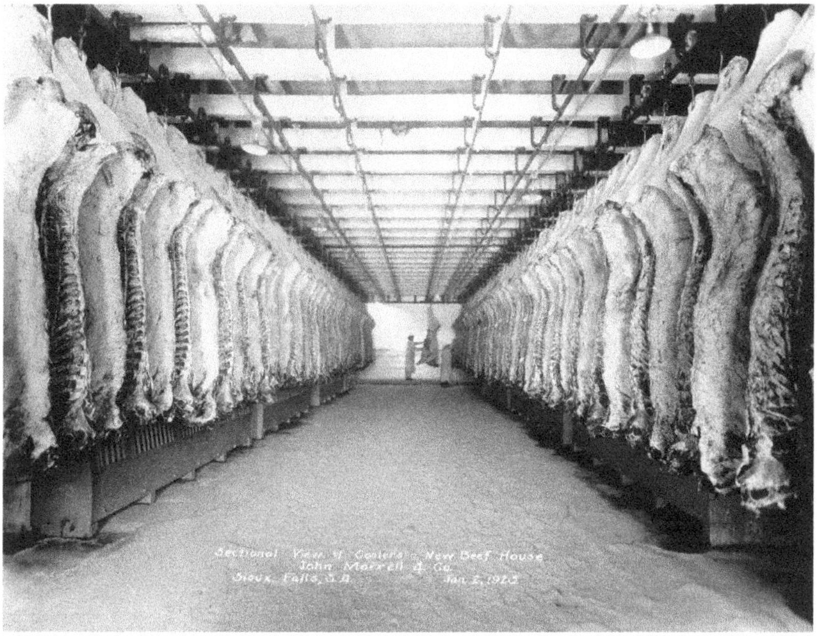

Figure 12. "Sectional View of Coolers, New Beef House, John Morrell & Co., Sioux Falls, SD, Jan. 2, 1925." Morrell Meat Packing Company Collection, collection MSC 0100, Special Collections and Archives, University of Iowa Libraries.

Subsequently, the Morrell album also depicts spaces for pork slaughter and processing.[48] The final few images in the album carry photographs of Morrell's various products—ham, bacon, lard, and luncheon meats as well as subsidiary products like eggs and butter—cut out and pasted onto the neutral white background of the paper. Ending in this way, the album provides what appears to be a snapshot of the corporation, showing the newest and most progressive aspects of Morrell's midwestern operations and giving an overview of its characteristic products, at a particular moment in time. In the album, the raw material animal, the human worker, and the machine come together in the ideal industrial space, constituting an industrial whole dedicated to generating the branded commodities that appear at the end. The photo album represents, and promulgates, a cohesive and coherent statement of Morrell's corporate identity. The album functions, in a way, as a self-fulfilling prophecy: in recounting its own history, Morrell offers a controlled narrative of orderly power emanating from an entity important enough to have its history recounted.[49]

A selection of images from this album did circulate in printed format. In 1924, Morrell began publication of *Morrell Magazine*, which circulated

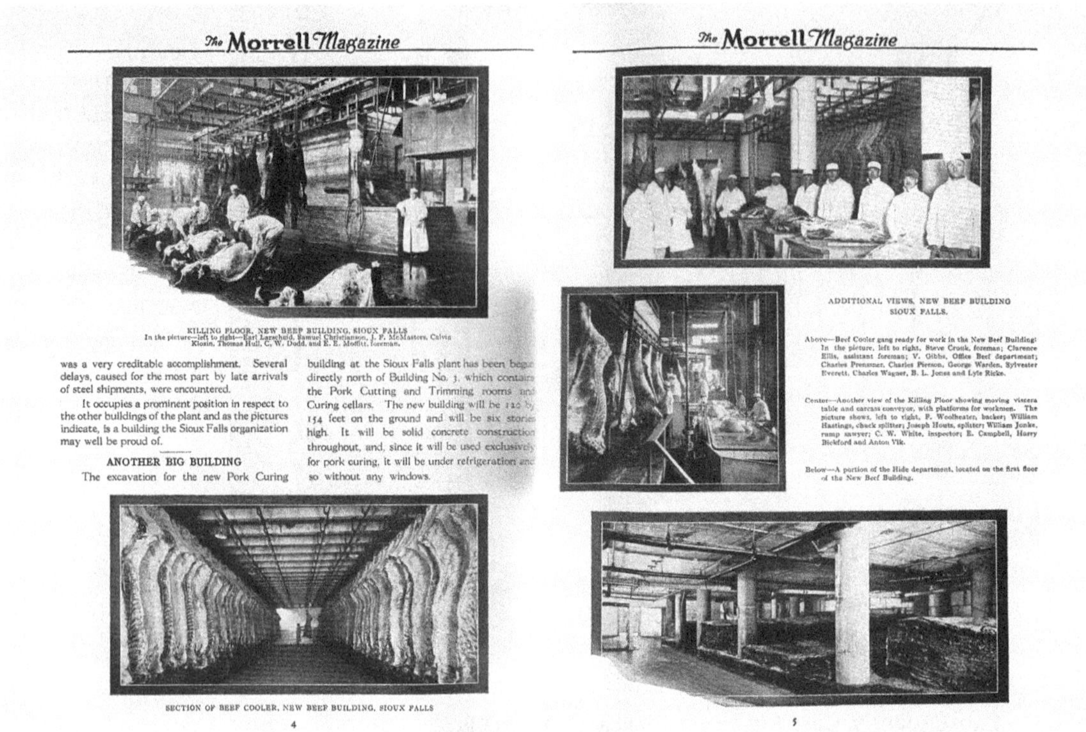
The Morrell Magazine

KILLING FLOOR, NEW BEEF BUILDING, SIOUX FALLS
In the picture—left to right—Earl Larschuid, Samuel Christianson, J. F. McMasters, Calvin Kloxin, Thomas Hull, C. W. Dodd, and E. E. Moffitt, foreman.

was a very creditable accomplishment. Several delays, caused for the most part by late arrivals of steel shipments, were encountered.

It occupies a prominent position in respect to the other buildings of the plant and as the pictures indicate, is a building the Sioux Falls organization may well be proud of.

ANOTHER BIG BUILDING

The excavation for the new Pork Curing building at the Sioux Falls plant has been beg directly north of Building No. 3, which contain the Pork Cutting and Trimming rooms an Curing cellars. The new building will be 120 by 154 feet on the ground and will be six stories high. It will be solid concrete construction throughout, and, since it will be used exclusively for pork curing, it will be under refrigeration and so without any windows.

SECTION OF BEEF COOLER, NEW BEEF BUILDING, SIOUX FALLS

4

The Morrell Magazine

ADDITIONAL VIEWS, NEW BEEF BUILDING
SIOUX FALLS.

Above—Beef Cooler gang ready for work in the New Beef Building: In the picture, left to right, Steve Crouk, foreman; Clarence Ellis, assistant foreman; V. Gibbs, Office Beef department; Charles Prenssner, Charles Pierson, George Warden, Sylvester Everett, Charles Wagner, B. L. Jones and Lyle Ricke.

Center—Another view of the Killing Floor showing moving viscera table and carcass conveyor, with platforms for workmen. The picture shows, left to right, F. Woolheater, backer; William Hastings, chuck splitter; Joseph Houts, splitter; William Jonke, rump sawyer; C. W. White, inspector; E. Campbell, Harry Bickford and Anton Vik.

Below—A portion of the Hide department, located on the first floor of the New Beef Building.

5

Figure 13. "New Beef Building," 4–5. University of Iowa Libraries.

to the company's workers. Part of a move in the 1920s toward a welfare-capitalist model of management, the magazine sold a vision of Morrell to its workers grounded in an evangelical Christian model of hard work. As Wilson Warren notes, Morrell's embrace of welfare capitalism came at a point in the company's history when there were deepening divisions between management and workers, as it became increasingly difficult for an average worker at Morrell to rise above a subsistence-level wage.[50] *Morrell Magazine* incorporated profiles of the company's founders, features focused on long-term employees and retirees, and platitudinous essays from company leaders. The publication contributed to Morrell's attempts to engineer a company culture at a point when, as Warren points out, any naturally occurring one had largely atrophied.[51] The feature article "New Beef Building, Sioux Falls Plant" came in the eighth issue of the magazine, published in February 1925.[52] It presents several of the beef building photographs from the album, but no images of the pork facility (fig. 13). The photograph of the killing floor and the perspectival image of the beef cooler both appear, as do images of the dressing line and the hide cellar. Captions with the images include the names of the workers depicted, as the publication attempts to align itself with worker interests.

Workers at some Morrell & Co. facilities in the 1920s complained bitterly of poor working conditions, and the company's refusal to improve facilities led to poor morale.[53] Featuring the new facility in this article, Morrell countered worker dissatisfaction with evidence that improvements were ongoing. The article touts the facility as "fireproof, sanitary, and well lighted."[54] But it also notes that "for the past few years these quarters have been entirely inadequate, and further growth was possible only by the erection of a new building."[55] Profit, not concerns for worker safety or health, ultimately drove the expansion. In this instance of internally circulated imagery, the display of photographs of Morrell's new facility to the company's workers comes off as so much peacockery: management attempting to show its largesse to workers, but in the process showing instead its stacked hand.[56]

Selling Slaughter to Sell Machines: The Allbright-Nell Company Catalog

In addition to meatpacking companies proper, the meat industry encompasses various auxiliary companies that supply specialty products for the meatpacking industry. Suppliers play key roles within the industry, their products often vitally important to the ongoing practice of industrial-scale animal slaughter but their position somewhat distanced from the actual practice. Some such companies have made extensive use of visual imagery in their catalogs and other printed matter. Many of these materials provide franker and more systematic depictions of animal slaughter than those produced and circulated internally by meatpacking companies themselves. In such documents, these auxiliary companies demonstrated their importance to the industry by showing what a vital role their products played. They also assumed an interpretive position, taking an intermediary role between the large meatpacking companies and the smaller operations, such as butcher shops and smaller-scale slaughter operations, that they also served. Their visual materials thus connected various branches of the meat industry, assisting in internal circulation of information.[57]

The Allbright-Nell Company, or ANCO, attempted such a positioning of itself and its products with a hardbound, book-length catalog of meat industry equipment, which it released in 1923.[58] As a manufacturing company, ANCO occupied an auxiliary position within the meat industry. But because it made many of the principal and most emblematic slaughterhouse machines, including the massive hog wheels, it played a central role in elevating the meatpacking industry to an exalted status as an epitomizer of progress and modernity.

The ANCO book functions primarily as a promotional document, but it incorporates a lot of content that is not immediately commercial in nature. Indeed, the company devotes the entire first half of the volume, nearly forty pages, to a sequential account of slaughter processes, illustrated with photographs, for various animals. This section begins with a linear and quite thorough account of the slaughter of hogs, followed by briefer but still revealing accounts of the slaughter of cattle and sheep. According to the book's introduction,

> In 1920 a new department was established for the purpose of making and collecting photographs not only of our own machinery and equipment, but also of general packing house operations. It is largely due to the ANCO Photographic Department that this present catalog is possible. However, we particularly wish to call attention to the immediately following pages, in which the modern methods of slaughtering hogs, cattle, and sheep have been illustrated step by step just as the different operations are performed under actual packing house conditions. This group of pictures has been selected from hundreds of photographs made by the ANCO Photographic Department all over the United States, and is the result of several years' work. We are inserting these pictures in this catalog, and hope they will prove interesting and helpful to our customers.[59]

The company had come to view the collection of photographs as a necessary, even crucial, step in understanding how its products were being used in packinghouses around the country. The catalog claims to deploy the photographs as a service to their customers; but the incorporation and sequencing of the images also permits the company to display and thus to reinforce its own expertise in manufacturing—positioning its products as essential contributors to the successful practice of industrial-scale meatpacking.

The company deployed images that could clearly demonstrate its level of access to and show the distribution of its products throughout meat-packing facilities and all the way along the slaughter and disassembly lines. The photos do not solely advertise ANCO's presence in the packinghouse; indeed, many of the photos show workers using hand tools, not ANCO's large machines. The photographs are direct in depicting the steps required to capture, restrain, slaughter, and bleed an animal. But at the same time, they are not confrontational, and they do not editorialize. They certainly do not poeticize or lament the animals' deaths. Animals appear as the industry regarded them, as raw material. The photos also show the labor of meatpacking, and workers themselves, dispassionately. If the pictures had glorified workers, they might have appeared excessively sympathetic to labor. And if the images had dwelled too long on the bloody nature of the work, they might have disrupted the dispassionate, mechanistic vision of slaughter that ANCO, as the creator of the machines, hoped to promote. The photos in the ANCO catalog offer a carefully, studiedly neutral vision of animal slaughter. In their dryness they deliberately exclude any messaging beyond the rationalism of industrial production.

The photographs in the ANCO catalog come mostly two per page, each accompanied by a brief caption given directly below the image. The sequence begins with hog slaughter. In the first image, hogs, already inside the packinghouse space, enter the confined area where they will be caught by the hind leg and shackled to the wheel hoist (fig. 14). The animals, though live, appear in shadow, a pair of them nearly the same shade of gray as the wheel and the ground below it. While a bit of retouching—a bit of highlighting—has been added on one hog's face to sharpen the features, overall the animals are allowed to fade into and become part of the background. Resembling the machine, they become mechanistic themselves.

The human worker, in contrast with the pigs, wears a coverall that has been heavily retouched to appear brilliantly white and unstained, an alteration repeated throughout the book. Certain other elements of the image have also been retouched, including details of the wheel hoist panel and chain, added to enhance the visibility of this ANCO product. The effect, here and throughout the catalog, is of a cleanly efficient space in which living beings and organic matter are sublimated not only to the machines but also to the capitalist logic that drives the entire packing process.

In the second, lower image on this page, the hogs dangle upside down

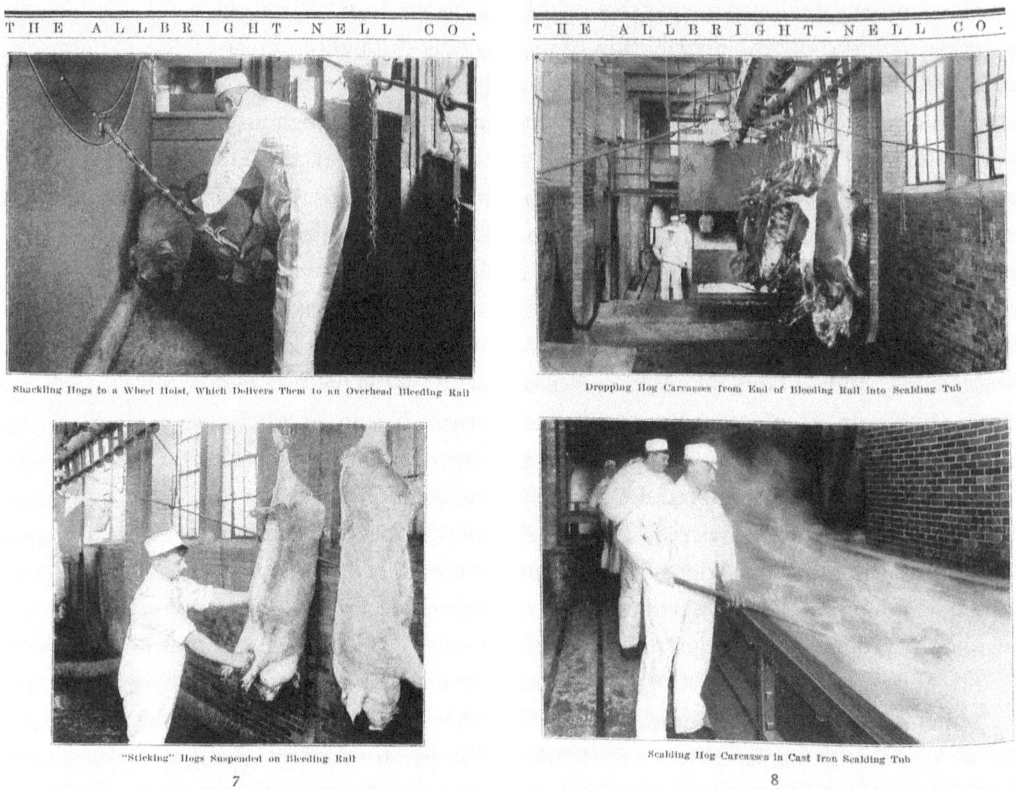

Figures 14 and 15. ANCO, *ANCO Machinery*, 7 and 8. Trade Catalog Collection, Special Collections, Chicago Public Library.

by one leg, passing in front of the sticker, who plunges a knife into each animal's neck to sever an artery. While the point of the knife's entry into the hog's neck is obscured by the animal's leg and thus not visible to the camera, the image nonetheless captures the precise moment at which the animal ceases to live. In an industry for which the animal functions not as a "being" but as the fundamental, and indeed the sole, raw material, bleeding constitutes a point both central and not wholly assimilable to the rationalism on which the industry pivots. Though bleeding is the central act required to render the hog an object, rather than a creature, it also requires that the industry recognize, at least for the moments it takes to drain away the blood, that meat production pivots on removing life from something that had possessed it.

Again, however, the photo presents the moment as dispassionately—as bloodlessly—as possible. The image includes only two hogs, one being stuck and one awaiting its turn. For clarity of the photograph, the other hooks that stretch away to the worker's left, toward the scalding bath, have been left empty. If they were not empty, the viewer would see a line of hogs, all bleeding out, the violent organicism of the process belying the desired effect of cool matter-of-factness. As it is, blood is present only here and there: a black line on the neck of the hog as the worker sticks it, a sticky shine on the floor in the background. Certainly the retouchers have removed from the man's smock and shirt any trace of blood. Normally his clothing would be spattered with it, even up to his hat; but in this image, his uniform virtually gleams white.[60]

The next two photographs depict hogs being first dropped into, and then pushed along through, the scalding bath that will loosen their bristles (fig. 15). The first of these shows, in the foreground, a row of hogs hanging, still bleeding. The ground below them, in the foreground, appears black with blood, and the brick walls in foreground are spattered. The image and the overall sequence benefit from photographic monochrome: blood and shadow appear nearly one and the same. In the lower photo on the page, workers stand with poles and push along the bodies of the hogs through the scalding bath. There is some retouching here on the men's uniforms, but the retoucher has also sharpened and clarified details on the water tank itself, appropriate since the tank was an ANCO product, advertised in the second half of the book.

The sequence continues, images often showing men and machines working in concert with one another to process animal bodies. In one image, discussed at the start of this book, an anonymous worker with his back toward the viewer watches a row of pigs rising out of the scalding tub toward an ANCO dehairing machine (see fig. 1).[61] The ANCO catalog treats most workers as it does this man, who appears sublimated or even awed by the machinery of slaughter.

US government inspectors, however, constitute notable exceptions to this rule. The first inspector appears, a few pages into the catalog, to inspect hog carcasses, ensuring the animals had been healthy enough to become healthy meat. Standing next to a packinghouse worker, the inspector waits for the worker to sever a hog's head, then examines glands inside the head. Because the inspector's presence connotes cleanliness, quality, and compliance with federal food safety guidelines, the photo sequence not only

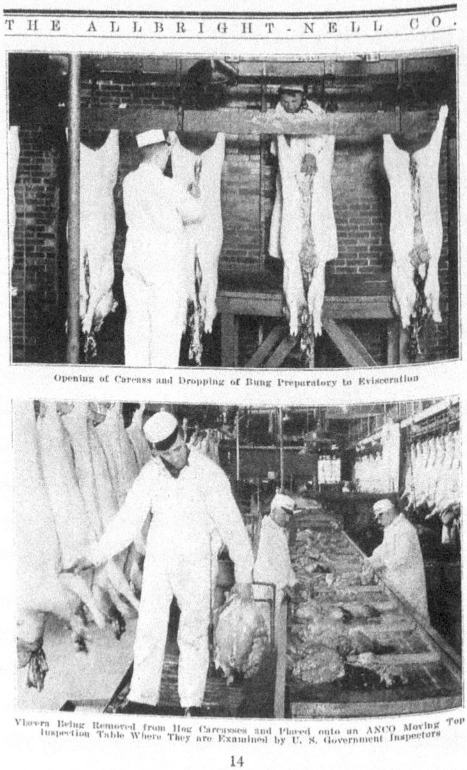

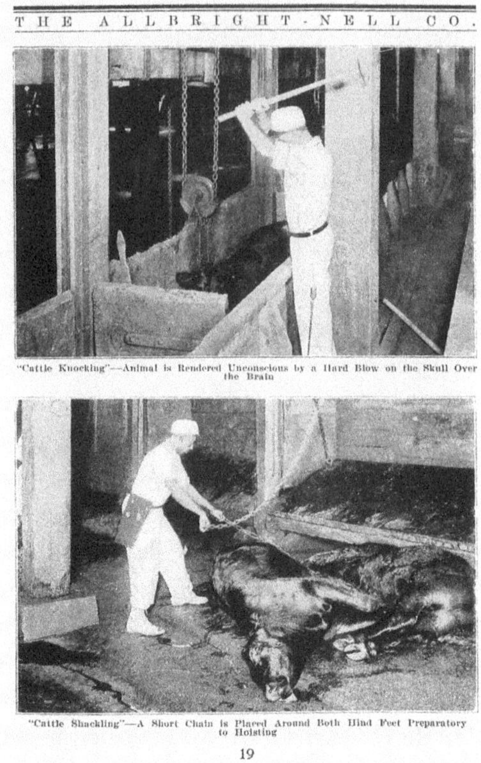

Figures 16 and 17. ANCO, *ANCO Machinery*, 14 and 19. Trade Catalog Collection, Special Collections, Chicago Public Library.

acknowledges but emphasizes his role. Some photos demonstrate both the packinghouse's compliance with food safety law and ANCO's role in ensuring it: one image shows "Viscera being removed from hog carcasses and placed onto an ANCO moving top inspection table where they are examined by U.S. government inspectors" (fig. 16). Here, again, the black-and-white tonalities diminish the photograph's potential to provoke disgust. The picture becomes, by virtue of the presence of the inspectors and with the assistance of monochrome, an image connoting purity and control, rather than filth and pollution.

After following the process of pork through the stages of dressing and splitting and, finally, into the cooler to chill as hanging half-carcasses, the photo sequence in the ANCO catalog shifts to examine how cattle are processed into beef. On the first page, the photographs show cattle being first

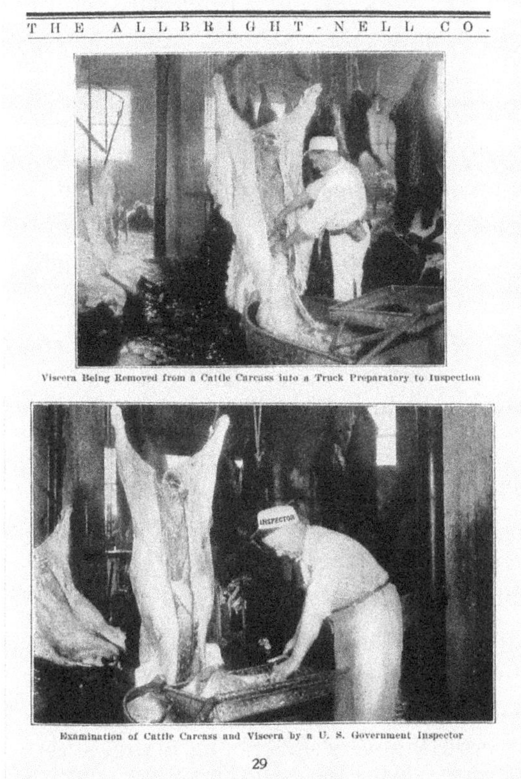

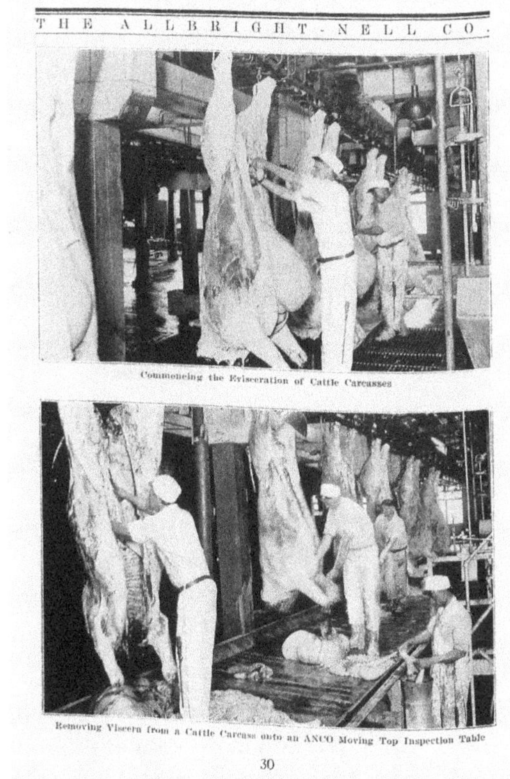

Figures 18 and 19. ANCO, *ANCO Machinery*, 29 and 20. Trade Catalog Collection, Special Collections, Chicago Public Library.

stunned with a blow to the head, then shackled by a hind leg for elevation in a head-down position (fig. 17). Next the animals are bled and denuded of skin; and then the heads are severed, a delicate process since the muscles and glands of the head must be kept intact for government inspection. Again, inspectors appear repeatedly; and even when they are not present in an image, their proximity is emphasized. One pair of images, appearing together on a page (fig. 18), first depicts "Viscera being removed from a cattle carcass into a truck preparatory to inspection." A white-clothed, white-hatted worker, tool pouch at his belt, removes a mass of guts—shining with wet, but indistinct and undifferentiated in monochrome—from a hanging carcass into a metal barrow. Below this image, the second picture on the page shows "Examination of cattle carcass and viscera by a U.S. government inspector." Here a different man takes the place of the first to examine the carcass and the viscera. His uniform appears superficially similar, but he carries no tool pouch, and his clothes, particularly his pants, have been retouched to remove any dirt or blood. Most obviously, the word *inspector* has been carefully hand-lettered on his hat. The photograph opts for obvious retouching over a more direct, unaltered realism, gesturing toward the importance of inspection as a representation of purity, safety, and trustworthiness in the meat industry. Later images maintain this emphasis, and show—as in the pork sequence—how ANCO literally supports the job of inspection, depicting inspectors examining viscera on what the captions assert are ANCO-branded conveyor tables.[62]

In media created for public consumption, the presence of non-white workers in packinghouses was sometimes suppressed.[63] But in the industry-targeted ANCO catalog, non-white workers have been photographed matter-of-factly, working alongside white workers, not consigned to the background or asked to stay out of the camera's range. In a pair of images in the beef sequence, Black workers remove and handle viscera (fig. 19), and later in the sequence, other non-white workers split and wash carcasses. Created for circulation within the meat industry, the book apparently saw no need to prevaricate about the constitution of its workforce. ANCO was engaged in selling its equipment by demonstrating its expertise, showing how indispensable its products were to the overall functioning of large-scale packing establishments. It was not, unlike popular advertising imagery, engaged in selling an image of the industry overall to a buying population assumed to be largely white, and thus it did not, at least in this way, have to pander to popular prejudices. As the workforce existed, so it seems to appear in the pages of the ANCO catalog.

By contrast, the catalog depicts not a single female worker. Women in the packinghouse worked mostly on parts of the line that dealt either with chilled meat or with sausage, trimmings, and by-product. Since those aspects of processing do not appear in this catalog, women by extension do not appear. This again contradicts trends in media generated by the meat industry for public, rather than internal, consumption: women workers' presence was often emphasized disproportionately in advertisements, public displays, and public tours of packing plants, with images of smiling young women workers—always white women—deployed to lend a homey, nonthreatening appeal to packinghouse spaces and to signature products like bacon and hot dogs. The ANCO catalog, however, had no need to

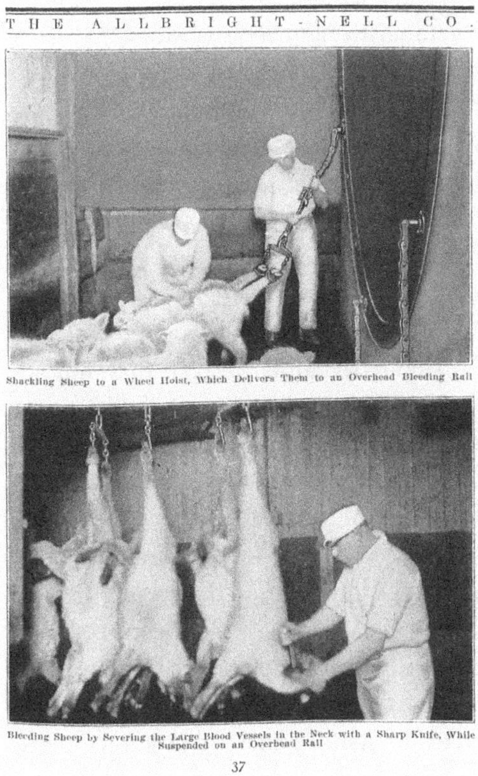
THE ALLBRIGHT-NELL CO.

Shackling Sheep to a Wheel Hoist, Which Delivers Them to an Overhead Bleeding Rail

Bleeding Sheep by Severing the Large Blood Vessels in the Neck with a Sharp Knife, While Suspended on an Overhead Rail

37

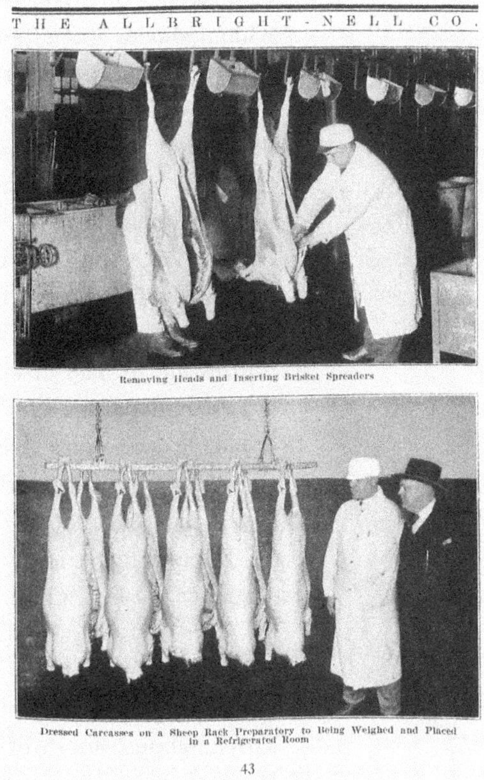
THE ALLBRIGHT-NELL CO.

Removing Heads and Inserting Brisket Spreaders

Dressed Carcasses on a Sheep Rack Preparatory to Being Weighed and Placed in a Refrigerated Room

43

Figures 20 and 21. ANCO, *ANCO Machinery*, 37 and 43. Trade Catalog Collection, Special Collections, Chicago Public Library.

entice its viewers to consume meat, only to purchase machines. Packing thus comes across in its pages as a mechanistically masculine profession.

Still engaged in demonstrating its relevance and necessity to the meat industry as a whole, the ANCO catalog shifts from the slaughter of cattle into an overview of sheep slaughter. Having covered the slaughter of hogs and cattle so thoroughly, the catalog moves more quickly here, lingering only where sheep slaughter differs markedly from that of other animals. Sheep are shackled in pairs, for instance, because they do not struggle as much as other animals; so one image shows how this is accomplished with a wheel hoist, while the photograph below it on the same page depicts the sticker severing the artery of an animal (fig. 20). Both images have been heavily retouched. In the photo of the hoist, the chains have been sketched in heavy-handedly, and the face of one of the sheep has essentially been painted back on. In the other, the hat and clothing of the worker have been whitened with little attention to naturalistic light and shadow.

In the latter image, a second alteration has also been made: the knife in the worker's hand has been outlined slightly, to distinguish it from the dark patch on the sheep's neck where it is being plunged into the artery to bleed the animal. This image, more than most images depicting slaughter processes, clearly shows the specific act that kills the animal, and the specific worker responsible for that act. Indeed, the images that have led up to this one—which have emphasized the docility and passivity of sheep, their apparent submission to slaughter—seem designed to prepare the viewer for this image. The sheep's submission to slaughter suggests that they accept slaughter; that they are essentially designed for slaughter.[64] The act of bleeding, shown so clearly here, seems not only natural but inevitable, the situation for which their docile, willing bodies were always (and will always be) destined.

The final image in the sheep sequence, and thus in the overall slaughter sequence, depicts a human figure unlike any other in the catalog (fig. 21, lower image). As with the pork and beef sequences, the sheep sequence ends with a display of fully dressed carcasses on a rack. Two men stand next to the rack. One, in a white butcher's overcoat and white hat, appears to be a packinghouse floor worker. Next to him stands a man in a dark three-piece business suit and a black top hat, his dress marking him as an outsider to the kill floor. A member of the sales force, an upper-level manager, or even an owner of the facility, the man eyes the carcasses with an appraising air. The caption of the image does not mention him, reading only "Dressed carcasses on a sheep rack preparatory to being weighed and placed in a refrigerated room." But his presence serves as a marker, denoting the end of the slaughter sequence and the start—on the very next page—of the equipment sales portion of the catalog. The man stands in for every buyer, every person in every packinghouse responsible for determining what equipment is to be purchased and how to purchase it—the catalog's true target audience. Here at the end of the slaughter sequence, the man not only controls the capital necessary to purchase equipment but represents capital itself, enabling the flow of money that both sustains and is sustained by this flow of blood.

Only subsequent to these extensive slaughter sequences does the ANCO catalog finally devote space to the equipment it carries, from hog wheels and tanks for scalding baths to machines for splitting hooves and wash-

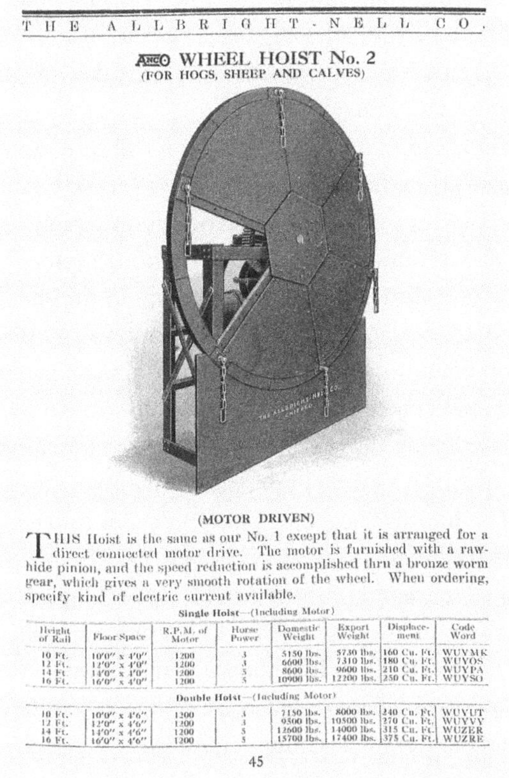
THE ALLBRIGHT-NELL CO.

ANCO WHEEL HOIST No. 2
(FOR HOGS, SHEEP AND CALVES)

(MOTOR DRIVEN)

THIS Hoist is the same as our No. 1 except that it is arranged for a direct connected motor drive. The motor is furnished with a raw-hide pinion, and the speed reduction is accomplished thru a bronze worm gear, which gives a very smooth rotation of the wheel. When ordering, specify kind of electric current available.

Single Hoist—(Including Motor)

Height of Rail	Floor Space	R.P.M. of Motor	Horse Power	Domestic Weight	Export Weight	Displacement	Code Word
10 Ft.	10'0" x 4'0"	1200	3	5150 lbs.	5730 lbs.	160 Cu. Ft.	WUYMK
12 Ft.	12'0" x 4'0"	1200	3	6600 lbs.	7310 lbs.	180 Cu. Ft.	WUYOS
14 Ft.	14'0" x 4'0"	1200	5	8600 lbs.	9600 lbs.	210 Cu. Ft.	WUYPA
16 Ft.	16'0" x 4'0"	1200	5	10900 lbs.	12200 lbs.	250 Cu. Ft.	WUYSO

Double Hoist—(Including Motor)

Height of Rail	Floor Space	R.P.M. of Motor	Horse Power	Domestic Weight	Export Weight	Displacement	Code Word
10 Ft.	10'0" x 4'6"	1200	3	7150 lbs.	8000 lbs.	240 Cu. Ft.	WUYUT
12 Ft.	12'0" x 4'6"	1200	3	9500 lbs.	10500 lbs.	270 Cu. Ft.	WUYVY
14 Ft.	14'0" x 4'6"	1200	5	12600 lbs.	14000 lbs.	315 Cu. Ft.	WUZER
16 Ft.	16'0" x 4'6"	1200	5	15700 lbs.	17400 lbs.	375 Cu. Ft.	WUZRE

45

Figure 22.
ANCO, *ANCO Machinery*, 45. Trade Catalog Collection, Special Collections, Chicago Public Library.

ing sausage casings. Here, the primary image-making mode of the catalog shifts. While some of the images of these machines are photographic, the large majority are prints (fig. 22). The shift in format enables the catalog to present each machine as an ideal, rather than a real, object. Surfaces in these images appear just shiny enough to connote newness but not so shiny that fine detail cannot be seen by the prospective buyer. Where a cutaway is necessary to demonstrate how a machine functions, the print images allow for that. In the few instances where a photograph is used, it demonstrates something specific, such as how a particular machine can function satisfactorily in a very small floor space.[65] But the majority of the images, isolating their machines on the page and showing them at the ideal angle, depict all the devices at roughly the same size. Scale becomes impossible to determine. Nearly every machine, no matter how large, appears in the catalog three inches high. Extracted from context, every piece of equipment becomes an object and a commodity. Purchasing via the print image, one buys not a particular machine but rather its platonic ideal, the device and its archetypal functionality fused into one.

Both halves taken together, the ANCO catalog deploys photography strategically and with sophistication. In the first half, images show scenes from specific slaughterhouse interiors, depicting real workers operating real machines for slaughter and disassembly of real animals. Although retouching interferes with the sense of indexicality in some places—in the sheep sequence, for instance, where some images border on the cartoonish—other images in the sequence, and even other areas in the same image, retain the base characteristics and appearance of photographs, their fundamental ontology not really threatened by these scattered areas of retouching. The reader derives a sense that ANCO fully understands the processes of industrial-scale slaughter. The photos legitimate ANCO's claim to have a proper, rightful place within packinghouses, showing how the company's equipment already exists in real spaces and contributes to real packinghouse operations. In the second half, photography retreats, appearing only when necessary; otherwise, it gives way to prints that show—and encourage the buyer to imagine—an ideal space, populated by ideal machines. In this way, the catalog encourages even slaughterhouse equipment buyers to engage in building packinghouses in the air; dream packinghouses furnished with the latest technologies, unsullied—for the moment—by animal effluvia or by the demands, vagaries, and insufficiencies of flesh-and-blood human laborers.

Throughout the ANCO catalog, images help to abstract both scale and a sense of impact. Everything carries roughly the same weight: a pull stroke from a knife, the blow of a sledgehammer, the rotation of a hog wheel, the hog wheel itself. The catalog is astonishingly thorough in its depiction of the processes of slaughter, but such directness does not equate to condemnation. These images give the lie to the assumption that any photographic image of industrial animal slaughter is inherently a critique. Photographically illustrated materials may just as easily uphold hegemonic relationships within the packinghouse. Absolute visibility begets absolute banality: if everything is of equal importance in the slaughterhouse, then perhaps nothing is really all that important, and certainly not worth getting upset about. By rendering slaughter processes perfectly visible and reducing slaughter machines to consumable objects, these materials attempt to render slaughter, even at scale, perfectly rational as well—a technocratic enterprise capable of feeding the world through the power of logic.

Perspective and Power

Pictures of living meat animals assist in reinforcing their meatly nature; and pictures of slaughter itself represent the absorption of the animal body into the flow of the production line. Subsequent to slaughter, the animal body is seen to have achieved its true, highest purpose. The ideal condition for the meat animal is the condition of carcasshood. In showing animal bodies as carcasses, photographs reinforce and extend this perception: achieving carcasshood, the meat animal transcends its base, fleshly existence, moving toward cleanliness, rationality, and even a divine geometry.

Photographers have imaged all stages in the dizzyingly complex disassembly process, but certain machines and certain views tend to appear far more often than others. Among these is a perspectival view down a massive refrigeration space, or cooler—a scene of carcasses, dressed and halved but not portioned, hanging to chill in tidy, evenly spaced rows that recede toward a vanishing point. The coolers themselves are central to packinghouse operations, their vastness a necessity to accommodate the vast number of animals that pass through each day, carcasses entering the cooler at one end still warm, passing on rollers along the length of the room, and exiting at the other end a day or so later, thoroughly chilled. Perspectival photographs of the coolers elevate them into something greater still, spaces that symbolize the triumph of cold capitalist logic over heat, blood, sweat, and offal—all the organic mess that industrial-scale slaughter is designed not only to strip away but to conceal from consumers.

Art history has long recognized the symbolic potential embedded in the linear perspectival view, a view common in Western art since it was codified during the early Renaissance. Taking advantage of the sense of command built into the perspectival view, as well as the senses of transparency, truthfulness, and even inevitability conveyed by the photographic medium, perspectival views of packinghouse coolers naturalize an illusion of human mastery over the animal. Indeed, such images do not just depict but help to generate that chilly logic.

The development of the linear perspectival system marked one of the crucial advancements in the visual culture of the Renaissance period.

Refined in the early Renaissance and codified by Leon Battista Alberti, linear perspective constituted a major step in the turn toward naturalistic and humanistic representation of the world, becoming a hallmark of the Western visual tradition. Perspectival images appear to organize themselves before the viewer specifically for her benefit, giving her the impression of having the one perfect stance that allows her to achieve the "right" view of the scene. Linear perspective has become so fundamental to the Western tradition that it is difficult to see it as a construction. Viewing a scene in perspective, the viewer unconsciously assumes "a conquering gaze from nowhere."[66]

But artists and critics have long recognized the artificialities and manipulations inherent in the perspective system. As Erwin Panofsky notes in *Perspective as Symbolic Form*, the linear perspectival system

> makes two . . . assumptions: first, that we see with a single and immobile eye, and second, that the planar cross section of the visual pyramid can pass for an adequate representation of our optical image. In fact these two premises are rather bold abstractions from reality, if by "reality" we mean the actual subjective optical impression. For the structure of . . . a purely mathematical space . . . is quite unlike the structure of psychophysiological space.[67]

Linear perspective, Panofsky goes on to say, marks "a triumph of the . . . human struggle for control [and] a consolidation and systematization of the external world."[68]

This privileging of the viewing subject cannot help but have ideological effects. By "settl[ing] [the center] in the eye," film theorist Jean-Louis Baudry notes, the linear perspective system anoints the viewer as "the active center and origin of meaning."[69] Further, the rationalistic basis of the perspectival system, as well as the machines like film and movie cameras that have helped to propagate it, "serve to conceal not only their use in ideological products, but also the ideological effects which they may provoke themselves."[70] The content of the image, and thus of the message being conveyed, is highly variable, but its effectiveness depends on the viewer identifying with the camera, taking the camera's physical position as her own position and the ideological point of view that it represents as her own ideology, at least for the duration of the viewing experience.

Photographs of packinghouse cooler interiors appear even more often than images of the hog wheels that so fascinated viewers and packinghouse visitors. Where the hog wheel was an impressive juggernaut, metonymically representing packinghouse operations as a whole, the cooler was—and remains—the space where flesh becomes meat. Writing about a modern-day packinghouse, Timothy Pachirat describes the cooler as a "tundra separating the humid jungle of the kill floor from the controlled manufacturing environment of [meat] fabrication."[71] In a cooler, Pachirat notes, space is "divided from above [by] long, horizontal metal rails [that] hang suspended from the ceiling and stretch in parallel lines."[72] To make images of the cooler, a photographer would stand at the end of one of the aisles created by these parallel overhead rails, aiming the lens down the length of the cooler. In the resulting photographs, the repeated forms of carcasses hanging from the rails create a pleasing staccato rhythm as they recede into the distance. For photographers making images in packinghouses, the idea to create a perspectival image of the cooler, or of the staging areas just outside the cooler, suggested itself repeatedly as an ideal means of conveying a sense of power, order, stability, and rightness: a natural choice, growing out of the Western pictorial tradition. Such perspectival views served to legitimate the business of slaughter on an industrial scale and, indeed, to connect industrial-scale slaughter to larger hegemonic, world-making practices.

In the first decade of the twentieth century, meatpacking giant Libby, McNeill, and Libby published a souvenir atlas.[73] An example of advertising ephemera of the sort often published by companies at the time and for a long time thereafter, the atlas was simultaneously functional and promotional.[74] It incorporated dozens of full-color maps of continents, nations, and US states and territories, before an illustrated section focused on its own, massive Chicago facilities. "Libby's Food Products Circle the Globe," asserted the cover, atop an image of a globe supported—or, rather, chained, in an unintentionally ominous fashion—between two pillars bearing a large letter *L*.

Inside the cover, under the heading "Good Foods Make Strong Nations," Libby asserts the geopolitical relevance of its own products. Heady with progressivist colonialism, the text asserts, "Of all the factors making for the development of the world, the civilizing of races, and the spread of knowledge, none has been as potent as that of good food. And when wholesome food came to be prepared in practically imperishable form, nature was conquered."[75] Libby's foods both "feed armies, and—most important, carry the possibilities of a fuller life to the peoples of the world." Nurture, in the form of nutrition, provides the key to civilization and geopolitical dominance: "The millions in Asia are halting, because they are starved. The rapidly increasing dominance of the American is due to the fact that we are the best fed people on Earth. . . . Where the American goes, his Libby's foods go with him."[76] The atlas itself offers a fascinating snapshot of a globe under the full influence of colonialism. Among the world maps, choices with regard to what the atlas highlights—an entire page, for instance, dedicated to the Philippine island of Luzon—reflect Americans' preoccupations at the time with regard to their nation's position on the world stage.[77]

Subsequent to these maps and a number of pages of infographics, the section offering "a short trip through Libby's kitchens: where foods are cooked daily, sufficient to feed 500,000 men" incorporates a series of six captioned photographs alongside brief explanatory texts, to give a sense of the scale of the company's endeavors. The section, three pages long and incorporating just a few images, is by no means a comprehensive view of Libby's operations. But in choosing its images, Libby reinforces its nascent brand: although it was still, in the first years of the twentieth century, a full-scale slaughtering and packing company, the company's emphasis had already shifted toward canning and preservation.[78] Of the six photographs in this final section of the atlas, three directly depict canning, while a fourth focuses on huge barrels used to cook product in preparation for canning, and a fifth shows the creation of one of the company's key products, beef extract. Only the sixth depicts fresh meat; but, tellingly, it does so in the form of a perspectival view of one of the company's beef coolers (fig. 23). In this "mammoth cooler," the caption declares, "long lines of dressed beef await their turn for the cutting machines and cooking rooms."[79] The caption implies that the cooler is one among many the company possesses, each similarly filled; the viewer is given to understand that this scene only hints at the company's exponentially greater capacity and scale. In the context of the atlas, and alongside photographs of Libby's canning operations, the perspectival view of the cooler stands in for all the raw material necessary to sustain Libby's global distribution of canned meats.

Perspectival photographs of meat coolers could be marshaled when a company, or even the meat industry itself, faced heated critique or otherwise felt itself to be under threat. In the face of accusations of chaos, perspectival photos of coolers function as assertions of order and rectitude. John O'Brien's 1907 *Through the Chicago Stock Yards* includes roughly three

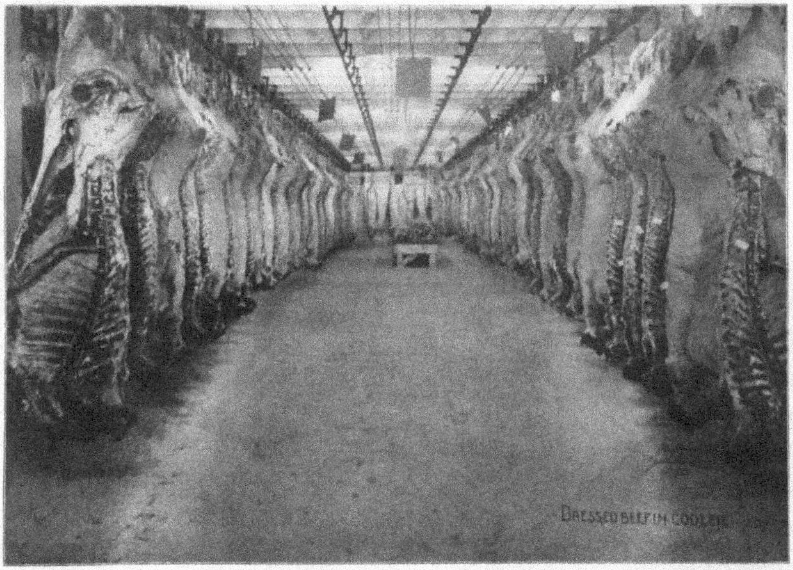

ONE OF THE MAMMOTH COOLERS WHERE LONG LINES OF DRESSED BEEF AWAIT THEIR TURN FOR THE CUTTING MACHINES AND COOKING ROOMS.

Figure 23. "One of the mammoth coolers [. . .]," in *Libby's Atlas of the World*, 38. Advertising Ephemera Collection, John W. Hartman Center for Sales, Advertising & Marketing History, David M. Rubenstein Rare Book and Manuscript Library, Duke University.

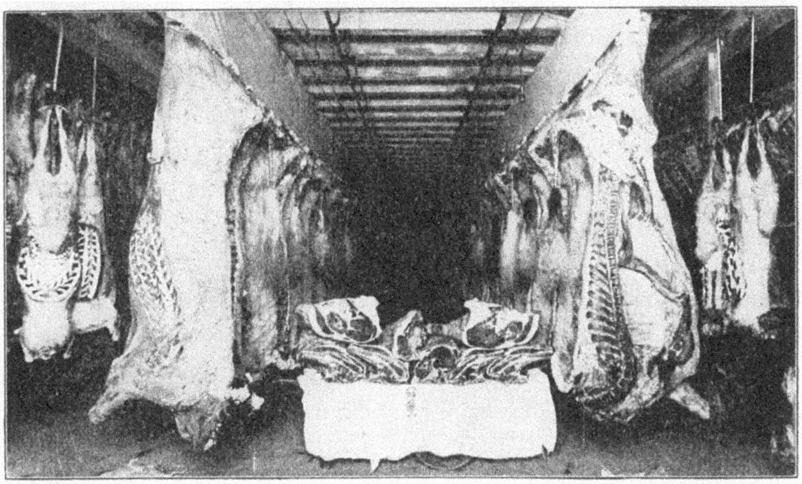

Christmas display in beef cooler.

Figure 24. "Christmas display in beef cooler," in O'Brien, *Chicago Stock Yards*, 37.

dozen photographs, of which no less than five offer perspectival views of either the coolers themselves or the staging area immediately outside: two images of the beef cooler; one inside and one just outside the pork cooler; and one of the sheep cooler.[80] Responding in part to Sinclair's critiques in *The Jungle* and the massive public outcry that Sinclair's book had occasioned just the year before, O'Brien's guide tries to bring meat production back down to something like intimate, human scale. One perspectival image shows a "Christmas display" in the beef cooler (fig. 24), reflecting a tradition, passed down from craft butchery, of cutting meat in decorative fashion for display in butchers' shop windows at Christmastime.[81] The image implies that packinghouse-scale slaughter preserves and grows out of the traditions of craft-scale butchery—an implication belied, in the image, by the very size of the cooler itself, extending away behind the small display in the foreground. Similarly, an image depicting hog carcasses being readied for chilling carries the caption "Finishing touches—ready for the pork cooler," implying that the human "touch" is applied to every carcass, that each one is treated with the utmost care and concern before it is deemed fully ready for chilling.[82]

Ultimately, however, the book cannot help reveling in just how large the meatpacking operations at Chicago really were. Captions below several of the perspectival images make clear that each image shows just "a section" of a larger room—that the infinitude implied by the perspectival recession before the lens is multiplied beyond the borders of the frame. The last of these images depicts not only the hanging carcasses of sheep but also a group of workers.[83] They appear to have ample room to maneuver and otherwise perform their duties within the open space before the camera. Again, the rails nearest the camera have been cleared of product specifically for the occasion of the photograph to create an open space that would not normally exist. Typically the cooler would be much more crowded, the workers performing their jobs not only in extreme cold but also in tight quarters. The perspective view, then, in this instance performs double duty: depicting the packinghouse as an orderly space created to foster and support abundance and showing it as a more humane, slower-paced workplace than it actually was.

Nearly twenty years later, Morrell & Co., too, hired a photographer to celebrate and commemorate the expansion of its beef processing facilities in Sioux Falls, South Dakota. As described earlier in this chapter, the Morrell album included a series of images depicting the spaces through which its product moved during the passage from live animal to carcass to packaged merchandise. One rigidly perspectival view of the cooler, published in *Morrell Magazine* (see fig. 12; fig. 13), stands as a centerpiece, presenting the facility in triumphal fashion as a symbol of the company's growth, prosperity, and modernity. Even partially emptied of its stock—as this cooler, too, has been partially cleared out to make a more pleasing image—the facility appears packed with surplus product. Good food, the photo implies, has been made readily available through the application of scientific principles and economic rationalism. The image becomes a very symbol of abundance and plenty, resulting from the triumph of human ingenuity over base animality. The cooler and its contents here epitomize the natural order.

Within industry publications, the meat industry routinely turned to perspectival views of its coolers to represent itself *to itself* as a locus of order, control, and cleanliness. B. Heller and Company was a supplier of additives and spices to meatpacking companies. It also published pamphlets and periodicals for circulation within the industry, among them its monthly magazine, the *Scientific Meat Industry*. The December 1905 issue carries a feature about Chicago's Schwarzschild & Sulzberger Company, or S&S, illustrated with several photographs.[84] The story celebrates the recent completion of S&S's new Chicago facility, and accompanying illustrations include three perspective views: one each of the pork and beef coolers and a third showing the inside of the company's "traveling sales office," a rail car used to sell the company's stock at locations all along the railway (fig. 25). Indeed, every one of the interior photographs in the article is a perspectival image of one of the company coolers. The photos give the impression that the entirety of the plant looks precisely this way: orderly, logical, clean, and wholly comprehensible from the viewer's point of view, as all linear perspective images are.

The house periodical of the American Meat Institute, the *National Provisioner*, also circulated (and still circulates) to an audience of industry insiders, incorporating stories and advertising designed to appeal to packinghouse managers and executives, butchers, and others with a vested interest in the meat industry.[85] Within its pages, advertisements and features alike are often heavily illustrated. Themes and trends emerge: at mid-century, for instance, advertisements for cooler refrigeration systems routinely incorporated perspectival photographic views of carcasses hanging in rows.

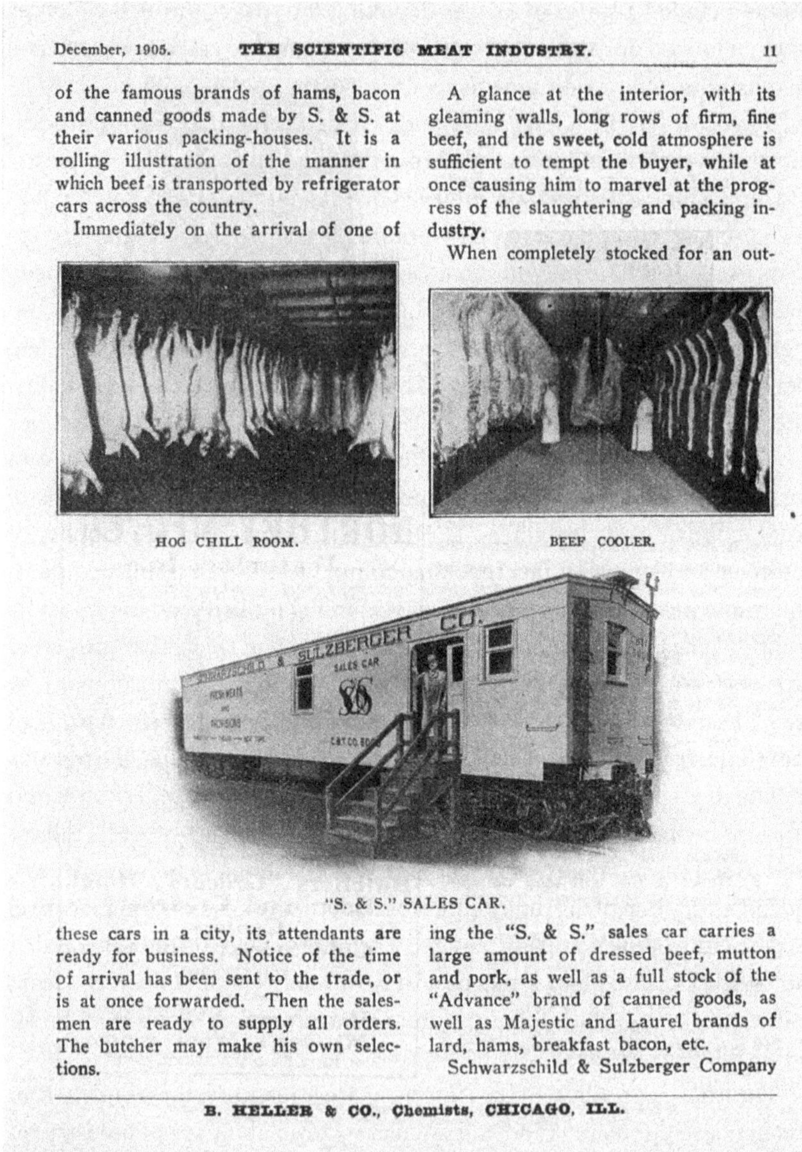

December, 1905. **THE SCIENTIFIC MEAT INDUSTRY.** 11

of the famous brands of hams, bacon and canned goods made by S. & S. at their various packing-houses. It is a rolling illustration of the manner in which beef is transported by refrigerator cars across the country.

Immediately on the arrival of one of these cars in a city, its attendants are ready for business. Notice of the time of arrival has been sent to the trade, or is at once forwarded. Then the salesmen are ready to supply all orders. The butcher may make his own selections.

A glance at the interior, with its gleaming walls, long rows of firm, fine beef, and the sweet, cold atmosphere is sufficient to tempt the buyer, while at once causing him to marvel at the progress of the slaughtering and packing industry.

When completely stocked for an outing the "S. & S." sales car carries a large amount of dressed beef, mutton and pork, as well as a full stock of the "Advance" brand of canned goods, as well as Majestic and Laurel brands of lard, hams, breakfast bacon, etc.

Schwarzschild & Sulzberger Company

HOG CHILL ROOM.

BEEF COOLER.

"S. & S." SALES CAR.

B. HELLER & CO., Chemists, CHICAGO, ILL.

Figure 25. B. Heller, "Schwarzschild and Sulzberger Company," B. Heller and Company Collection, Hanna Holborn Gray Special Collections Research Center, University of Chicago Library.

The tidy and well-managed carcasses testify, in their uniformity and abundance, to the effectiveness of the systems themselves. A 1949 ad, "Hydro-Thermal Cold Coil Refrigeration," features a photograph of halved and quartered carcasses, their rows converging at the end in shelving units that are themselves lined with tidy rows of primal cuts (fig. 26).[86] The camera has been positioned high in the space, so it can effectively capture not only the rows of carcasses but a good view of the ceiling, where the refrigeration system features prominently in the image. The text in the advertisement emphasizes the low rate of "shrinkage" (net weight loss due to desiccation) generated by this cooling system—and, by extension, the increase in profit to the packinghouse. "Think what you'll save . . ." the text urges. In the context of the ad, the perspectival photograph thus testifies to the refrigeration system's ability to preserve not only an abundance of flesh but an abundance of revenue.

In another *National Provisioner* ad, this one for the Marlo Coil Co., the row of hanging carcasses appears within a photograph manipulated to look like a drop of water (fig. 27).[87] Here, too, the text emphasizes the potential profit to the packinghouse. The carcasses are doubly contained in this image: within the refrigeration space and within the confining shape of the printed water droplet. The ad performs acts of layered signification, much like those described by Roland Barthes: the hanging carcasses in the packinghouse, already complexly legible as a collective sign of the company's effectiveness, efficiency, and solvency, become within the water droplet a source of potential profit or loss, to be protected from the threat of desiccating dry air and a concomitant drain on company returns.[88] "Marlo Refrigeration Units," the ad assures the reader, "guard against spoilage and loss of color . . . assure maximum profits by extracting heat—not moisture (weight) from meat."[89]

An ad for Gebhardt Cold Air Circulators similarly deploys a perspectival image of a refrigeration room in one of its advertisements (fig. 28).[90] In the image, rows of carcasses stretch away toward a back wall where a group of executives and salesmen in street clothes and hats confer, oblivious to the camera's presence. The camera angle again includes much of the ceiling, where the outlines of the refrigeration system housings have been retouched, their outlines sharpened and their shiny, gleaming stainless steel surfaces touched up. A second image, below and overlapping the first, shows an oblique perspectival view of another workspace, again with

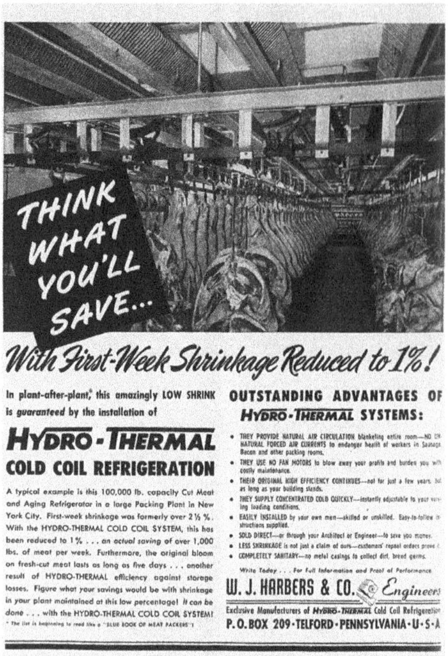

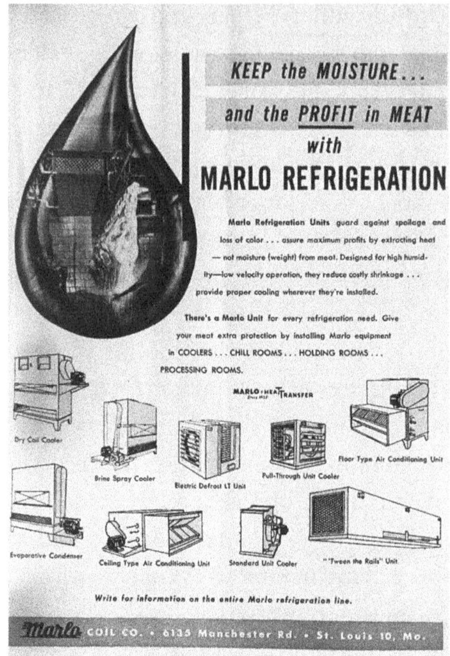

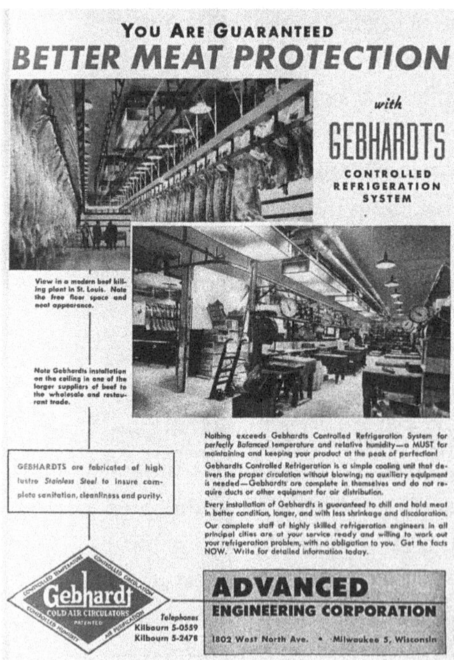

Figure 26. (top left)
"Hydro-Thermal Cold-Coil Refrigeration" advertisement, *National Provisioner* 121, no. 5 (July 30, 1949), 12. Parks Library, Iowa State University.

Figure 27. (top right)
"Marlo Refrigeration" advertisement, *National Provisioner* 121, no. 15 (October 8, 1949), 5. Parks Library, Iowa State University.

Figure 28. (left)
"Gebhardt Cold Air Circulators" advertisement, *National Provisioner* 121, no. 6 (August 6, 1949), 15. Parks Library, Iowa State University.

emphasis on the ceiling and the cooling system. Text above both images emphasizes health, safety, and stability of the product: "You are guaranteed BETTER MEAT PROTECTION." Below the first image, a smaller caption notes that this is a "View in a modern beef killing plant in St. Louis. Note the free floor space and neat appearance." The amount of floor space, of course, has little to do with refrigeration and cannot be attributed to it; and the open appearance of the space is again due to a missing row of carcasses, removed to reveal the refrigeration units in the ceiling, the focus of the advertisement. But the perspectival image seems such an intuitive way to represent a cooler that these machinations are rendered almost invisible. The hyperengineered perspectival image becomes, counterintuitively, the natural vision of industrial-scale meat production.

Writing of the Chicago meat industry in the early twentieth century, Weems notes that "the program that emerged for visualizing The [Union Stock] Yards marked an attempt to simultaneously comprehend, celebrate, and disarm the massive-scale slaughtering system, and to rationalize its diminishment of living bodies (and beings) as inviolable wholes."[91] Perspectival views, Weems continues, played a key role in this visual program, marking "the apex of the stockyards narrative . . . [and] the point at which the animal body is fully and irrevocably rendered from animate being to industrial product."[92]

But for all their cold rationalism, their chilly logic, these perspective photographs also held an appeal for viewers that might be called emotional, or at the very least personal. In viewing a perspectival photograph, the viewer assumes that point of view, and the command it implies, as her own. The perspectival view is an embodied view, placing the viewer within the space and organizing the view around her. Photographs of packinghouse coolers convey abundance, surplus, and ready supply, generated by a cold logic. However, the spaces shown in the photographs also constitute the final stop on the animal's journey from flesh into meat. Seeing images of the bloodier processes of slaughter and dressing, the viewer may cringe because she is like the animal, because she perceives the flesh and the blood as being like her own flesh and blood. Here in the cooler, however, creaturely essence and bodily heat are chilled away. The images invite viewers to identify not with the animal but with the camera, with the machine—and perhaps, by extension, with the larger machinery that is the packinghouse overall. All that has come before, such photos imply, has led inevitably, inexorably, to this. Not only the animal's death but its conversion into a unit, its subsumption into the packinghouse system, has been prefigured from its birth. The passage from creature to commodity is a linear, unidirectional passage. It follows that linear perspective should be called upon to visualize that passage.

Packinghouses distributed these photos in various ways. Some went into company reports, to be seen by stockholders, while others, like the ones in the *Scientific Meat Industry* or the *National Provisioner* magazines, went to an audience of people mostly within the industry broadly defined, like butchers and grocers. Still others appeared in company publications for wider distribution and even in local newspapers. As such, photographs commissioned by meat industry entities did not just mirror an existing reality; they helped the industry to concretize and then to propagate—to render, in Shukin's dual conceptualization of that term—a reality in which order and logic dominated, in which sleek, clean machines triumphed over messy organic substance.[93] Packinghouse photos converted nature's chaos into consumable, edible order, helping to sell industrial-scale slaughter as a model of machine-age progress. Photos like these deny the bedlam described in texts like *The Jungle,* replacing the mental images conjured by such texts with visual images in which disorder has no place. They do not reflect a reality so much as make one, satisfying consumer desire for a product with a spotless provenance, a flawless (even bloodless) lineage. "Of course, that view of infinite vision is an illusion, a god trick," as Donna Haraway notes.[94] But it has been an extraordinarily successful one: perspectival views of packinghouse interiors aid the meatpacking industry in establishing its almost-divine right to treat animal bodies as base matter, to make order and sustenance for the human body politic out of animal chaos.

While the meatpacking industry is in some ways very much like any other corporate interest, the nature of its business sets it apart. Its traffic in flesh and blood sometimes requires it either to address questions that verge on the metaphysical or to labor mightily to deny the existence of such questions altogether. Photos have played a critical role in this self-imaging, assisting companies in justifying and naturalizing both industrial-scale slaughter itself and the power dynamics that made it possible. Packinghouses epitomize technocracies, micro-universes in which animal bodies—and the bodies of human laborers—are sublimated to the overpowering rationale of the disassembly line. Within the packinghouse, metal dominates over flesh, the inorganic over the organic. Packinghouses represent a triumph of rationalism. If the process is messy and imperfect—if the vagaries of individual animal bodies and animal behavior and human bodies and human needs sometimes gum up the works and slow down the line—this represents not a fundamental flaw in the system but rather just a slight deviation, a momentary departure from the ideal logic on which the packinghouse operates. In such a technocracy, the mechanical eye of the camera constitutes the perfect witness: it can be trusted to see through the gore, to look past the distractions of flesh and blood, to the cool rationalism beneath. Indeed, the industry's desire to present itself and its technologies as embodying an ideal logic, as epitomizing progress, has made photography the ideal medium—a medium that can propagandize without seeming to propagandize, that can convert the editorial into the natural. Even where they confront the bloodiest aspects of industrialized meatpacking, through selective framing and other almost-invisible manipulations photographs can impart a sense of necessity, balance, order, beauty, even higher purpose to the repetitions of slaughter at scale.

CHAPTER TWO

Realizing Corporate Visions

Torkel Korling and Arthur Siegel

AMERICAN MEATPACKING COMPANIES had to rely on the photographers in their employ to help them express their visions for a glorious future of industrial-scale meat production. The majority of the time, these photographers' names and any details of their lives would go unrecorded. Some might work for the meatpacking company itself; many large American corporations maintain photographic departments.[1] Others might work for hire, whether as freelancers or as employees of a larger photographic or marketing firm.[2] Authorship of photographs, like that of the text in promotional materials, fell most often to the corporation itself, not to any single individual.

Occasionally, however, meatpacking companies hired photographers for larger projects, ones in which photography itself played a more substantial or even a primary role as carrier of meaning and narrative. In these instances, the photographer's name and participation might be recorded or even publicized, their authorship imparting a sense of authority or quality to the finished product. Indeed, an individual photographer's participation might carry, as it does with the projects profiled in this chapter, an expectation that the photographer would bring to the project their own sensibility, vision, and style. The finished product thus constitutes a collaboration between author and client, enacting—updating, for an industrial-capitalist age—the age-old relationships of mutualism, symbiosis, and negotiation between artist and patron that have characterized so much art production in the Western world since the early modern era.[3]

The projects profiled in this chapter, though still corporate promotions like those examined in chapter 1, appear to have distinct authorial visions that set them apart, to some extent, from the albums discussed in the prior chapter. They call into question the idea of individual vision and highlight the extent to which an individual style, a distinct aesthetic or ethos, may be assimilated to, sublimated by, or deployed on behalf of corporate messaging.

The first photographer profiled here, Torkel Korling, worked for two separate companies over the course of the 1940s: the Rath Packing Company of Waterloo, Iowa, in 1941; and Oscar Mayer & Co., probably in 1944–1945. Korling's project for Rath resulted in a mass-produced and mass-distributed book about the company's facilities in Waterloo; his work for Oscar Mayer, on the other hand, seems never to have been published and exists in a sole album copy. A well-known freelancer who had worked for *Life* magazine, Korling brought something of that magazine's messaging about American identity and prosperity to his work for his corporate clients.

The second photographer, Arthur Siegel, also worked for Oscar Mayer, hired by the company in the 1950s to create a mass-distributed booklet profiling its operations at its headquarters in Madison, Wisconsin. Although he was photographing the same Oscar Mayer facilities that Korling had depicted only about a decade before, Siegel's aesthetic sensibility and the subjects to which he turned his attentions differ widely from Korling's. Even working for the same clients, even in a corporate arena, these photographers clearly had leeway to work from their own consciousnesses and histories. These projects serve as case studies for an exploration of the uses of expressive photography in service of corporate ends—the usefulness of aesthetic vision for promotional purposes.

Torkel Korling: Rath Packing Company and Oscar Mayer & Co.

Torkel Korling undertook at least two projects for meatpacking companies in the 1940s. A Swedish-born, Chicago-based documentary photographer and naturalist, Korling photographed both natural and industrial subjects with equal facility throughout his career.[4] His freelance work for *Life* magazine, perhaps his highest-profile photography, included a well-regarded series of images of Frank Lloyd Wright's brand-new Johnson Wax Building in Racine, Wisconsin, published in 1939.[5] On the strength of such work, Korling received his first meat industry commission, for the Rath Packing Company.

To celebrate its fiftieth anniversary in the meatpacking business in 1941, the Rath Packing Company put on a weeklong celebration and commissioned a book to memorialize its history and show off its state-of-the-art Waterloo plant.[6] To manage it all, Rath hired advertising and public relations firm Young & Rubicam. Titled *Waterloo Packer: The Story of the Rath Packing Company*, the book included text by the Chicago-based writing team of Geoffrey Parsons and Robert Yoder and images by Korling. Parsons worked as a Chicago correspondent for the *New York Herald-Tribune*, Yoder for the *Chicago Daily News*; and both wrote, sometimes separately and sometimes as a team, for the *Saturday Evening Post*. As for Korling, a Young & Rubicam executive wrote to a member of the Rath staff that the photographer "specializes in pictures that tell a story, not just portraits or snappy compositions."[7] Since Korling had worked on assignment not only for *Life* but also for other corporations assembling commemorative volumes, executives of Rath and of Young & Rubicam found him an especially apropos hire.[8]

Modeled on *Life* magazine, the book told the Rath company's story in photo-essay format, Korling's photographs functioning as a primary carrier of the narrative.[9] Taking the reader from farm to factory to supermarket, *Waterloo Packer* seemed to offer an intimate look at the company's operations. But the institutional book and all the fiftieth-anniversary festivities were actually carefully planned publicity, soft advertising designed to augment the company's name recognition and market share. Co-opting *Life*'s photo-essay format, *Waterloo Packer* presented a controlled vision of industrial-scale slaughter.

Encapsulated within Rath's commemorative publication and its fiftieth-anniversary celebration overall are many themes that recur in meatpacking imagery: questions of sequencing within and circulation of photo-texts; forms of institutional and industrial storytelling and identity construction; narratives of meat making as inflected by gender, race, and culture; the nature of meatpacking labor; and constructions of animality and animal agency in an industry that pivots on rapid, organized, and infinitely repeatable slaughter. In commissioning an institutional biography and embracing photography as a means of telling its story, the Rath company created a powerful argument for the relevance and continued growth of industrial-scale meat production as an economic and caloric engine of mid-century America.

Within *Waterloo Packer*, a feeling of peaceful nostalgia dominates at the start, as Korling's first photograph introduces readers to the notion of Iowa farms as what the caption calls "the stockroom, the source of supply" for meatpacking plants.[10] The following page maintains the sense of yearning for the past: the first animals the reader encounters are not meat animals but draft horses, hitched to a planter into which a farmer pours seed corn (fig. 29). In 1941 draft horses were almost obsolete on modern farms, so although the text introduces the Rath company as a modern and prosperous present-day organization, Korling's photograph conveys nostalgia. The subsequent two-page spread clarifies why the book has begun with plant rather than animal agriculture: because "Corn makes hogs—and Iowa makes both" (fig. 30).

The book presents Rath's location in Iowa as a key asset. Ample corn makes for tasty, tender pork, and Iowa's system of good paved roads permits quick transfer of animals from the farm to the packinghouse. Though the text is frank about the conversion of live animals to "superior meat —tender and firm at the same time," the captioned images introduce pigs as "passengers" on the journey from farm to packinghouse, implying the animals' willing participation in the process. In the subsequent spread, farm trucks pull up to the plant to unload their cargo. The animals move into Rath's enclosed "hog hotel," where they experience comfort and protection from the weather while they await slaughter (fig. 31).

After a lesson on Rath's history and profiles of its executives, *Waterloo Packer* takes readers into the meatpacking plant itself. The book elides the process of slaughter itself: previously shown living, hogs reappear already dead and drained of their blood (fig. 32). They now begin the process of disassembly, of becoming meat. In the leftmost image of the sixth spread, a man strips off of one carcass a shiny black substance—a kind of tar, into which the carcasses are dipped after slaughter and which, when peeled off, takes most of the hog's hair with it, reducing the amount of shaving and scraping necessary to clean the hog's skin. The caption under the next image attempts to revive and revitalize the animals, minimizing the fact of their deaths with a joke: "Slaughtered hogs are the liveliest." The caption notes that carcasses move on the overhead rails at a speed of roughly twenty feet per minute, a speed that permits the slaughter and handling of six hundred hogs per hour. Other images in this spread depict the disassembly and chilling processes from various angles. The spread overall shows

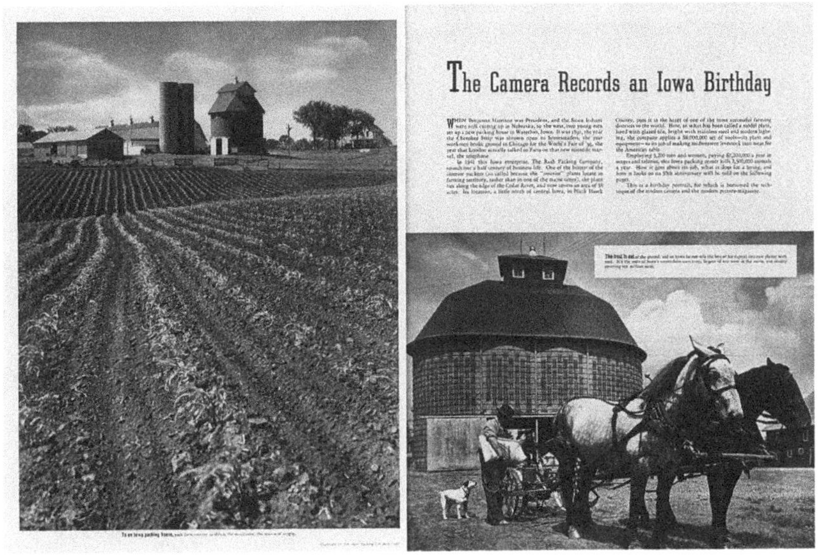

Figure 29.
Waterloo Packer, first spread, with text by Geoffrey Parsons and Robert Yoder and images by Torkel Korling. Special Collections and University Archives, Iowa State University Library.

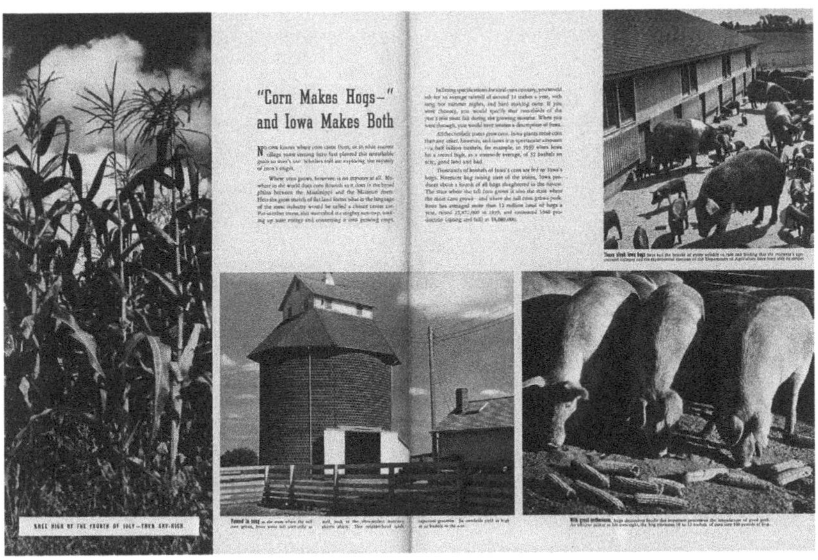

Figure 30.
Waterloo Packer, second spread, with text by Geoffrey Parsons and Robert Yoder and images by Torkel Korling. Special Collections and University Archives, Iowa State University Library.

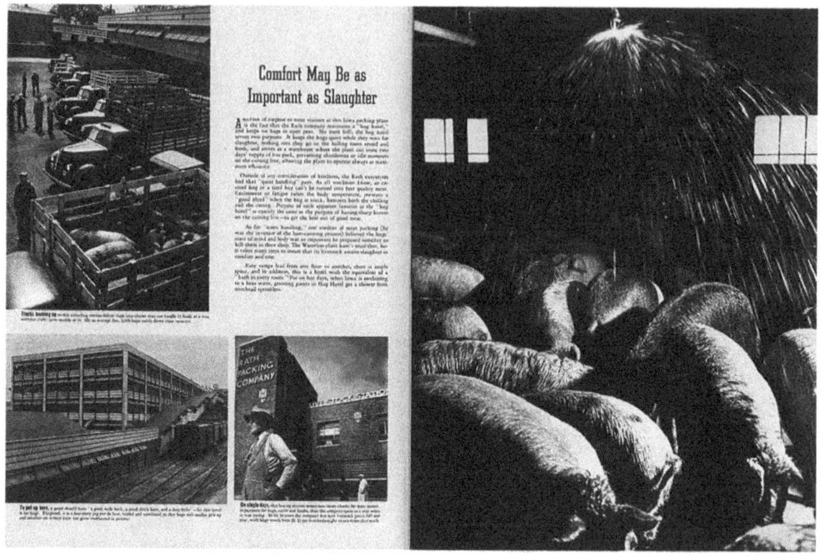
Comfort May Be as Important as Slaughter

Figure 31. *Waterloo Packer*, fourth spread, with text by Geoffrey Parsons and Robert Yoder and images by Torkel Korling. Special Collections and University Archives, Iowa State University Library.

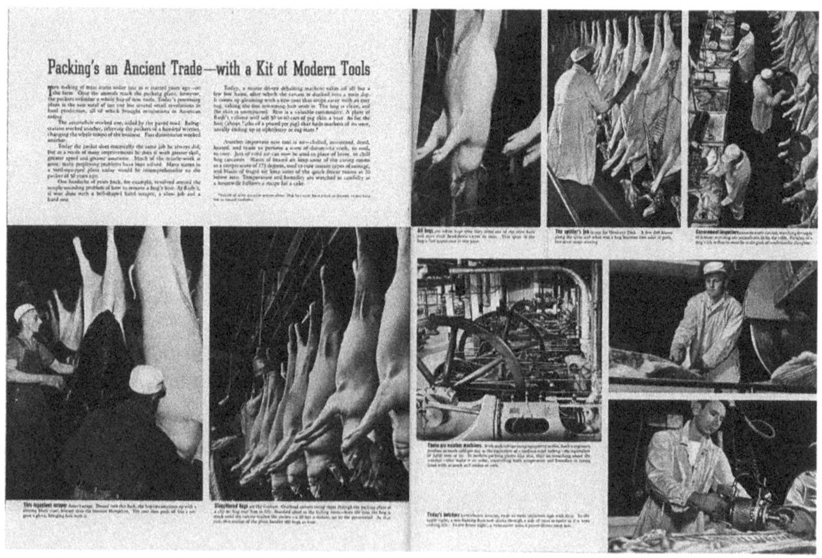
Packing's an Ancient Trade—with a Kit of Modern Tools

Figure 32. *Waterloo Packer*, sixth spread, with text by Geoffrey Parsons and Robert Yoder and images by Torkel Korling. Special Collections and University Archives, Iowa State University Library.

how the speedy conversion of hogs into meat relies on both skilled labor and heavy mechanization.

The subsequent spread introduces readers to the disassembly line, where 350 workers "outnumber" the hog, reducing it to cuts of meat and by-products. Then the book follows for a few pages one particular cut, the belly, as it is first smoked into bacon by male workers and then sliced, weighed, counted, and packaged by female workers (fig. 33). Bacon and ham serve as the public face of meatpacking, demonstrating how the industry traffics in deliciousness, achieved through ancient means gone high tech. The book's final spread (fig. 34) focuses on distribution of the Rath company's products, shown first in boxes on conveyors, then being loaded into trucks, and finally unboxed and presented to the woman consumer by her local butcher.

Waterloo Packer constituted just one component of a huge fiftieth-anniversary celebration for the Rath Packing Company, months in the making and days in the unfolding. Conceived and implemented almost entirely by executives of the Bureau of Industrial Service, the public relations arm of Young & Rubicam, the Rath celebration was designed explicitly to attract the attention of journalists and newspaper editors and, through them, the general public.[11] From 1940 through 1941, Rath executives poured a tremendous amount of effort and time into planning their anniversary celebration. A Bureau of Industrial Service employee, Torrey Stearns, managed the project and proposed a series of events and productions.[12] Among other events, there would be an employees' day; a "farm day" for area farmers; and a banquet for politicians, shareholders, local businessmen, journalists, and other worthies. Each of these would feature the same program of entertainment, which would also function as a point of interest and a way to generate popular curiosity about the events.

The explicit purpose of all this was a soft sell: each anniversary event was designed to be newsworthy, inviting coverage by local and even national news outlets that would put the company name in the public eye in places other than the advertising section of the paper. Stearns's final publicity proposal for the anniversary celebration details, over several pages, a series of press releases that his company would write for Rath. Every event of the anniversary week would receive its own press release, and in addition to brief releases, Young & Rubicam would also compose a series of extended features. These would be sent to national news outlets as well as to local

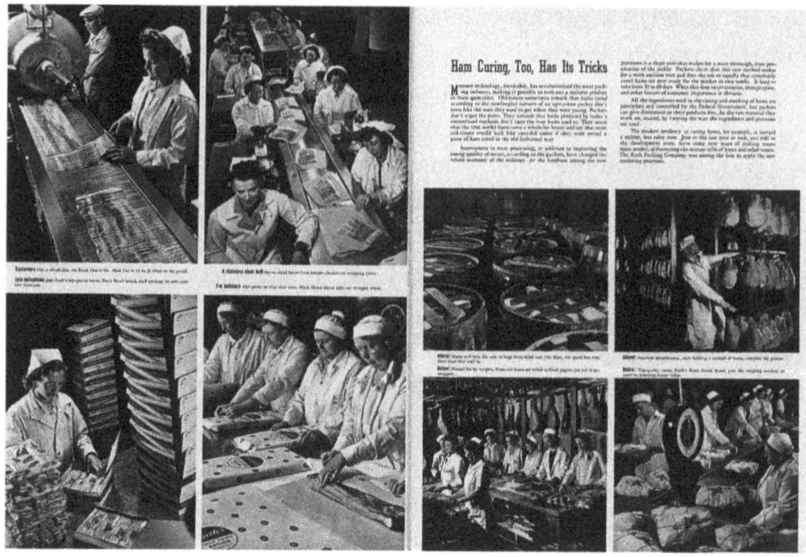

Ham Curing, Too, Has Its Tricks

Figure 33. *Waterloo Packer*, ninth spread, with text by Geoffrey Parsons and Robert Yoder and images by Torkel Korling. Special Collections and University Archives, Iowa State University Library.

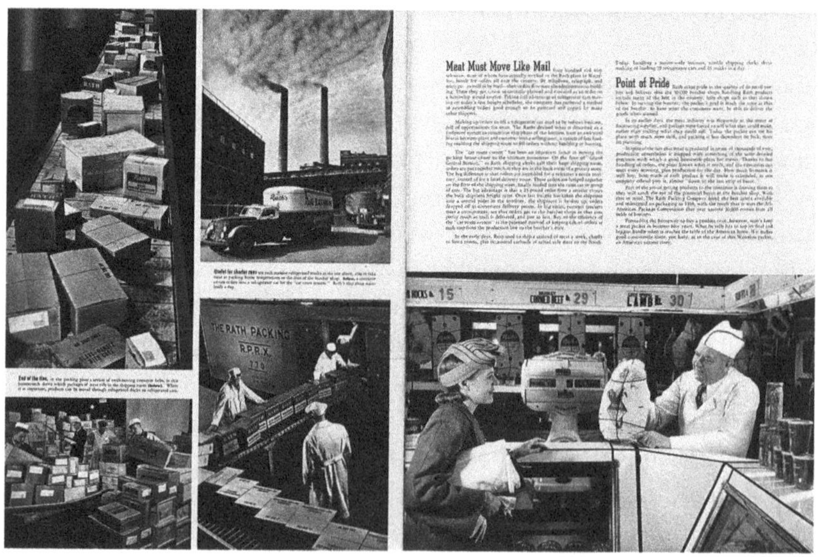

Meat Must Move Like Mail

Point of Pride

Figure 34. *Waterloo Packer*, eleventh and final spread, with text by Geoffrey Parsons and Robert Yoder and images by Torkel Korling. Special Collections and University Archives, Iowa State University Library.

outlets in every market where Rath products were present, especially those cities in which Rath, had a branch operation.[13] Information about Rath would blanket local and national news. Rath should, Stearns proposes, accommodate the press by setting up a press headquarters at its Waterloo plant during the anniversary week, where "reporters can come to obtain information and copies of stories. Typewriters should be available in this room."[14] The proposal emphasizes that Rath would not be buying this coverage: "News and editorial space is not for sale. . . . However, newspaper and magazine editors are always interested in NEWS about outstanding accomplishments of business firms, and this plan provides them with NEWS about The Rath Packing Company."[15] Good news about Rath's "outstanding accomplishments," Stearns argues, would increase the company's prestige, its name recognition, and ultimately its sales.[16]

Like *Waterloo Packer*, these events were explicitly designed to reflect attention back onto the company they celebrated. The entertainment program, for one, brought the company a great deal of public attention, though initially it had presented some interesting problems. After considering various options, Young & Rubicam and the Rath company had settled on a group of entertainers from Texas and Mexico who performed a Mexican-themed variety show that included elements of a traditional bullfight. As the show evolved, they began to see its potential to be one of the more newsworthy events of the anniversary week. There was some concern, however, about the possibility of negative publicity surrounding the bullfight. Although the packing industry faced little opposition in the 1940s from groups opposed to the slaughter of animals for meat, bullfights did draw the ire of animal welfare groups.[17] On reading a proposal for the entertainment program, Stearns wrote to Byron Benson, advertising manager of the Rath company, that he was "wondering if there wouldn't be a danger of attracting S.P.C.A. opposition" and worrying that the bullfight might actually be illegal.[18] In response, Rath's vice president and director of sales, R. A. Rath, decided that the bullfight should be advertised explicitly as bloodless.[19] Via telegram, R. A. Rath ordered Stearns to "soft pedal the bullfight in your releases except in minor way and improve Panamerican Mexican angle."[20] The entertainment, ultimately called the "South of the Border Pan-American Cavalcade and Bloodless Bullfight," thus grew to become part of an overall goodwill endeavor, potentially generating not only news interest but trade relationships. The mayor of Waterloo was encouraged to send a letter (actually written by employees of Young & Rubicam in consultation with the US Department of State) to the Mexican ambassador in Washington, DC, inviting Mexican president Manuel Ávila Camacho to the Rath banquet.[21] The president did not attend, but ultimately the spectacle and the diplomacy it represented were among the most heavily reported features of Rath's anniversary week.[22]

Another anniversary week event that drew a great deal of publicity involved the local Native American population, specifically members of the Meskwaki Nation (Sac and Fox Tribe) of central Iowa. Conceived not by the advertising agency but by Rath executives, the event would culminate in the induction of company president J. W. Rath as an honorary member of the Sac and Fox Tribe.[23] In early 1940, Benson was dispatched to the tribal land in Tama, Iowa, to discuss the idea with tribal leaders. In his report on the visit, Benson writes that he had pitched the idea of the tribal induction on the basis of the Rath company's "service to the tribe": the Rath brand name "Black Hawk" and native headdress logo had been inspired by Chief Black Hawk, a great Meskwaki leader of the nineteenth century. Benson continues,

> In view of the fact that Mr. J. W. Rath had originally proposed the name "Black Hawk" as the brand to be adopted for the finest quality products which we process, it seemed to [me] that it would be very fitting on the occasion of the 50th anniversary, for the tribe to honor him with the induction into the tribe and to give him the Indian name of "Black Hawk." In this manner it would associate their tribe with our products and would give them extensive publicity in newspapers and through other channels.[24]

Tribal leaders ultimately agreed to the Rath company's proposal. They had some requirements, to which the Rath company acceded: tribal members had to be properly compensated for their time, and their expenses reimbursed; and the tribe would construct a native village, to remain in place throughout the weeklong anniversary celebration, from which tribal members could draw income supplemental to their compensation.[25] The Meskwaki Nation thus had a sustained presence throughout the anniversary week. Tribal participation culminated in an evening induction ceremony, held during the main banquet, in which J. W. Rath received a traditional headdress. This event, too, generated the desired publicity, receiving prominent coverage in a number of Iowa newspapers.[26]

All the anniversary events, including *Waterloo Packer*, pivoted on core assumptions about racial, cultural, and gender identities. Non-white workers made almost no appearance in the book, despite constituting nearly 10 percent of the Rath workforce in 1941, a number that would climb dramatically in the coming decade.[27] The company's manufacturing plant itself appears, in the pages of *Waterloo Packer*, almost exclusively white and male, with white women present to satisfy specific, well-defined secondary roles. Within the book's pages, white men farm grain, raise pigs, and transport them to the packinghouse, where other men, also mostly white, slaughter and dress them, all under the supervision of the white male executives profiled at the book's heart. White women at the packinghouse perform select jobs, such as receiving and sorting trimmings or wrapping bacon and hams, but these duties come after, away from, and secondary to the heroic-scale acts of slaughter, dressing, and quartering required to still and apportion the animal body. Where a job is primarily performed by women—in the pages of *Waterloo Packer* as in the meat industry more broadly—it is generally less critical and thus lower paid. Some jobs performed by women also incorporated a public-facing component: at the Rath plant, as at facilities throughout the industry, the clean, efficient, heavily mechanized bacon- and ham-wrapping rooms were often used as points of pride on public guided tours of meatpacking plants. The bacon-wrapping room shown in *Waterloo Packer* (see fig. 33), typical for the industry, employed young white women to serve as the meat industry's friendly, accessible public face.[28] Aside from them, the only other woman featured in the book is the consumer, who appears in the book's final image (see fig. 34). Young, white, and fashionably dressed, she smilingly takes the butcher's advice to purchase Rath products.

The entertainment program, too, leaned heavily on racialized performances. The Mexican-themed entertainment programming was rolled into an overall program of international, interracial goodwill, and with the company president's honorary "induction" as a member of the Meskwaki Nation, the company and its chief executive donned a racialized costume.[29] Through the anniversary celebrations and commemorative booklet, white male executives from Rath and Young & Rubicam manufactured a space within which meat production was performed as a primarily white male endeavor, with femininity and non-white masculinities made lesser than, sublimated to, or placed in service of the accomplishments of white men.[30]

As Young & Rubicam executives acknowledged in their communications, the company targeted the female consumer almost exclusively with these promotional efforts, aiming to increase the company's prestige among "housewives who see the news stories about the anniversary." But if the company performed *for* women, very little of the performance involved women themselves.

Beyond gendered and racialized inflections, depiction of workers overall in *Waterloo Packer* was determined in part by the company's relations with organized labor at the time. By 1941 the Waterloo plant was the largest single meatpacking facility in the nation.[31] Like all large-scale American meatpacking companies, Rath practiced on a vast scale, the interior of the packinghouse essentially a factory for the disassembly of live animals into their component parts. At Rath in 1941, the hog kill and cut lines processed an average of six hundred and sometimes as many as one thousand hogs per hour. Although some positions in Rath's packinghouse demanded considerable skill, particularly those centered around slaughter and skinning, much of the work required relatively little training; and even the skilled workers performed just a few operations repeatedly, rather than the dozens of operations required of a butcher working on a small scale.

In the present day, the meatpacking industry identifies animal welfare organizations as primary loci of friction and sources of negative publicity. But in 1941 the major source of friction in the meatpacking industry, as in many other American industries, was organized labor. One union in particular, the Packinghouse Workers' Organizing Committee (PWOC; later renamed the United Packinghouse Workers of America, or UPWA) was having increasing success recruiting workers at Rath and other area meatpacking companies.[32] In Waterloo and nationwide, the PWOC's growing popularity was due in large part to its progressive attitude toward racial inclusivity: the union made enrolling both white and non-white workers a priority, reasoning that a powerful collective bargaining position depended upon cross-racial solidarity.[33] Meatpacking companies watched this activity closely, Rath included: in early 1941, the PWOC had begun recruiting among Rath workers, with some success, and Rath management kept itself apprised of the union's growth in part by acquiring copies of the union's own newspaper.[34]

But *Waterloo Packer* deliberately, assiduously, avoids such issues. Not only does it enact a state of racial sanitization on its workforce, it takes pains to depict labor as fully assimilated, and fully sublimated, to the mechanics of the packinghouse as a whole. Captions may laud workers' skill, but the book does not name or individualize them, as it does with the company's executives. They appear as highly competent functionaries of the packinghouse, human tools of the industry. The question of their employment, the quality of their work, its difficulty, and the toll it takes on their bodies: such issues never arise in *Waterloo Packer*.

Animals, too, appear in the book's pages fully sublimated to the industrial machine. Within the sequence, pride of place is given to the company's hog hotel (see fig. 31), where pigs rested for several hours or overnight after arriving on packinghouse grounds. In interviews conducted by Stearns of Young & Rubicam—documents that then became Parsons and Yoder's primary sources of information as they composed the text of *Waterloo Packer* —multiple respondents mentioned this structure as a central symbol of Rath's modernity and prosperity. One hog buyer, Bert Hoffman, talked of how much better it was to warehouse the hogs indoors than outdoors, where overheating could sicken them and reduce the quality of their meat.[35] R. A. Rath touted it as key to the quality of the company's pork, because it permitted the company to rest hogs in a sheltered, cooling environment prior to slaughter, drastically reducing their stress and thus improving their meat.[36] He and other employees clearly believed that the hog hotel stood as a symbol of the company's forward-thinking embrace of the latest scientific research. Indeed, the hog hotel seemed to symbolize the company's very success itself: that the company needed to warehouse so many hogs demonstrated not only its own success but the success of large-scale slaughter and even of capitalism itself, with its promises of abundance and surplus.

This spread, including Korling's image of hogs massed under a cooling shower of water, was designed to speak to the Rath company's concern for porcine well-being as it might affect the quality of their final product. For the time period, such a celebration of industrialization, of the triumph of the capitalist ideal, was perfectly appropriate. But from the point of view of even a few years later, the layout and text call up some chilling associations. In focusing on a structure built to warehouse living things for the purposes of slaughter, an architecture of waiting to die, the layout calls to mind the mass murder of the Second World War, already well underway in 1941.[37] The image originally connoted comfort, but after the fact, the photograph becomes something else entirely: a dirge, an unintentional lament.

Waterloo Packer took pride of place among the events and productions for Rath's fiftieth anniversary. It cost far more to produce than any other line item in the anniversary budget; and on its completion, Rath mailed copies of it to executives from a number of other major food production firms, some of whom replied with their congratulations and comments on the quality of the book.[38] Korling's images, the major draw, served to express the company's vision through the photographer's lens. Stylistically, the photographs exhibit Korling's own extensive experience with industrial commissions, his ability to express the corporate vision via powerful, well-constructed compositions. Korling received a shooting script from Rath, which the final publication shows he followed quite faithfully.[39] But the vision of Rath's production lines comes across as an idiosyncratic, singular vision, and an unvarnished one. Korling's documentary style, here, serves the ends of the meat industry, conveying that his attitude of admiration for the achievements of American industry should be shared by everyone who comes across the book.[40]

After receiving its own copy of *Waterloo Packer*, Oscar Mayer & Co. expressed its admiration by hiring Korling, probably around 1944–1945, to undertake a similar project.[41] Although no mass-printed book resulted, the Oscar Mayer company archive at the Wisconsin Historical Society includes a cohesive album, with ninety-two high-quality photographs bound in a leather ring binder.[42] Maintained intact and in order, the album in some ways conveys an organized vision of Oscar Mayer's operations at its Madison, Wisconsin, plant.[43] But in other ways it is enigmatic: while the photographs are numbered on their mats, corresponding captions have not been maintained alongside the pictures, and the album contains no other text. Although the album may have been intended for publication when it was created, the images do not seem to have appeared in any published context. Narrative structure in the album must be inferred from the photos themselves.

That said, connections do suggest themselves. In creating the Oscar Mayer album, Korling followed the model set by his earlier assignment for the Rath Packing Company. Oscar Mayer's album bears many similarities to Rath's *Waterloo Packer*. The Rath publication takes the reader from farm to packinghouse to supermarket, tracing the journey of Rath's products from their animal origins to packaging and sale. The Oscar Mayer album addresses many of the same subjects and follows a similar trajectory; and some individual images bear close resemblance to counterparts in the Rath publication. But the Second World War had arisen in the intervening years since the 1941 publication of *Waterloo Packer*. The Oscar Mayer album contains clear and repeated references to the war, and in particular to the plant's contributions to the war effort—one of its underlying themes, and a key way in which this album differs from Korling's earlier production for Rath.

The Oscar Mayer album opens with a series of exterior photographs of the company's main facility in Madison. The initial shot takes a distant view, and subsequent images move in closer. The company's wartime service banner, listing the number of former employees serving in the military during the ongoing Second World War—1,241, at the time—hangs above a second banner reading "Our job is MORE MEAT. It's fightin' food!" Several of these photographs emphasize the collegiality of workers from different parts of the plant, conveying a sense that at Oscar Mayer, everyone dedicates themselves in kind to the war effort. In one image a group of workers leaves the plant for the day, the variety of their dress—mechanics' overalls, dungarees, the white overcoats worn by butchers and inspectors—attesting to the variety of their positions. Another image depicts a white-coated worker from inside the plant talking with a delivery driver, the latter leaning proprietarily on his truck.

After these establishing shots, photos in the album close in more tightly on the plant itself, revealing in the process Korling's familiarity with the modernist machine aesthetic by then fairly widely embraced by American photographers.[44] The picture numbered "No. 10" in the Oscar Mayer album, in particular, bears a more-than-passing resemblance to Charles Sheeler's famed 1927 image of crossed conveyor belts at the River Rouge Ford Plant, down to its focus on a loftily distant water tower, photographed from below so it appears flattened against the sky (fig. 35).[45] In creating this homage, Korling exercised his artistic agency and showed off his timely knowledge of aesthetic trends. More than just an expressive fillip, however, the image ultimately serves his client's purposes quite nicely: American meat making seems to absorb some of the mythos already clinging to American automaking.

The Oscar Mayer album then moves inside the plant. The interior tour begins with impressive machinery, incorporating images of some of the dynamos that kept the plant running; but the album then shifts to emphasize the human element. The first people Korling shows inside the packinghouse are not factory workers but farmers, chatting and laughing in their stained, well-worn overalls and battered work boots as they wait to receive payment for the animals they are delivering (fig. 36). After these photographs, a pair of images at the end of this human-focused sequence support the notion that the album originally showcased Oscar Mayer's support for the war effort: both show a man in military uniform, sitting with other men at a table. He is probably an employee leaving the plant for active military duty, and these pictures show people sitting down with him to say goodbye.

This human interlude functions as a transitional point from scenes of the plant exterior to images of its interior operations, including the animal material in which it dealt. Only with the twentieth photograph in the album does the viewer encounter pigs, which first appear hanging head-down from an overhead rail. Like *Waterloo Packer*, this album elides slaughter itself: already dead and stripped of their hair, the pigs' bodies gleam uniformly white. The lone worker in the image checks a pig's ear for the tag with which inspectors will track it as it passes through processing. Although he wears a stained apron and heavy rubber gloves, in the photo he does not perform any specific cut or otherwise advance the pigs' process from animal to meat. The pigs seem simply to have arrived here without human intervention, their deaths literally foregone. Elision continues in the next image (fig. 37). Carcasses appear split and partially dressed, but the labor of splitting and dressing has been largely bypassed. Several men stand on an elevated grate. The grate is covered in gore, but in the monochrome of the photograph it looks like white foam, not the bloody mess it actually was. The actions of the man closest to the camera—the last few steps in the dressing process—stand in for the entire course of "hot side" activities in the Oscar Mayer plant, from slaughter to gutting and cleaning.

Subsequent to this image, carcasses appear only chilled and cut, no longer animals but objects—or rather, series of objects: hunks of flesh, cuts of meat. The disassembly process seems largely a machine-driven affair. In one picture, portions of pork pass under a sawblade and are separated into cuts entirely by machine, human workers consigned to the background. In two subsequent images, workers become, like sailors, metonymically identified entirely by their hands, all their labor concentrated in one part of the body (fig. 38).[46] Their knives and their chain-mail gloves further this dissociation of hand from body: with their metal coverings, the men themselves become machines.[47]

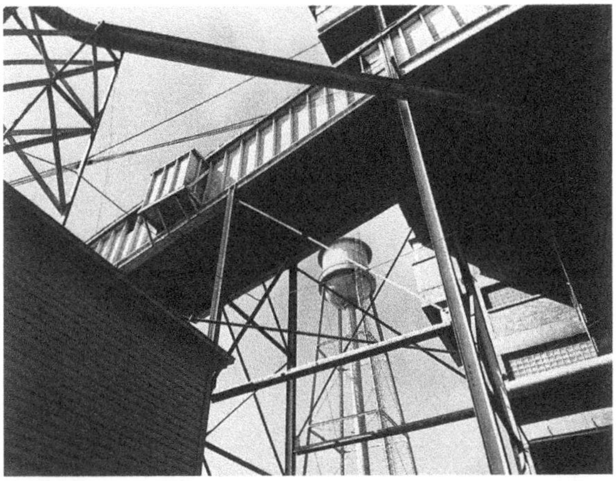

Figure 35.
Torkel Korling, Exterior with water tower, Oscar Mayer Plant, Madison, WI, ca. 1944–1945. Image no. 10, Oscar Mayer photograph album, Special Collections and Archives, Wisconsin Historical Society (accession number M2016-042).

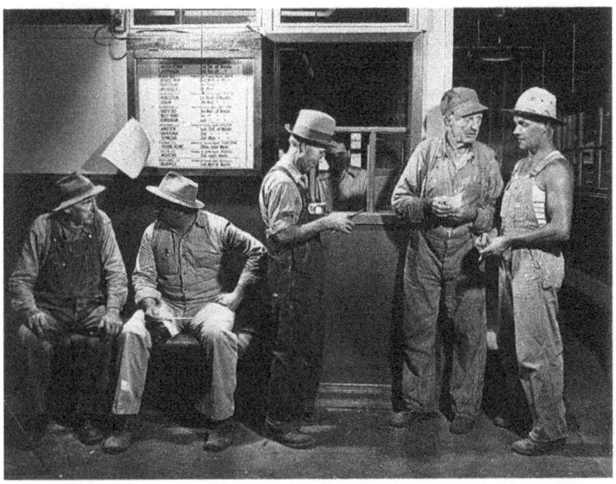

Figure 36.
Torkel Korling, Pay booth, Oscar Mayer Plant, Madison, WI, ca. 1944–1945. Image no. 17, Oscar Mayer photograph album, Special Collections and Archives, Wisconsin Historical Society (accession number M2016-042).

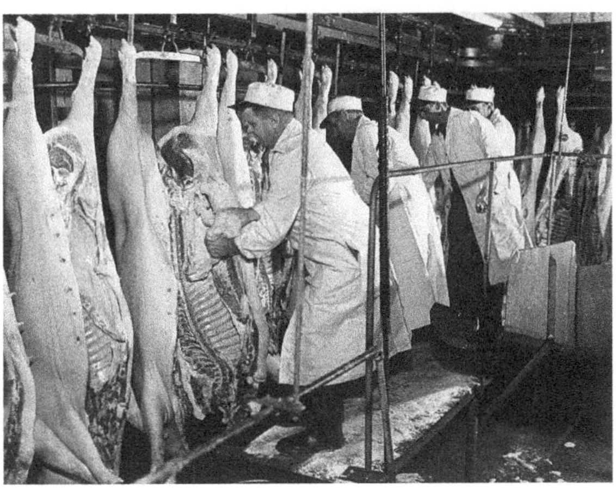

Figure 37.
Torkel Korling, Pork kill facility, Oscar Mayer Plant, Madison, WI, ca. 1944–1945. Image no. 21, Oscar Mayer photograph album, Special Collections and Archives, Wisconsin Historical Society (accession number M2016-042).

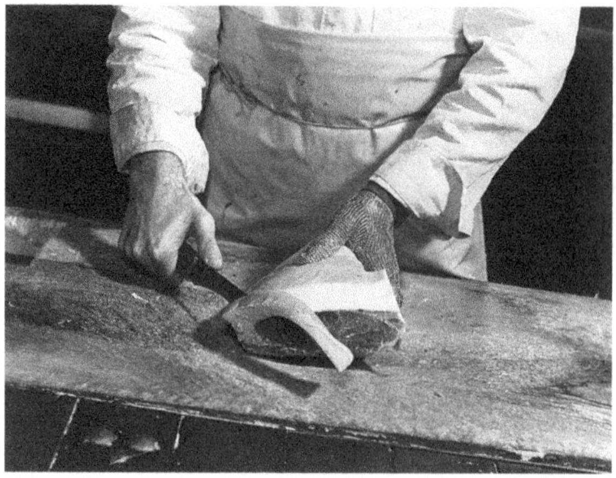

Figure 38.
Torkel Korling, Pork cut line, Oscar Mayer Plant, Madison, WI, ca. 1944–1945. Image no. 26, Oscar Mayer photograph album, Special Collections and Archives, Wisconsin Historical Society (accession number M2016-042).

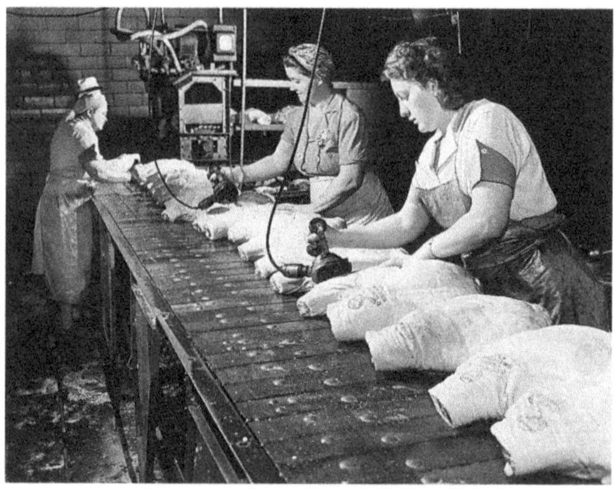

Figure 39.
Torkel Korling, Pork processing, Oscar Mayer Plant, Madison, WI, ca. 1944–1945. Image no. 28, Oscar Mayer photograph album, Special Collections and Archives, Wisconsin Historical Society (accession number M2016-042).

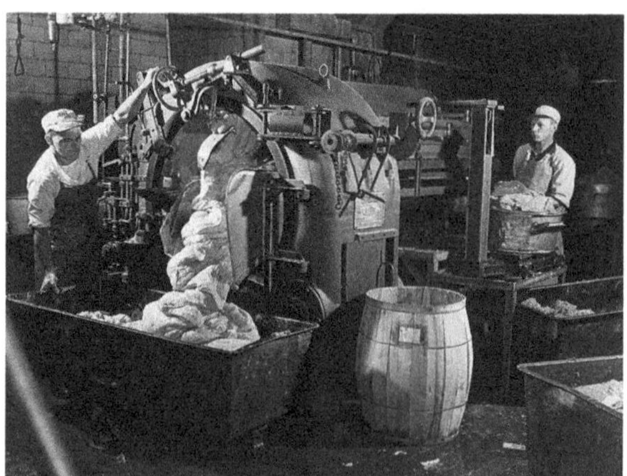

Figure 40.
Torkel Korling, Pork processing, Oscar Mayer Plant, Madison, WI, ca. 1944–1945. Image no. 36, Oscar Mayer photograph album, Special Collections and Archives, Wisconsin Historical Society (accession number M2016-042).

In this album and in the Rath book, as in many representations of the meat industry, certain cuts of meat, particularly bacon and ham, are given pride of place. Bacon often functioned as a kind of standard-bearer cut of meat for the entire industry: company displays at international exhibitions often featured bacon over other cuts, and public tours of meatpacking companies often spent more time in the bacon-branding, bacon-slicing, and bacon-packaging areas of the plant than in any other. Well-chilled, carefully portioned, and no longer bloody, the meat was, at this point, far removed from its animal source. The work of turning cuts of meat into bacon, ham, and other processed and preserved cuts also tended to be fairly strictly gendered. The more demanding work, particularly the salting of bacon, the hanging of hams and sides of bacon in smokers, and the operation of the smokers, was done by men. Much of the labor of branding, slicing, portioning, and packaging was mechanized, by contrast, so while the labor involved a great deal of repetition (and hence the potential for repetitive stress injuries, a constant in meatpacking plants), it was, by comparison with labor in other parts of the packinghouse, relatively clean and physically undemanding. Concomitantly, bacon- and ham-processing rooms often employed women—particularly, as Roger Horowitz notes, young white women. When visitors toured packinghouses, and when bacon processing went on display at international expositions, the women themselves went on display, too.[48]

The Oscar Mayer album upholds these industry norms. In a couple of images, tidily dressed white women process raw hams and sides of pork, labeling them and preparing them to hang (fig. 39); while in other images, crews of men salt and stack slabs of bacon and hang cuts of meat in the smoker. Work that involves lifting, hauling, and hanging meat falls to men, while Korling depicts women searing labels into cuts of meat and placing them on hangers, the latter task evoking domestic tasks particularly clearly. The photos demonstrate that even in a workplace where women's employment was a norm, "natural" gender orders and assigned gender roles continued to pertain.[49]

These pictorial assertions of natural order complement the album's overall message regarding the equally "natural" hierarchy of human over animal, the foregone conclusion that animal bodies were no different from any other raw material. In converting raw animality into consumable meat, by extension, the operations of the packinghouse proved just as necessary as those of any other heavy industry to the war efforts as well as to the advance of economic and industrial progress more broadly. Even aesthetic decisions in the photos uphold these assertions. The carefully framed images deploy slashing diagonals, rhythmic repetitions, and sculptural machine forms to pleasing effect, conveying a sense of order, professionalism, and cleanliness. The album implies that even during the course of its regular daily business, in the processing of its showpiece cuts and the production of its most prominent branded products, Oscar Mayer and its employees contributed to the war effort, helping to uphold a democratic world order.

Like other industry-sponsored depictions of the meatpacking industry at mid-century, this album emphasizes a sense of near-endless productivity, conveying the plant's ability to feed the world. Repetition of forms accomplishes this in some images, while size and volume accomplish it in others. In a sequence focusing on lard processing and meat grinding, these semiliquid substances pour out of immense grinders, folding like fatty ribbons into gargantuan wheelbarrows (fig. 40). Human operators appear subordinated to the machines, feeding them bits of pork fat and residual flesh with pitchforks. Other machines then portion the substances into boxes and cans, human beings intervening little in the packaging processes. The dominant theme in many of these images is simply "muchness": piles of metal boxes, rack upon rack of bologna. The theme persists even in images showing Oscar Mayer's in-house ice production: workers pull ice cubes from huge molds, then stack them into massive walls of ice inside a cavernous, hangarlike space (fig. 41).

Up to this point, the album has treated the creation of meat as a largely mechanical process, with human workers as facilitators. Animals have appeared as carcasses, the base material out of which Oscar Mayer constitutes its products. Now, finally, the album shows live animals. This sequence presents the animal experience essentially in reverse, starting with images closest to the packinghouse itself and then moving outward and away. Two images show pigs inside an enclosed space at the packinghouse, where they are kept cool by showers of water—their last stop before the slaughtering pen (fig. 42). Here Korling's photos closely resemble those he created for *Waterloo Packer* in 1941, again suggesting that he used the earlier commission as a model for this one (fig. 31).[50] Next, the Oscar Mayer album shows an outdoor pen at the packinghouse, and then a sequence focused on the animals' point of origin: the farm. The album shows a farm depopulated, then

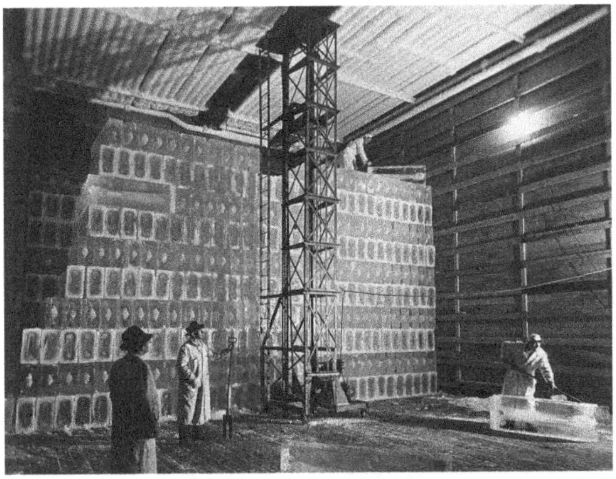

Figure 41.
Torkel Korling, Making ice, Oscar Mayer Plant, Madison, WI, ca. 1944–1945. Image no. 56, Oscar Mayer photograph album, Special Collections and Archives, Wisconsin Historical Society (accession number M2016-042).

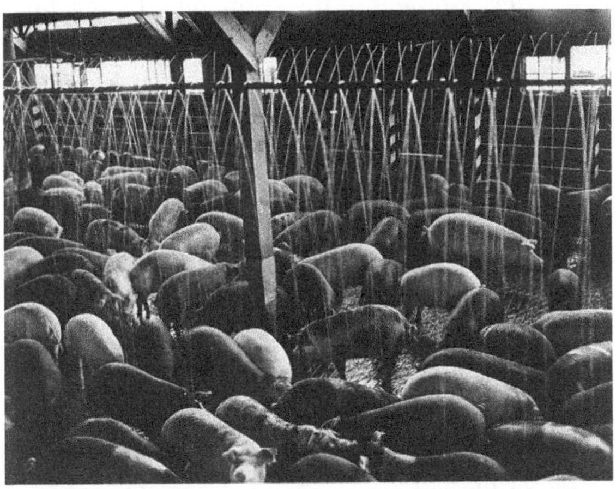

Figure 42.
Torkel Korling, Pig confinement facility, Oscar Mayer Plant, Madison, WI, ca. 1944–1945. Image no. 58, Oscar Mayer photograph album, Special Collections and Archives, Wisconsin Historical Society (accession number M2016-042).

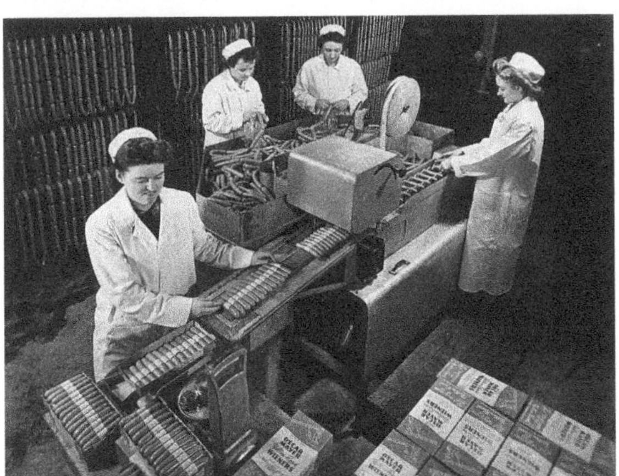

Figure 43.
Torkel Korling, Packaging hot dogs, Oscar Mayer Plant, Madison, WI, ca. 1944–1945. Image no. 78, Oscar Mayer photograph album, Special Collections and Archives, Wisconsin Historical Society (accession number M2016-042).

focuses on cattle, then human workers here and there doing farm labor, and finally pigs. Subsequently a series of images shows farms at a distance, always with a road in the foreground, gesturing at the process of animal transport.

Returning to the meatpacking plant, the album lingers where Oscar Mayer felt its strengths lay, presenting image after image of processed meats: bacon, ham, and the company's signature product, hot dogs. One photo depicts the production of hot dogs as an interaction among three elements: humans—here, tidily dressed women, aware of and prepared for the presence of the camera—machines, and the products themselves (fig. 43). Hot dogs hang in the background, linked and ready to package; and they also appear coming out of the company's "Cartridge Pack" packaging machine, bound in sleeves and ready to sell. Other photos focus on the machine alone. A prized development at the company, the machine performed a series of tasks that once would have required several workers: unlinking franks, lining them up neatly, and packaging them.[51] This automation rendered human labor at many of these jobs unnecessary, and the few women workers left appear like the machine's handmaidens, attending it, feeding it, scooping up what it spews forth continually. Other photos focus on the machine alone, fetishizing and glorifying it like an industrial altar.

Clean and shining, the "Cartridge Pack" machine introduces the notion of intellectual, rather than physical, labor associated with the meat industry. The result of human ingenuity, the machine serves as an ideal bridge image between the production-focused earlier part of the album and the last dozen or so images, which feature Oscar Mayer's research laboratories, sales and accounting department, and home economics division. Two photos show food scientists and technicians working in the company's research and development division (fig. 44). They manipulate beakers, dials, vials, and solutions in a bright, open space, their white coats and their work duties those of laboratory scientists, not butchers. These images, along with others showing workers in shirts and ties taking telephone orders, present the clean, white-collar aspects of meat-making work. Presaging the postwar glorification of scientists and laboratory work in American corporate and popular culture, these images present industrial progress as the result of intellectual, as well as physical, labor.[52]

The last set of images (fig. 45) show the Oscar Mayer research kitchen, where the company employed home economists to test its products, developing recipes and suggestions for how they might be used in American

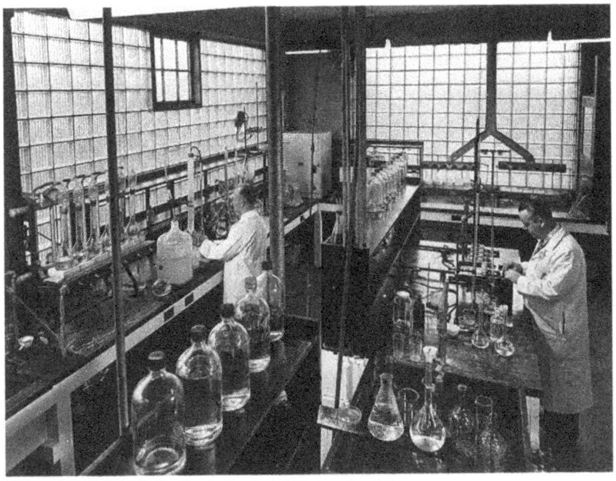

Figure 44.
Torkel Korling, Research and development, Oscar Mayer Plant, Madison, WI, ca. 1944–1945. Image no. 85, Oscar Mayer photograph album, Special Collections and Archives, Wisconsin Historical Society (accession number M2016-042).

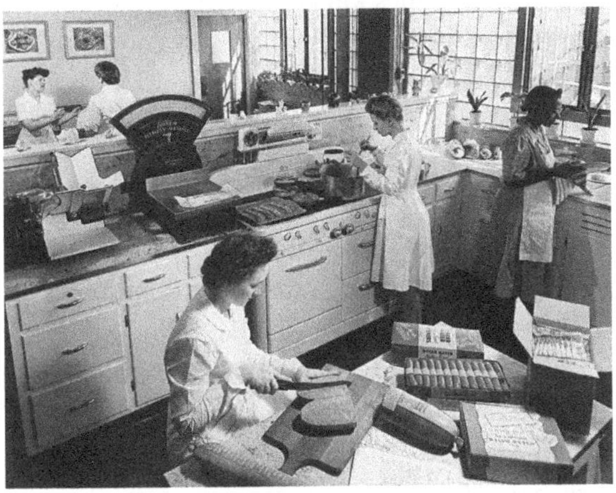

Figure 45.
Torkel Korling, Research kitchen, Oscar Mayer Plant, Madison, WI, ca. 1944–1945. Image no. 89, Oscar Mayer photograph album, Special Collections and Archives, Wisconsin Historical Society (accession number M2016-042).

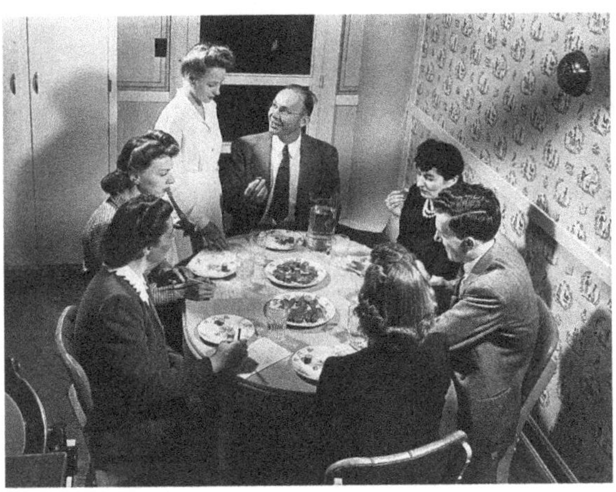

Figure 46.
Torkel Korling, Research kitchen, Oscar Mayer Plant, Madison, WI, ca. 1944–1945. Image no. 92, Oscar Mayer photograph album, Special Collections and Archives, Wisconsin Historical Society (accession number M2016-042).

middle-class homes. Although women worked in certain positions along the disassembly line and on other parts of the packinghouse floor, only in home economics divisions such as this one were female workers the norm.[53] This kitchen resembles a domestic space, but details mark it as a professional environment: specialized scales and other equipment stand on the counters, and the images also show several women sitting at desks in an office area, divided from the kitchen by a low wall.

Like the scientists, the home economists wear uniforms that distinguish them from packinghouse workers, most of them clothed in starched white dresses, unstained by blood or other meat effluvia. One woman, however, wears different clothing: a Black woman, standing near the sink, has the uniform of a domestic servant rather than the home economists' sterile whites. All of the women in the home economics kitchen perform traditionally feminine kitchen-related roles, but while the white women conduct tests and develop recipes, the Black woman washes dishes. Like a home kitchen, the experimental kitchen may have been a female-dominated space, but it was not an egalitarian one.

In the final images, one of the white home economists serves a meal —composed wholly of hot dogs—to an ersatz "family" of adults who sit around a breakfast table in a domestic-looking interior, making notes on their preferences (fig. 46). The home economist stands, stiffly and deferentially, at the shoulder of the man sitting at the head of the table, in the father's position. Enacting a role somewhere between mother, hostess, and waitress, she hovers while the group gives their impressions of the food, responding with wan politesse to the "father's" jocularity in the second image.

Serving such focus groups proved a necessary aspect of recipe testing, common enough that the company had constructed a dedicated space—this faux breakfast nook with its small table, vinyl chairs, and patriotic-themed wallpaper. In the context of the album, however, these images take on a second layer of meaning, consummating the album's narrative. Indeed, this is a key way in which the Oscar Mayer album differs from *Waterloo Packer*. Where that album concluded with Korling's image of a woman purchasing meat at a butcher shop, here the Oscar Mayer album remains within the confines of the company facilities, but manages, at the same time, to deliver the viewer into a domestic space. The album bypasses the typical commercial narrative of factory to consumer in favor of a cozier tale of factory to home (and factory *as* home), conveying pseudodomestic culinary bliss. The album's overall arc thus reinforces Oscar Mayer's public image as a provider not just of meat but of a specific, maximally convenient type of meat, ideal for a particularly American brand of rapidly modernizing domesticity. Consummating the album's underlying theme, Korling's images show how Oscar Mayer, through its market research and development, works to prepare for an anticipated shift from wartime economizing to postwar abundance.

From the early twentieth century, Oscar Mayer & Co. endeavored to set itself apart from other meatpacking firms, placing early emphasis on branding and promoting particular products, especially hot dogs, over and above other aspects of its production.[54] For this album, produced late in the Second World War, Oscar Mayer employed Korling to reinforce its brand identity and to connect its extant identity and business practices with the national interest. Already well-versed in the visual language of the machine age, Korling applied his skills to generating a coherent, cohesive visual narrative for Oscar Mayer. With his work for *Life* magazine, Korling had produced photo essays well-aligned with the magazine's overarching foci on American identity and American productivity. With his work for the Rath company's *Waterloo Packer*, he shifted seamlessly to a format more directly (though not necessarily more explicitly) tied to the company's promotional interests. The Oscar Mayer album represents a kind of apex of this arc. Though probably intended to run alongside an explanatory text, or at the very least a series of captions, the album reads clearly and cohesively even in their absence. The photographer's individual style, interests, and skills coalesce here, Korling himself becoming a medium for Oscar Mayer's corporate interests. Even in the absence of text, Oscar Mayer sings, through the photographer, a pictorial hymn to its own productivity and patriotism.

Modern American Meat: Arthur Siegel for Oscar Mayer & Co.

At mid-century, Oscar Mayer & Co. engaged in a complex messaging program that incorporated both hard-sell advertising tactics with softer public relations campaigns to shape the company's image and impart a sense of "corporate personality" without overtly selling products.[55] Commemorative albums and books constituted a key part of Oscar Mayer's public relations strategy, and the company continued to commission them into the postwar period.[56] One volume, with images commissioned from commercial and art photographer Arthur Siegel, appeared in 1958 to mark the company's seventy-fifth anniversary. Much like the other volumes analyzed in this chapter, the book, titled *A Tradition and a Company: Oscar Mayer & Co.*, functioned not just—not even primarily—to commemorate the company but to publicize it and situate it as stable, trustworthy, and familiar.[57]

A Chicago-based photographer, Siegel's work ranged widely across fine art photography, documentary photography, and corporate commissions such as the one for Oscar Mayer.[58] Trained as a sociologist during his undergraduate studies, Siegel was attuned to interpersonal dynamics and social roles as they played out in organizations and societies, an attention that reveals itself in his public and industrial imagery.[59] The modernist aesthetic that suffuses his work derives in part from his—for the time, highly unusual—photographic training at Chicago's New Bauhaus. An institution founded in 1946 by Hungarian expatriate and refugee László Moholy-Nagy, the New Bauhaus organized itself on principles articulated at its namesake German design school, the Bauhaus, earlier in the century. Like its predecessor, the New Bauhaus embraced new image-making technologies like photography at a time when most art and design schools still ignored them. The exposure to European modernism and experimentalism Siegel received at the New Bauhaus helped to define his career. During the Great Depression, Siegel worked for the federal government's storied Farm Security Administration, making documentary photographs under the guidance of the legendary Roy Stryker.[60] Subsequently he taught photography at the successor school to the New Bauhaus, the Chicago Institute of Design.[61]

Siegel's corporate commissions, like the one for Oscar Mayer, benefited from his wide-ranging experience and experimentalism, allowing him to photograph all sorts of people in all sorts of places. His distinctive style, marked both by his experience and by his own sensibility, likewise emerges even in his corporate work. Sensitive to human expression and experience as well as to the aesthetic appeal of the machine in an industrial age, Siegel's photography manages to seem both cold and warm at once, both rational and humanistic. As a choice to photograph the productions and outputs of American industry, then, he would have been hard to beat.

A Tradition and a Company carries an image of the company's headquarters on the cover and portraits of the company's three founders on its first page. Facing a text detailing Oscar Mayer's founding, the book presents a picture of the company's flagship products, packaged and branded hot dogs, rolling out of a machine. From there, the book offers an overview

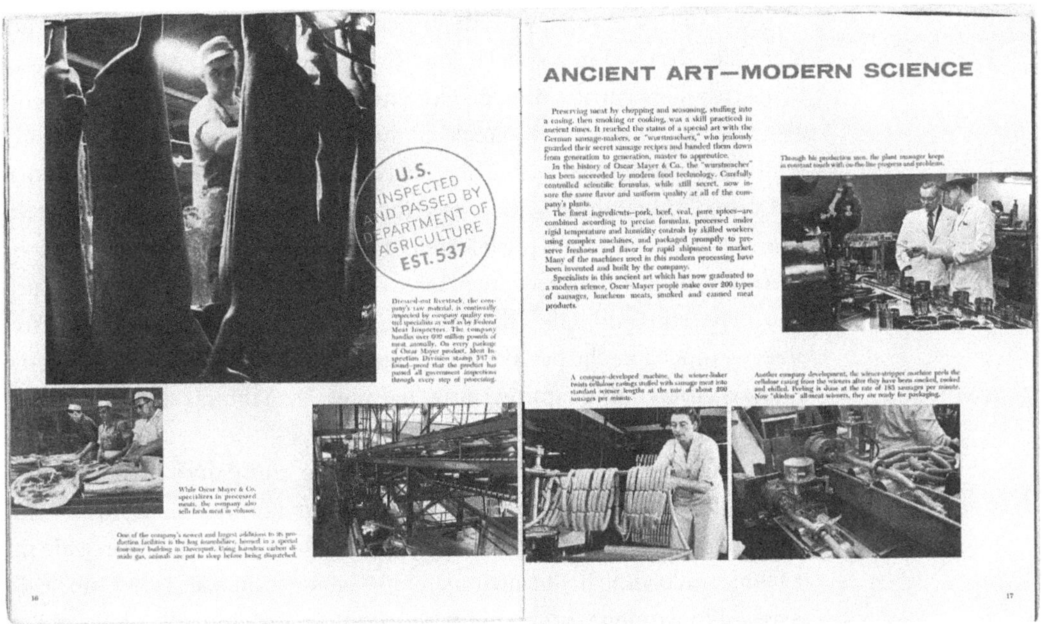

ANCIENT ART—MODERN SCIENCE

Preserving meat by chopping and seasoning, stuffing into a casing, then smoking or cooking, was a skill practiced in ancient times. It reached the status of a special art with the German sausage-makers, or "wurstmachers," who jealously guarded their secret sausage recipes and handed them down from generation to generation, master to apprentice.

In the history of Oscar Mayer & Co., the "wurstmacher" has been succeeded by modern food technology. Carefully controlled scientific formulas, while still secret, now insure the same flavor and uniform quality at all of the company's plants.

The finest ingredients—pork, beef, veal, pure spices—are combined according to precise formulas, processed under rigid temperature and humidity controls by skilled workers using complex machines, and packaged promptly to preserve freshness and flavor for rapid shipment to market. Many of the machines used in this modern processing have been invented and built by the company.

Specialists in this ancient art which has now graduated to a modern science, Oscar Mayer people make over 200 types of sausages, luncheon meats, smoked and canned meat products.

Dressed-out livestock, the company's raw material, is continually inspected by company quality control specialists as well as by Federal Meat Inspectors. The company handles over 600 million pounds of meat annually. On every package of Oscar Mayer product, Meat Inspection Division stamp 537 is found—proof that the product has passed all government inspections through every step of processing.

While Oscar Mayer & Co. specializes in processed meats, the company also sells fresh meat in volume.

One of the company's newest and largest additions to its production facilities is the hog immobilizer, housed in a special four-story building in Davenport. Using harmless carbon dioxide gas, animals are put to sleep before being dispatched.

Through his production men, the plant manager keeps in constant touch with on-the-line progress and problems.

A company-developed machine, the wiener-linker twists cellulose casings stuffed with sausage meat into standard wiener lengths at the rate of about 200 sausages per minute.

Another company development, the wiener-stripper machine peels the cellulose casing from the wieners after they have been smoked, cooked and chilled. Peeling is done at the rate of 183 sausages per minute. Now "skinless" all-meat wieners, they are ready for packaging.

16

17

Figure 47. Oscar Mayer, *Tradition and a Company*, 16–17, with photographs by Arthur Siegel. J. Walter Thompson Company Archives, Corporation Vertical Files, John W. Hartman Center for Sales, Advertising & Marketing History, David M. Rubenstein Rare Book and Manuscript Library, Duke University.

of the company, presenting it as a communitarian and benevolent organization: kind and protective toward its workers, beneficent toward and collaborative with farmers, meticulous in observing food safety protocols for its consumers' health.

Siegel's photographs constitute the primary content of the booklet, each accompanied by an explanatory caption. But the booklet offers information strategically, appearing to show much while revealing little—less, even, than Korling's books. Photographs of the interiors of the company's processing facilities, for instance, are sparse, generic, and cryptic. Just one page (fig. 47, *left*) includes images of the disassembly or cut floor, showing a man dressing carcasses (which hang in shadow and so close to the camera that they appear as undifferentiated, detail-free silhouettes) and another crew breaking down sides of fresh meat into smaller cuts. The caption accompanying this image notes that "While Oscar Mayer & Co. specializes in processed meats, the company also sells fresh meat in volume."[62] This text positions meat processing as a kind of side activity for the company rather than as the core activity that enables the subsequent production of all the company's products.

The same page carries another image showing the company's "hog immobilizer, housed in a special four-story building in Davenport [Iowa]. Using harmless carbon dioxide gas, animals are put to sleep before being dispatched."[63] Looking down along the machine from above, Siegel presented it as a series of pleasing diagonals, supported by a jumble of supports and struts. Although close looking at the image reveals a man standing at the base of the machine, and pig bodies confined within it making their way up the conveyor, at a glance it comes across as a generically industrial structure. The image stills and reifies the machine, presenting it as a fixed and passive object—much like the passive voice used in the soothingly anonymous, deliberately forgettable caption (how "harmless" can the gas and the machine really be, given their function?).

The facing page (fig. 47, *right*) touts the company's invention of a wiener-linking machine. Lest the emphasis on automation seem to come at the expense of human labor, Siegel shows a woman worker alongside the machine, unloading it. She no longer links sausages herself, a job historically assigned to women workers; now she tends the machine and delivers it of its products, midwifing the hot dogs into existence. Nearby on the same page, company managers, in suits under their white coats, peer down into cans that stand open, ready to receive product. Subsequent pages show workers peering at dials and tending machines, standing in rows at assembly line stations, and performing other cryptically industrial tasks, all in service of helping hot dogs and sausages to continue, unceasingly, on their way through the plant and out the door. All of these workers, whether male or female, white-collar or blue-collar, appear to have been relieved of physically demanding work by advances in automated manufacturing: they watch, monitor, observe, or inspect rather than actually make.

Subsequently the booklet devotes several pages to managerial functions and depictions of company leadership. In the images, men (and they are all men, apart from the secretaries) sit around conference tables, smoking pipes and cigarettes, surrounded by papers and ashtrays and coffee cups —the detritus of male mental labor.[64] They talk and argue—folding their arms, leaning back in their chairs, deploying what one caption terms "body English" to sell their ideas and move the company forward (fig. 48). The company's owners and "policy makers" are depicted in individual portraits; another section shows product testing by female home economists who bustle about the company's model kitchen, preparing samples to offer to

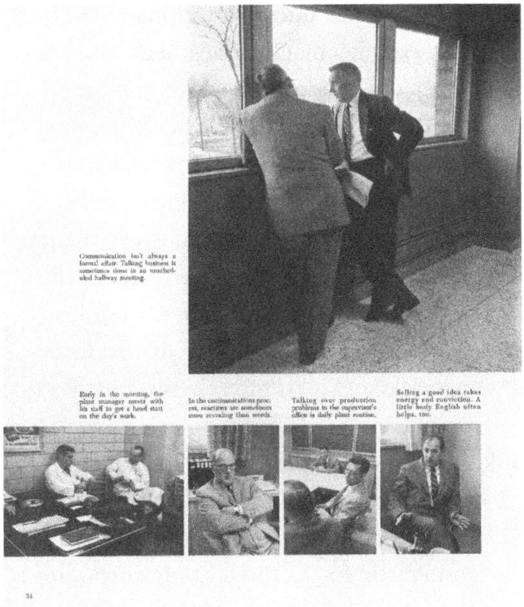

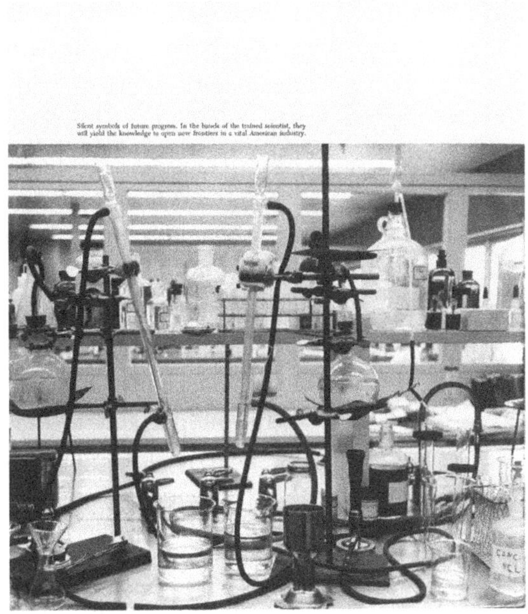

Figures 48 and 49. Oscar Mayer, *Tradition and a Company*, 34 and 48, with photographs by Arthur Siegel. J. Walter Thompson Company Archives, Corporation Vertical Files, John W. Hartman Center for Sales, Advertising & Marketing History, David M. Rubenstein Rare Book and Manuscript Library, Duke University.

male taste testers. A section depicting sales and promotion is dominated by consumers—women and children—along with "Little Oscar," the company's mascot. Drawn in advertisements with childlike features, "Little Oscar" was actually embodied for decades by a succession of men with dwarfism, who wore a chef's uniform and traveled with the company's iconic Wienermobile.[65] The booklet concludes by focusing on the Research and Development Department, shifting focus to the company's future and concluding with an image not of meat, or of hot dogs, but of a laboratory, shown—in iconic 1950s fashion—as an intriguing jumble of flasks, beakers, and tubing: "Silent symbols of future progress," the caption asserts, which will, "in the hands of the trained scientist . . . yield the knowledge to open new frontiers in a vital American industry" (fig. 49).[66]

Throughout the booklet, Siegel's images carry a formalism that was both Siegel's own style and a mode suited to the rationalist production model practiced and promoted by his client, Oscar Mayer & Co. The photographs lack, even avoid, the heroizing gestures toward labor that appear in Korling's publication for the Rath company or in other better-known examples such as Lewis Hine's 1932 *Men at Work*, a paragon of machine-age photography.[67] Such celebrations of labor and human-machine collaboration, born from the early twentieth-century machine aesthetic and the democratic populism of the Depression era, made less sense in the corporate culture of the late 1950s and early 1960s.[68] Siegel's Oscar Mayer booklet instead emphasizes the company as a totality, a machine made of machines, ready to produce product at high volume to support the efficient, capitalist production of the American postwar era. Everything and everyone in the Oscar Mayer universe—animals, machines, farmers, line workers and managers, inspectors and their food safety protocols, home economists, salespeople, and even executives—serves the company and its products. Everything and everyone, up to and including Siegel himself, is sublimated to the hot dog and the brand.

Siegel and Korling both took corporate commissions with regularity. Neither man would have regarded working within the bounds and restrictions set by their clients as a novelty. The commissions provided work, fairly generous terms of compensation (especially for Korling, doing corporate work during the Great Depression), and perhaps an interesting challenge. In employing these photographers to image their facilities, production lines, and managerial cultures, Rath and Oscar Mayer sought to deploy the various visual languages of documentary photography to create texts that seemed, in their readers' eyes, documents in themselves, unvarnished representations of the company's operations. The photographers brought their distinctive aesthetic sensibilities to the table: Korling's, a celebration of the machine age honed by his work for *Life* magazine in the 1930s; Siegel's, a careful, rationalist formalism bespeaking the influence of the New Bauhaus on his training and even on his very aesthetic perception. But the terms of the arrangement implied that the corporate clients would lay claim to any element readers might find pleasing, including the aesthetics of the imagery itself. Indeed, it was hoped readers would conclude that the aesthetic element derived from the production process, not from how the process was imaged: that the role of the photographer was to help corporate clients bring the beauty of process to the surface, highlighting existing hidden harmonies—not to create art but to assist in aestheticizing, and so apotheosizing, industrial creation. In this regard, meat proved no different from any other product: its slick surfaces gleaming, it could symbolize the future just as well as any automobile or household appliance.

Purity, Harmony, Identity

Marketing Meat to Women

THE RATH PACKING COMPANY celebrated its fiftieth anniversary in the meatpacking business by releasing its commemorative book *Waterloo Packer*. As discussed previously, the book presented a controlled and carefully managed vision of American industrial-scale meat production as primarily a white male activity. Women workers performed only a select few jobs at the meatpacking plant, such as bacon packaging. The most prominently featured woman in the book was the female consumer depicted in the book's final spread (see fig. 34), purchasing meat from a male butcher.

Correspondence around the creation of *Waterloo Packer*, however, reveals that though women were marginalized *within* the book, in reality they were its primary target audience. Indeed, the entirety of the Rath fiftieth-anniversary celebration, including not only the commemorative book but the galas, dinners, performances, and other events, aimed to increase the company's name recognition among female consumers. As one planning document, written by a consultant from Young & Rubicam, notes,

> Publicity for The Rath Packing Company's fiftieth anniversary . . . would serve to establish the age and reliability of the company in the mind of the general public and thereby increase its prestige. . . . At the present time, the name "Rath" on meat products may signify little or nothing to housewives who have never used those products. It is just another brand name. They know nothing of the company or its history. Following the anniversary publicity, however, the name "Rath" would signify a company that has been in business for fifty years—and therefore must manufacture quality products—to all housewives who see the news stories about the anniversary. The products of the company would have increased prestige in their minds. They would be more inclined to try those products.[1]

Reaching "housewives" like the one depicted in the book's final spread was thus a key strategy in the circulation of *Waterloo Packer*. Rath distributed the book free of charge, and apart from a few gift copies for journalists and food-industry worthies, the company's salesmen delivered most of the fifty thousand copies to customers along their wide-ranging sales circuits throughout the US. Although most of their direct contacts—butchers, grocery store managers, and other merchandisers—would have been men, salesmen were explicitly instructed not to deliver the brochure to their customers' places of business but rather to "meet your customer away from business, to meet his family and to make sure that this brochure will reach his home."[2] Even in families that did business in meat, then, women's purchasing power was recognized as a driving force. *Waterloo Packer* posited meat *production* as a primarily male activity, with limited and controlled female presence; but Rath and its advertising consultants recognized that increasing meat *consumption* required an appeal to women, the people who most often prepared and, more important, purchased meat.

This chapter considers how, in promoting its products, the American meat industry has depended heavily on controlled visual access, much of it designed to appeal to women, much of it *including* women. Consciously permitting, or restricting, visual consumption of its products and its production processes has been critical to its rise as a major American industry. Controlled deployment of visual imagery has been a core means by which the industry has granted consumers selective access to the sources of their meat, highlighting some aspects of meat's path from farm to table while suppressing others. The visual and performative aspects of meat production contributed in crucial ways to the increase in meat consumption in the early to mid-twentieth century. Advertisements not only promoted products but demonstrated behavioral patterns of increased meat eating, designed to generate—and highly successful in generating—actual changes in consumers' behavior.[3]

Gendered performances within racially homogenized spaces constituted a central operational strategy of meat industry publicity between the beginning of the century and the mid-1960s. Promotional material posited the archetypal American consumer of meat, and thus the primary target of meat industry advertising, as a white woman, responsible for the domestic activities required to run a household. Advertising and public relations agencies conducted extensive studies to determine and to cater to the preferences and unspoken assumptions of real consumers, of course: from what time of day they preferred to shop, to the conclusions they drew about a cut of meat based on its color, to what direction they preferred to walk once inside the supermarket.[4] But in ads, companies generated a *vision* of the American woman in whom, they hoped, their consumer would see an idealized version of herself. Advertisements depicted her as middle-class, married to a man working a steady job with regular hours, and mother to at least one, but usually not more than two, children.[5] Sometimes she was pretty and young, though (usually) already confident in her skills as a homemaker; sometimes she was a cheerful, pleasant-looking matron, her reassuring presence a sign of quality in the product. And she was, with very few exceptions, white.[6] These norms extended not only to the fantasy women who appeared in advertisements but to the real women employed by the meat industry as home economists, whose work proved vital to expanding the perception of meat as a simple, everyday, fundamentally American foodstuff.

In itself, the gendered nature of meat industry promotion is no real revelation: the majority of advertising, particularly ads focused on food, home goods, and other products associated with domesticity, targets and has targeted women.[7] What matters here is the effect that such appeals had on the perception of the product overall. The gendered and racially homogenized nature of meat advertising helped to domesticate meat itself in the mid-twentieth century. Gendered appeals posited meat as an everyday food rather than a luxury food—a proposal that myriad twentieth-century women embraced wholeheartedly. Women's behavior, accordingly, constituted the primary means by which meat became (and by which it remains) a food consumed daily by Americans across wide spectra of class, race, and culture.

Sometimes, publicity and advertising materials directly—albeit not graphically—acknowledge the process by which animals become meat. This is the dynamic that plays out in *Waterloo Packer*, where a small army of people labors to bring meat to the white American woman's table. By far the dominant strategy in meat industry advertising, however, involved suppression of the fleshly animal nature of meat in favor of what we might term its "foodly" nature—how it not only could but *should* become a central feature of the American table: available across a range of price points, within the reach of cooks across a range of skill levels, central to everyday meals and feasts alike. Taking to heart, and taking literally, the adage that people don't want to know how the sausage gets made, meatpacking companies mostly avoided depicting their packinghouses or packinghouse workers in advertising. Instead, they more often leaned on other groups of employees. Scientists from their research and development divisions show up with some frequency in advertising, bringing with them implications of scientific rigor in meat production.[8] But the workers who appear most frequently by far are the home economists who worked for many companies. This phalanx of women workers served as the American "housewife's" representative and proxy within the industry. In advertisements, cookbooks, and other publicity materials, home economists channeled the efforts of the industry into recipes, presentations, and menus, translating the hard machinery of meat production into the soft stuff of the home environment.

From early in the twentieth century, not only the content but the very makeup of advertising itself incorporated deep assumptions about gender preferences. Elspeth Brown notes how, in the first two decades of the twentieth century, photography was used more often in ads designed to appeal to men. Perceived, in these early years, as a straightforward and fundamentally factual medium, unable to carry complex messaging, photography tended to be deployed in ads employing rationalistic, "reason why" pitches. Appearing in male-oriented business publications, these photographically illustrated ads functioned as "visual analogue[s]" to the logical arguments for why a particular item or brand was best. Ads designed for women in the early twentieth century were often more impressionistic, relying on traditional illustration to increase desire. Carrying associations of refinement and culture, drawn illustration added cachet for brands, as well.[9]

Advertisers soon came to recognize, however, that photography could be just as persuasive as any other medium—indeed, more so, because of the air of flat-footed, literal factuality that clung to the photograph. Writing in 1937, Paul Parker notes that since 1920 there had been "greater use of photographs and transparent wash drawings, as opposed to the ubiquitous opaque water color and black and white oil paintings of yesterday."[10] Marketing scholar Richard Pollay confirms that "early in the [twentieth] century [advertising] artwork was primarily pen and ink or engravings, as appropriate to the then prevailing printing technology. Original paintings in a variety of media were popular for the next two decades but were gradually displaced by photography, which was utilized in 60 percent of . . . ads by the 1930s."[11] Just like the picture magazine editors of the Depression era who quickly learned how to deploy photography liberally and effectively in service of a controlled and fundamentally boosterish vision of the modern Western world, advertisers saw the medium's potential to sell a fantasy of perfection that much more believable because it appeared wholly real.[12] This required some work, of course, as Parker notes: "I do not mean . . . to imply that the advertiser can merely show a photograph of his product and expect to convey the most strikingly effective communication. . . . The photograph must be retouched to remold the object so that it is realistic in [a] mnemonic or anticipatory sense; so that the 'should be' dominates the actual."[13] Traditional illustration and photography came to coexist in meat advertising, as in much other twentieth-century American advertising, working together to present a vision of a near future in which perfection was wholly achievable, available for the price of a purchase.

Purity, Harmony, and Identity

Within the meat industry, visions of perfection proved highly and increasingly desirable in the first decades of the twentieth century. There were certain key problems that every major meatpacker shared alike, and that advertising and publicity were often called upon to solve. One key concern was what one might term "*The Jungle* problem" (although concerns about meat's health and purity predated Sinclair's novel): the persistent need to assert the health, wholesomeness, and cleanliness of American meat products.[14] Publicity should not attempt to offer a preemptive denial of the conditions described in *The Jungle*; rather, publicity should behave as though such conditions had never existed and could not possibly exist. Meat advertising had to read simultaneously as sophisticated and naïve, appealing to the consumer's deepest physical and psychological cravings while never acknowledging that meat could ever be anything *but* pure and healthful. Meat products had to seem so clean, so pure, and so safe that the consumer would never even consider the alternative. Someone encountering *The Jungle* (or any other nightmarish account of American meat production) should read it as slanderous hysteria, a fever dream, even a kind of science fiction, representing a dystopia thankfully removed from modern American reality. It was a tall order.

Another concern shared by the industry overall focused on harmonious labor relations. Labor unrest occurred regularly in the meat industry over the course of the early to mid-twentieth century. Like other factories, meat-packing plants depended for their efficiency on the division of slaughter and disassembly processes into discrete, repetitive tasks; and as in other industries, the needs and goals of meatpacking management and labor tended to be at odds. Meatpacking management attempted to maximize efficiency and expand narrow profit margins by mechanizing as many operations as possible and deskilling the rest, reducing dependence on expensive skilled laborers. Labor, in turn, asked for dignified treatment, adequate pay, and health care sufficient to cover the injuries—from repetitive stress, accidents with knives and saws, and even threats from animals—common in what could be quite a dangerous work environment. Strikes in meatpacking plants could have real and immediate effects on the American food supply chain, highly undesirable at a time when the meat industry was endeavoring to make its products ubiquitous on tables in the US and worldwide.[15] Downplaying the difficult and bloody nature of much meatpacking work, conveying an image of paternal friendliness toward their workforces, and —above all—guaranteeing the American public access to an unending supply of healthy meat became central concerns of the meat industry and thus central drivers of its public relations and public imaging initiatives.

The very fact that the meat industry shared so many concerns and worries, that every meatpacking company had to combat so many of the same issues in its public imaging, hints at another key problem: identity. Every meatpacker had to negotiate the question of how to differentiate itself from other companies in the business in the minds of the meat-buying American public. But the very nature of the problem demanded that each company solve it in a way that was, or at least seemed to be, unique. We saw in the prior chapter how meat industry organizations used visual imagery in part to convey a sense of corporate identity. But how did that translate into company identity, brand identity, as it was projected to the public?

In solving this problem and establishing company identity in the minds of the public, every meatpacking company had to contend with issues inherent in the structure of the industry itself. In the early years of the twentieth century, a consumer going to her butcher would generally have very little control over the origin of the steak she bought. Such factors depended far more on behind-the-scenes dynamics—on the salesmen from whom her particular butcher purchased goods; on her city's access and connection to the vast rail and road networks that carried meat out of Chicago, Sioux City, Omaha, Madison, and other midwestern meatpacking hubs—than on her own preferences or her own ability to differentiate between one meatpacker and another. Consumers generally would not demand "a Swift steak" or "a roast from Armour." To the extent that consumers had a choice about what they purchased, they would choose by cut, or by what they perceived to be the quality of the cut relative to others, not by brand.[16]

As advertising increasingly placed (the illusion of) control over brand preference in the hands of the endpoint consumer, meatpacking companies found themselves faced with the question of how to appeal directly to buyers, how to make themselves and their products memorable and desirable. Many companies arrived at the same solution: displacement of brand identity away from fresh cuts—such as steaks, chops, or roasts—onto processed, packaged, preserved products. These might be edible products: canned hams, presliced bacon, or potted meat products like SPAM—the flagship pork product of George A. Hormel & Company.[17] Or they might be by-products like soaps and fertilizers, a category of product that has bolstered profit margins in the American meat industry from its earliest days. Slim profit margins on meat drove meatpackers to wring revenue out of animals however they could, treating animal bodies as sites for extraction of a host of products: hides and leathers, fats, soaps, fertilizers, bone china, and gelatins.[18] Branded meat products and by-products could, if successful, raise the fortunes of the entire company that manufactured it, carrying steaks and roasts with it along the way. Oscar Mayer may have made more than hot dogs; but in a fundamental way, hot dogs made Oscar Mayer.

Purity and Harmony: Armour and "The Ham What Am"

The American "Big Four" meatpacking companies, Armour, Swift, Morris, and Hammond, dominated the industry at the turn of the century.[19] The companies pioneered not only industrial meat production but also its advertising, and these common concerns are built into their publicity campaigns from earliest years. Founded by Philip Armour, a ruthlessly competitive man whose business extended well beyond meat production, Armour and Company grew in the later nineteenth century to become one of the biggest of the Big Four, battling Swift for supremacy even as the companies worked together to form an oligarchic pool, cooperating to fix prices and drive smaller competitors out of business.[20]

As early as the 1890s, when a series of public health crises drove passage of some of the first food safety laws targeting the meat industry, Armour exhibited across its advertising a persistent preoccupation with purity in its advertising for both food products and by-products, often expressed by association with whiteness.[21] In this it was hardly alone: advertising itself, as Anne McClintock shows, grew not only alongside but directly out of the colonialist endeavor, as colonial labor exploitation led to surpluses of cheap goods. Soap, in particular, became emblematic of this shift and was among the earliest products to be branded and marketed. "In order to manage the great soap show, an aggressively entrepreneurial breed of advertisers emerged," McClintock writes, "dedicated to gracing each homely product with a radiant halo of imperial glamour and racial potency. . . . Soap was credited not only with bringing moral and economic salvation . . . but also with magically embodying the spiritual ingredient of the imperial mission itself."[22] Armour, aggressively promoting branded by-products of its meat business, simplified the formula even further. Whiteness and purity themselves became universal signifiers across the company's brands: its soups were "White Label," its soap was "White Soap," its lard was "Simon Pure" (a common adage at the time, the term *simon-pure* means "absolutely pure" or "untainted") (fig. 50). From the start, the ads targeted women, the presumed recipients of the chivalric attentions of the "Armoured" knights shown in many of the company's ads. The cultural and racial hierarchies implied in the imagery of crusading warriors only aided in promoting the purifying qualities of its products.

After the publication of *The Jungle*, many of these associations were strengthened by the advertising subtext. The implicit point of comparison became not the taint of racial impurity but the threat of industrial production with its filth and contaminants. Sinclair's hellscape evoked Armour's production facilities in Chicago, among others, but the company's advertisements instead harkened to a halcyon era.[23] By 1911 Armour's Simon Pure had become "this wonderful *cream* of lard," a pure white substance implicitly rising to the top, above other products. One advertisement from

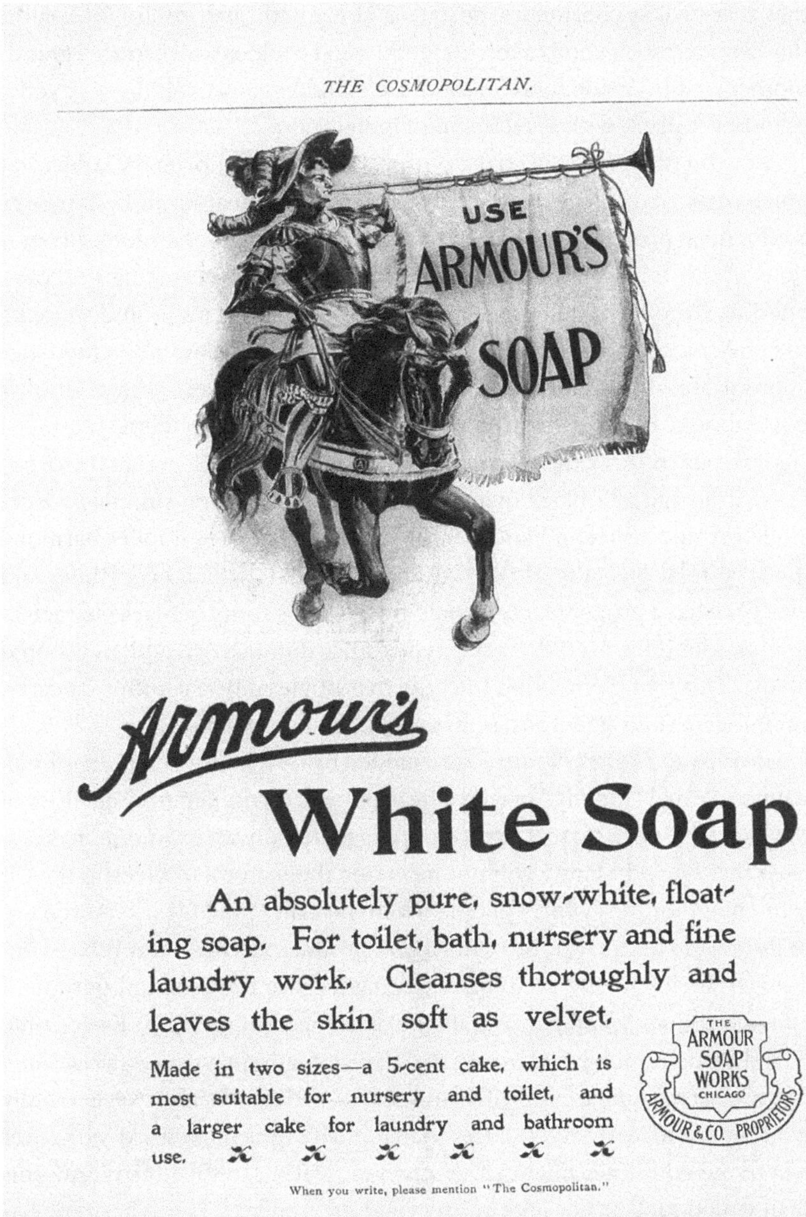

Figure 50. "Armour's White Soap" advertisement, 1897. Roy Lightner Collection of Antique Advertisements, 1871–2009, box 1, John W. Hartman Center for Sales, Advertising & Marketing History, David M. Rubenstein Rare Book and Manuscript Library, Duke University.

that year evokes consumers' nostalgia, along with their desire for health: the "wonderful pies and cakes, doughnuts and cookies which graced grandmother's table" made with Simon Pure could be consumed by everyone, "children and all, without thought of indigestion."[24]

But the company's primary output, and thus the primary subject of much of its advertising, was not by-products but meat itself. In Armour's ads for meat products, the emphasis shifts from purity to harmony. Promotions of the company's Star Ham deploy a complex messaging program, building from a combination of imaging, text, and repetition over years to convey a sense of the ham as both everyday and elevated, a foodstuff appropriate for elegant meals but within budget for an average family's pocketbook. From at least 1906, the company regularly deployed racial stereotypes to carry this messaging. Armour's ads for Star Ham incorporated an image of a Black man wearing a white uniform, smiling broadly while carving a slice of ham. Happily servile, his image implies harmony in social relations, a world of rigid hierarchies in which everything and everyone has a proper place. Donald Bogle, in his study of Black characters in film, identifies five "character types" that dominate in white-authored media. This figure embodies the type that Bogle calls "the tom . . . hearty, submissive, stoic, generous, selfless, and oh-so-very kind."[25]

In one 1906 ad, the figure is surrounded by text, addressed to an unseen woman named "Mandy," in which he expounds on the superior qualities of the ham (fig. 51). The text, written in exaggeratedly racist idiom, makes it clear that he and Mandy will not be eating the ham themselves; rather, it is intended for the (white) people whom he calls "the Folks."[26] At the top of the ad and in the text, he refers to the product as "The Ham What Am," a slogan the company used for years to advertise the product, both with and without an accompanying image of the man himself. An Easter time advertisement, published nearly two decades later in a 1924 issue of *Good Housekeeping*, depicts him with shiny skin, a wide smile, and exaggeratedly white eyes and teeth (fig. 52). He stoops, and his eyes tilt upward: no matter where the consumer might sit or stand in relation to the illustration, this man would look at her with subservient deference.[27] The stereotype has roots in long-standing American depictions of southern plantation life; but Armour's ads imply that for its white consumers, a genteel lifestyle no longer has to be the province solely of a landed elite. These advertisements encouraged middle-class white women, too, to imagine themselves the pos-

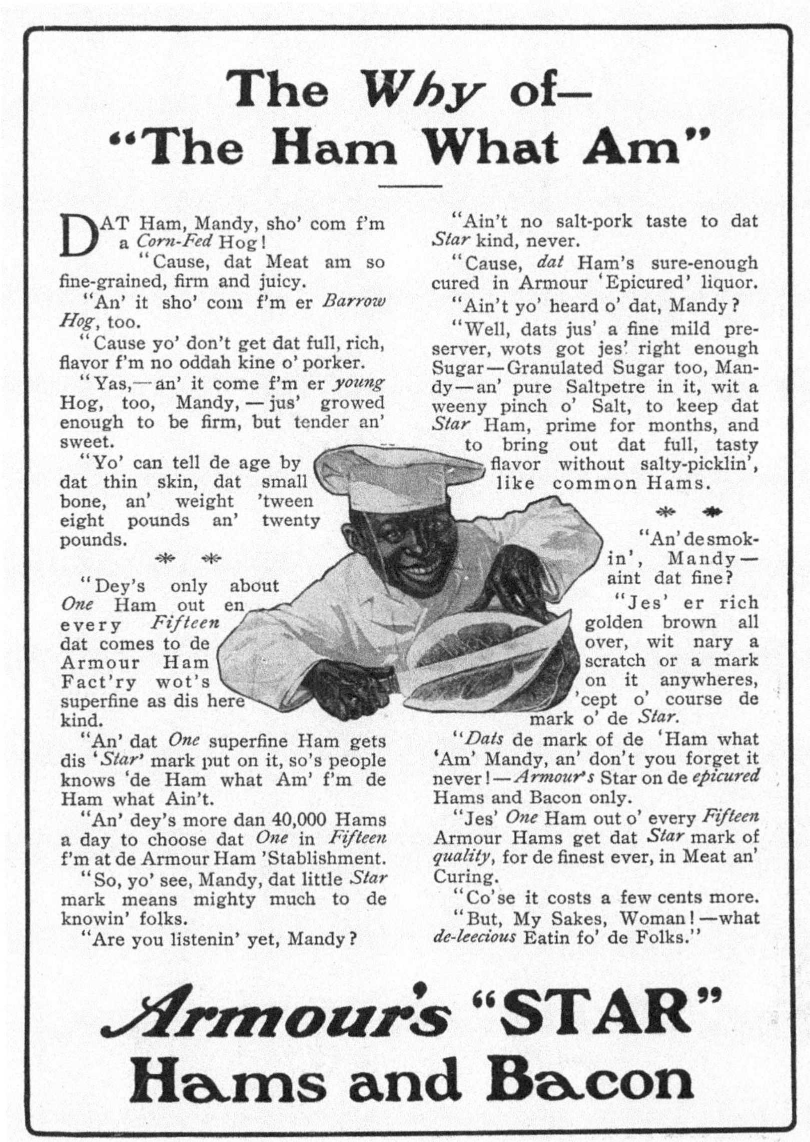

Figure 51. "Armour's 'Star' Hams and Bacon" advertisement, 1906. Roy Lightner Collection of Antique Advertisements, 1871–2009, box 5, John W. Hartman Center for Sales, Advertising & Marketing History, David M. Rubenstein Rare Book and Manuscript Library, Duke University.

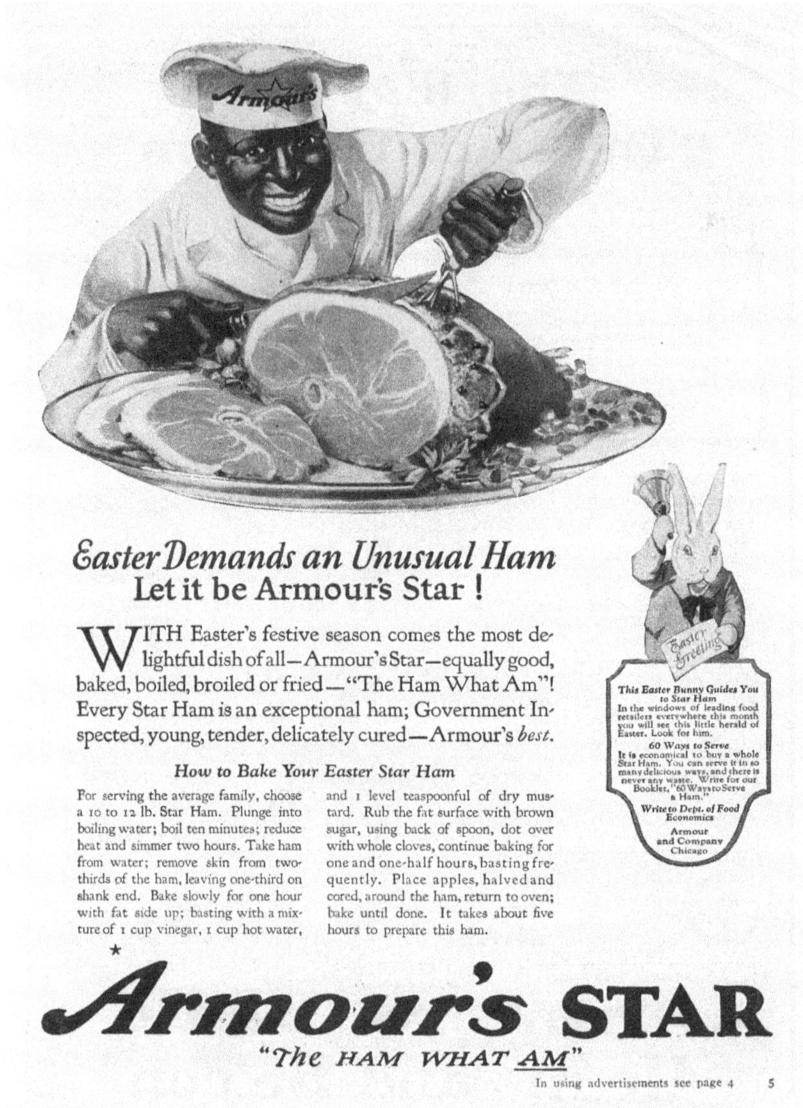

Figure 52. "Armour's Star" ham advertisement, 1924. Roy Lightner Collection of Antique Advertisements, 1871–2009, box 1, John W. Hartman Center for Sales, Advertising & Marketing History, David M. Rubenstein Rare Book and Manuscript Library, Duke University.

sessors and overseers of vast properties and happily, harmoniously servile people.[28] Through this smiling man who exists to serve, desiring nothing for himself, the audience can imagine that he serves them, that they are people who deserve to be served.

The catchphrase "The Ham What Am" proved powerful enough that it could evoke the figure's presence (and the concomitant status it imparted to the consumer) even in his absence. Another 1924 Armour ad carries an image of "A tasteful June meal": a pitcher of iced tea, biscuits, a potato casserole, peas, a jelly, and a ham on a platter, with one oversized slice of ham on a plate (fig. 53).[29] Below, an image of an entire wrapped, branded Star Ham sits alongside a text asserting that "you don't have to prepare elaborate dishes to gratify the average man. . . . Tempt him tonight with this 'easy to prepare' Ham Steak Dinner." The brand name and the catchphrase sit at the bottom of the ad: "Armour's Star: The Ham What Am." In the context of this ad alone, that folksy catchphrase with its Seussian syntax seems a bit out of sync with the assertions of elegance in the rest of the ad. But for the white target audience of the Armour ad, primed by decades of the company's repetitious messaging, the catchphrase alone was enough to trigger audience buy-in to the stereotype, and to the intimations of elegance—grounded in racial hierarchy—that it implied. Meat already occupied an elevated status among foods. Large cuts like hams and roasts were additionally rarified, the food of special occasions and holidays, worth saving one's pennies to purchase. These ads promised the consumer that she was worthy of having and serving such good food and of occupying the superior social status it connoted. By surrendering to this happy simpleton and the slice of meat he offered, the consumer could realize certain core satisfactions: thrift, everyday elegance, a happy and satiated husband.

Armour's early twentieth-century ads overall respond to a number of the core issues facing the meat industry. They convey purity through repetition of words associated with whiteness and cleanliness. They minimize the presence of labor—both the labor required to generate and package the product and the labor of the consumer after purchase of the product—and place emphasis on harmony and ease. Where work is represented, notably in the image of the man carving the Easter ham, the representation signifies both the elegance of the product and the (presumptively white) consumer's social and racial superiority over the harmoniously deferent servant. And we see in Armour's ads the emphasis on a processed, branded product,

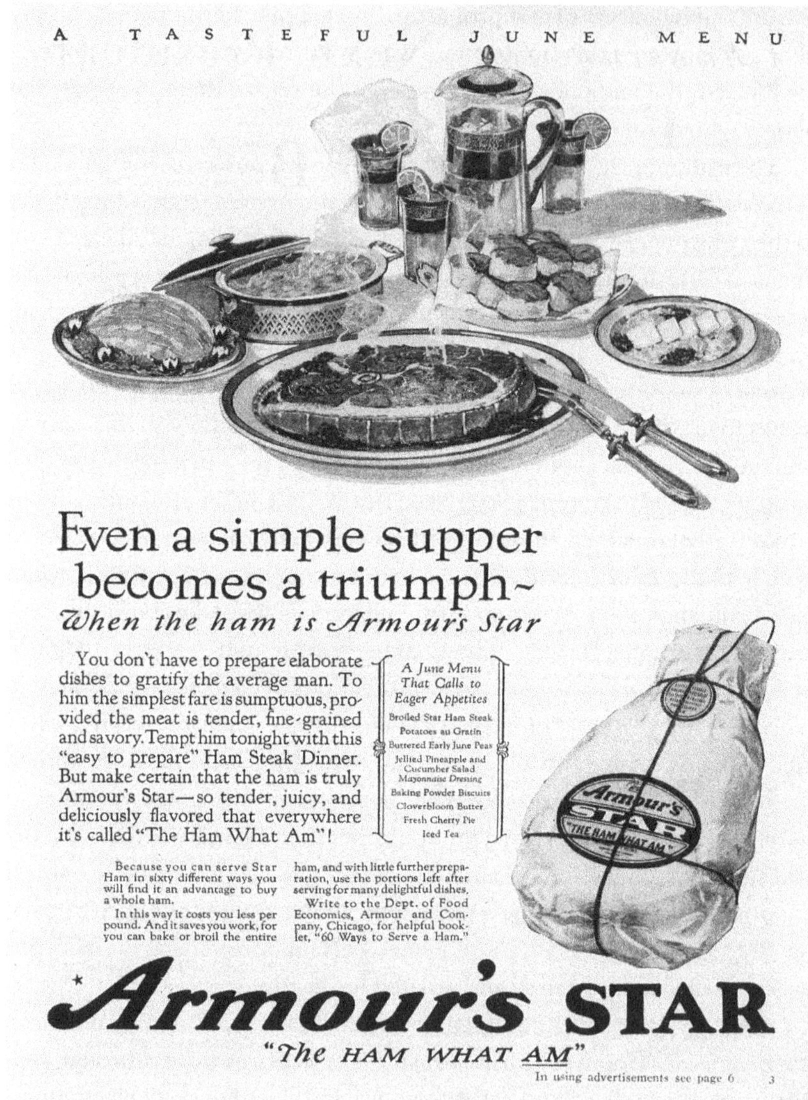

Figure 53. "Armour's Star" ham advertisement, 1924. Roy Lightner Collection of Antique Advertisements, 1871–2009, box 1, John W. Hartman Center for Sales, Advertising & Marketing History, David M. Rubenstein Rare Book and Manuscript Library, Duke University.

the Star Ham, so recognizable and reliable that it carries its own catchphrase. The features that characterize so many later meat industry publicity campaigns all appear here in Armour's early promotions, developed into a sophisticated network of messages, relatively stable and building on themselves across products, campaigns, and years.

Identity Reigns: Hormel and Oscar Mayer

The advertisements of George A. Hormel & Company span the period, starting in the early 1900s and expanding dramatically in the later 1930s and 1940s, when print advertising became a significant aspect of Americans' visual landscape. Founded in 1887 as a butcher shop in Austin, Minnesota, Hormel quickly expanded to become a full-scale slaughtering operation modeled on those in Chicago.[30] It initially lacked the reach of the Chicago firms due to its provincial location; but by 1910 it had begun to seek out national business, and its first national advertisements appeared in that year.[31] One 1911 notice promotes the company's "Dairy Hams and Bacon" as having been created from "porkers of *preference—only* the choicest selection of prime young milk- and grain-fed hogs raised in southern Minnesota's famous dairy section."[32] These were available by mail order, the ad also notes, "if your dealer does not stock Dairy Brand goods." The company's proximity to its animals' grazing grounds, its location in a region famed for (pure, white) dairy, and its distance from Chicago helped it to skirt "*The Jungle* problem" almost without trying.

This ad, like some other early meat industry promotional material, acknowledges the creatureliness of meat animals in a way later ads often do not, discussing the animals' diet and how it will impact the animals' subsequent quality once they, in turn, become diet for humans.[33] Such forthrightness tended to fall away as the century progressed. Indeed, by the late 1920s the company, though prospering as a meat-making enterprise, had begun to shift away from identifying as such in public, going all in for canned products to the extent that it came to seem almost more a canning company than a meatpacking company. It foregrounded its canned hams, beef broths and soups, and, by the later 1930s, SPAM. A hybrid of sausage and by-product, SPAM was immediately a "large-volume seller," driving the company's success in coming years: "A large share of the million dollar profit shown by Hormel in 1938, a year in which most packers took heavy losses, is evidence of the part played by this new product in expanding sales."[34] The product became within just a few years indelibly associated with cafeteria service "mystery meat," after American GIs and schoolchildren found themselves consuming the product, sometimes multiple times daily, during and after the Second World War. Despite widespread mockery, however, sales of the product grew throughout the 1940s, helping to cement consumer expectations that cheap, readily available meat products should feature prominently at every meal.[35]

One early SPAM advertisement, dating from the late 1930s, demonstrates how images and text converged in meat industry advertising to produce complex messaging about consumption, family, and the pleasant ubiquity of inexpensive meat. The ad incorporates three images of different recipes, showing the product's versatility: one image depicts a "SPAMburger"; one, a "SPAM 'n' salad"; and one, a tray of fixings for making "SPAMwiches" (fig. 54). At top, a fourth image shows a family of three outdoors—the father kneeling next to a portable grill on the grass to cook SPAM, while the mother bends before him to take a piece with a bun, and a little boy stands between them, his speech bubble reading, "We like SPAMburgers a lot!" The mother's enlarged and disembodied head also floats at the top right corner of the page, looking down over it all and noting the product's versatility. Text attributed to her as "Mrs. Havstad" notes how good tasting the product is and how often the family eats it in the summer. At the bottom, next to an image of a can of SPAM, is an inset box with an image of another woman, the company's home economist, "Mary Butler of the SPAM family," describing the product's wholesome-sounding ingredients. Various other images and text boxes appear there, too—a Good Housekeeping Seal of Approval and an official-looking stamp from the American Medical Association reading "Eat meat—it helps you stay fit."

Complex and overdetermined, the ad places certain demands on the viewer. The eyes are directed here and there. Two women, a consumer and a professional expert, testify to the product's quality. The product itself appears repeatedly: indoors and outdoors, alongside vegetables and bread, in its package and in slices. Yet even a cursory glance allows the viewer to apprehend the message that SPAM has a place, even many places, on the American table; that model American families (read: white, young, and attractive families) use and appreciate the product. Once a luxury item, meat here becomes an everyday, multiple times per day staple. Indeed, the

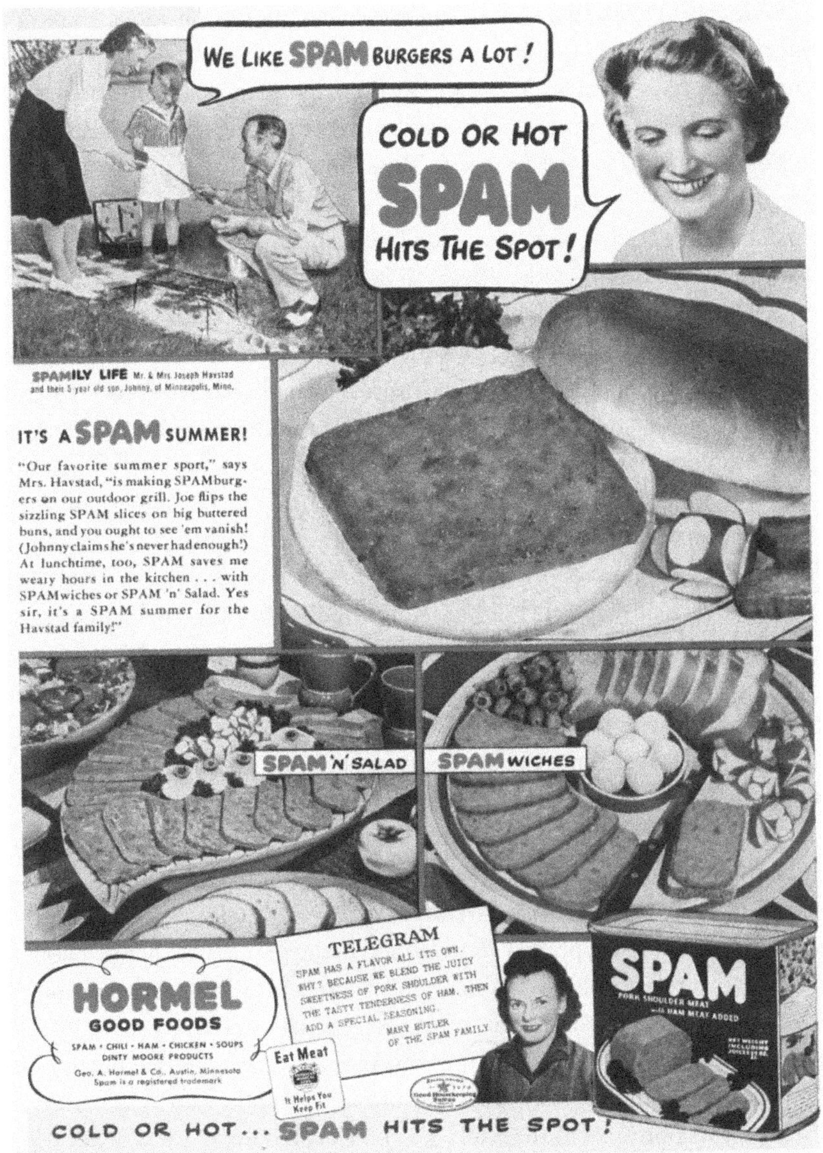

Figure 54. Hormel SPAM advertisement, ca. 1937. Roy Lightner Collection of Antique Advertisements, 1871–2009, box 5, John W. Hartman Center for Sales, Advertising & Marketing History, David M. Rubenstein Rare Book and Manuscript Library, Duke University.

product is not even sold *as* meat but rather as a branded product: the word *meat* appears only in the tiny text on the image of the can—"pork shoulder meat with ham meat added"—and in the American Medical Association stamp. Although Hormel ran (and still runs) a massive, full-scale slaughtering operation in Austin, Minnesota, the ad implies that it is primarily a canning operation. Production of *meat* would require labor from packinghouse workers; preparation of *meat* would place demands on the consumer's time. SPAM, on the other hand, appears to demand little or no effort from anyone. If meat industry advertising aims to minimize labor and maximize branding, SPAM ads represent an effective coalescence. Where once fresh meat had represented a seasonal pleasure, Hormel's ads for SPAM work to demonstrate that the opposite of luxury is not abstemious penury but rather an easy everyday abundance.

If Armour represents meat industry advertising in its infancy, and Hormel demonstrates the cementing of meat's "everydayness," Oscar Mayer constitutes in some ways a paragon, not just of meat industry advertising but of advertising overall.[36] Founded in Chicago in 1883, Oscar Mayer expanded rapidly from a sausage retailer into an industrial-scale meatpacking organization. It never abandoned its dedication to processed products, however, presciently cultivating the company name as a widely recognizable brand.[37] Over time, the company name came to epitomize American food itself, becoming not only a brand but a phenomenon. Although it makes a number of highly recognizable products, such as bacon, bologna, and other lunch meats, in the popular mind, Oscar Mayer is hot dogs, and hot dogs are Americana. Oscar Mayer in some ways took a tack precisely the opposite of Armour's prewar strategy, and in the process further developed Americans' relationship with meat. If Armour's prewar consumer aspired to an elevated social status, with all the large roasts and happy servants that accrued thereto, by the mid-twentieth century consumers had become so comfortable with everyday meat consumption that they no longer needed to justify its purchase to themselves as an elevated luxury. It was not even *meat* anymore, it was simply food: packaged and branded, recognizable and familiar.[38]

Targeting a woman consumer, Oscar Mayer's advertisements do not try to seduce her with the fantasy that she has a domestic staff, or even a single servant—why would she need one, after all? The Oscar Mayer family comes across in its advertising as self-sufficient, even self-contained:

an isolated unit, nuclear and stable, able to do for itself.[39] The American housewife of Oscar Mayer's postwar era does not aspire to be upper-class but more perfectly middle-class, more perfectly herself. As such, the company's advertisements do not try to persuade consumers that they *deserve* to have meat; rather, the company was wildly successful at convincing buyers that meat was *not* special but supremely democratic: available to everyone, within the reach of everyone, part of the American birthright.[40]

At mid-century, Oscar Mayer endeavored to make itself ubiquitous, to insert itself into the American psyche to the point that Americans regarded Oscar Mayer virtually as part of their birthright. To do this it drew on, depended on, and subtly manipulated the deep symbolism of Americanness, the American family, and an American childhood in ways remarkable precisely for their calculated innocence.[41] An advertisement, particularly one for an already wildly successful brand, would not attempt to tell the audience anything new about the American family; rather, it ought to draw on what audiences already feel and tell them things they think they already know.[42] If social mores change, advertising might reflect those changes; perhaps it might even aim to further and boost those changes. But it should not initiate change in itself.[43]

In Oscar Mayer's ads and other image productions of the late 1950s and early 1960s, the company forged and continually reinforced a public image through repetition and reuse of the same images and text in changing configurations. Black-and-white photographic illustrations coexist in some of these ads with larger color photos or illustrations from drawings, the real and the ideal working together to convey a sense of the company as truthful and trustworthy, a concrete and reliable presence in American consumers' lives. In others, drawn illustrations appear on their own, carrying the fantasy of lives made perfect and easy—ideally harmonious—through purchase of the right products.

Two images appear in a 1961 advertisement for Oscar Mayer bacon (fig. 55), designed to run in the glossy color rotogravure supplements found in weekend editions of newspapers (black-and-white versions exist as well).[44] At top, a small black-and-white photograph depicts a scene from the Oscar Mayer headquarters. In this image, a group of "bacon testers," identified in the caption as two vice presidents and the president of Oscar Mayer, sample bacon in a model kitchen. A sense of purity pervades the clean environment, at once kitchen and laboratory. Two unnamed women,

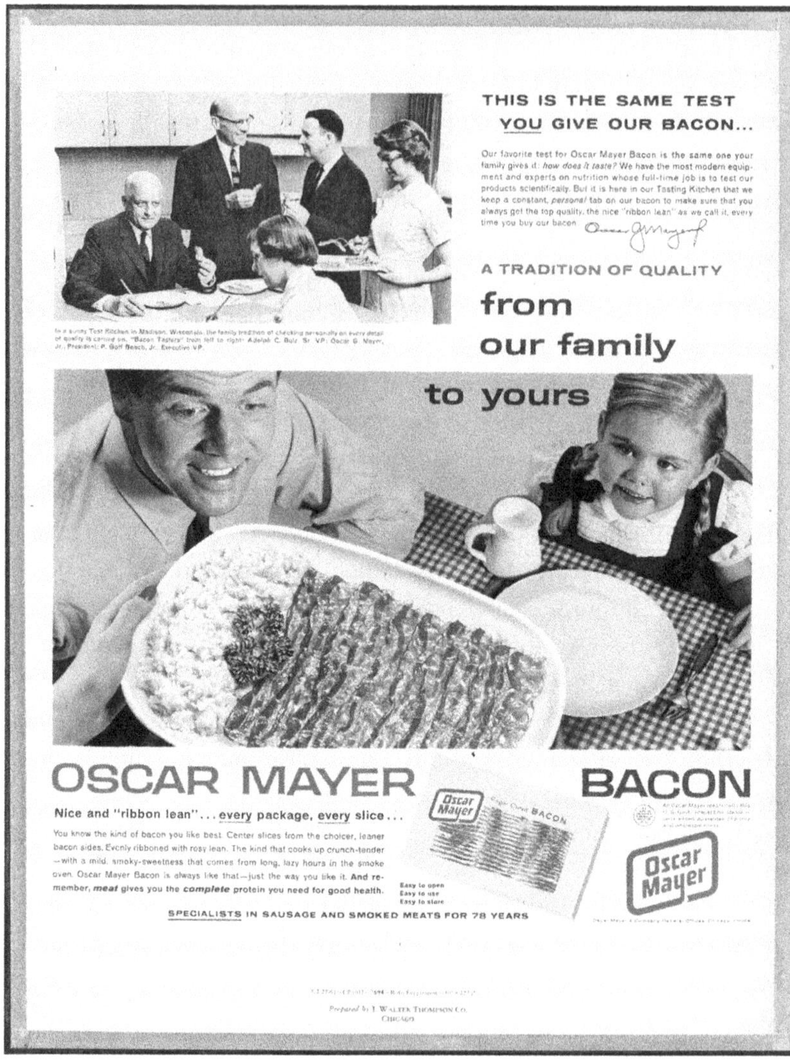

Figure 55. "Oscar Mayer Bacon" advertisement, 1961. J. Walter Thompson Company Archives, Domestic Advertisements, box OM1, John W. Hartman Center for Sales, Advertising & Marketing History, David M. Rubenstein Rare Book and Manuscript Library, Duke University.

both uniformed home economists, record the testers' reactions. Two of the men look at one another, while the third, seated, smiles toward the women. Their heads bowed as they record data, neither woman returns the gaze. They are, in one sense, working women, doing a job in a professional environment; but they also appear servile and subservient, subjects of (and subject to) the male gaze, unable to reciprocate. Like many home economists in meat industry imagery, their status is uncertain and unstable: their recipe development, product testing, and data collection form a valuable component of meat companies' public relations endeavors; but they also enact a heightened version of the American woman's role as household nurturer and manager. They demonstrate to American women not only the best way to make bacon but the best way to *be*, and the best way to serve.

As part of this same advertisement, large text nearby reads, "This is the same test you give our bacon. . . . A tradition of quality from our family to yours." Below, the primary illustration enacts and realizes this transfer of knowledge and product from corporate to private realms. In this photograph, a little girl of about five and a man, wearing a shirt and tie, look up with excitement as a platter of bacon and eggs, decorated with a sprightly sprig of parsley, descends toward a nicely set table. The blue of the tablecloth, the man's shirt, and the child's jumper contrast pleasingly with the yellow of the child's shirt, the same color as the eggs on the platter. Blue and yellow, in turn, work together to set off and show off the streaky-red bacon. The mother appears only as a single hand, holding the platter, one shiny thumbnail the same bright red as the meat. The man and child look up at the platter; the image looks down at them. The woman barely appears, and yet we feel her presence keenly because, the angle implies, *we* are *she*. Our point of view is her point of view; her manicured hand is our hand; and her family, by extension, is our family—just as the caption says it is. In the two images in this ad, the roles of women, vis-à-vis meat and the family, coalesce: the ad proffers a vision of the full scope of women's lives in the 1950s.[45] Although home economics could be a demanding and rewarding career for women at the time, the ad presents it as hardly a career at all. Rather, it is presented simply as mothering outside the home rather than inside it.[46] Again, photographs in advertising often functioned to embody a rationalistic or scientific viewpoint, while drawn illustrations carried emotionally affective content. This ad reinforces its messaging about women's roles and duties in maintaining a harmonious home by deploying both idioms together.

Meat industry advertising of the mid-twentieth century overwhelmingly presents the purchasing and preparation of meat as a woman's job, part of her role as caretaker and provider. The backyard barbecue, however, constitutes a notable exception, a place where gender roles with regard to food preparation and service have long been consistently inverted.[47] In a SPAM ad from the later 1930s, the topmost image depicts the husband grilling while the wife and the son stand nearby. This presentation, this stereotype, persisted (and still persists) both in meat industry advertising and in American culture more broadly.[48] "Barbecue culture offered a vision of the rugged and individualistic provider—the potent patriarch who sees to his family's needs and is rewarded with admiration and obedience."[49]

This is particularly prevalent in Oscar Mayer's summer advertisements for hot dogs. One ad from 1962 places the viewer in the position of the person who "mans" the grill, looking down at a pair of man's hands as they brush a shiny glaze onto the hot dogs. In another ad, an image at the top shows a woman and two children sitting to enjoy a picnic lunch (fig. 56). A huge male hand reaches into the picture, and into the middle of the family group, to deliver a hot dog, speared on a grill fork. The man's hand, like the woman's in the bacon advertisement, synecdochally evokes the man himself, completing the family picture and showing his (oversized) contribution to it.[50]

Another ad shows a man serving a hot dog to a woman, the latter appearing again only as a pair of disembodied hands (fig. 57). Here we occupy the position neither of the man nor of the woman but of an observer, watching the interaction. Ads including this image ran throughout the early 1960s. In the full-color iteration, which ran in *Life* magazine, text above the image reads, "If you like 'em juicy-plump and really fresh . . . Oscar Mayer wieners are for you!"[51] One can only look at so many ads for hot dogs before the phallic nature of the wieners asserts itself. This ad in particular seems deliberately semantically loaded: the man smilingly offers a wiener while the woman gently parts her bun with her perfectly manicured fingers, ready to receive it. Like Freud's cigar, the hot dog lends itself to subtextual readings. Oscar Mayer's print ads embedded themselves in the viewer's subconscious via incessant and ubiquitous repetition. If they could take advantage of

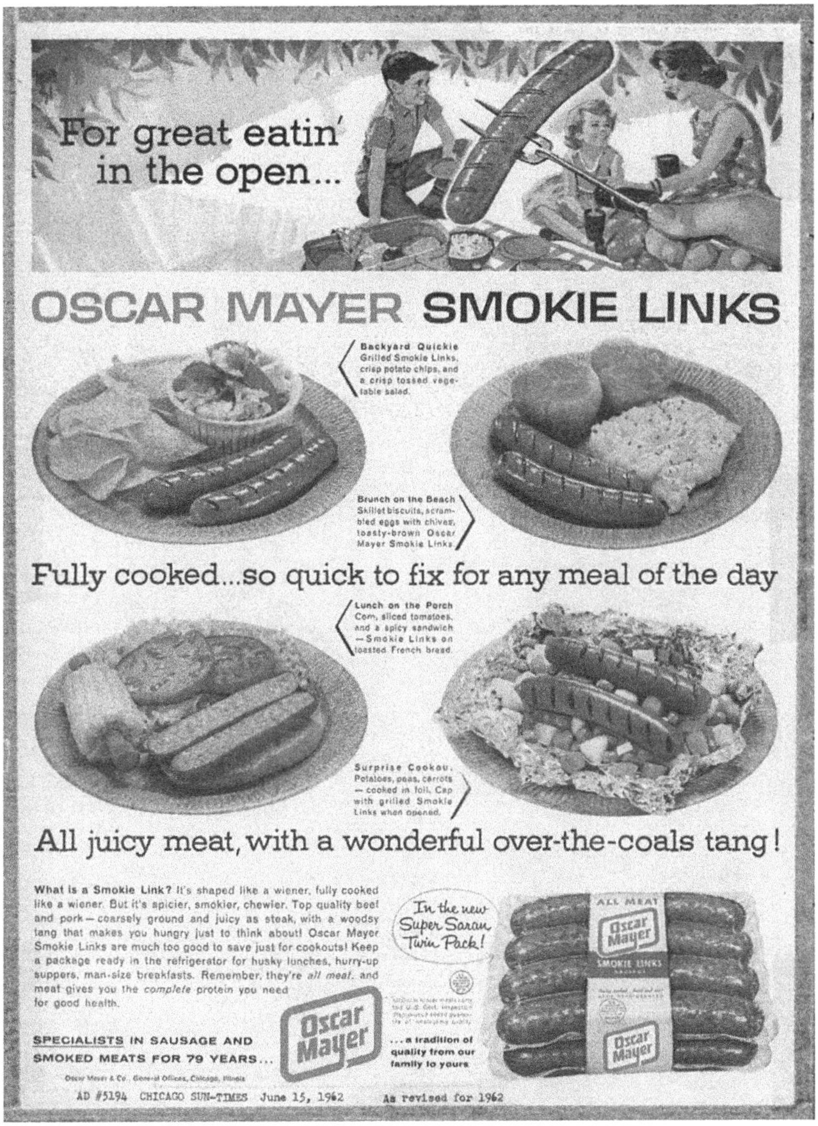

Figure 56. "Oscar Mayer Smokie Links" advertisement, 1962. J. Walter Thompson Company Archives, Domestic Advertisements, box OM1, John W. Hartman Center for Sales, Advertising & Marketing History, David M. Rubenstein Rare Book and Manuscript Library, Duke University.

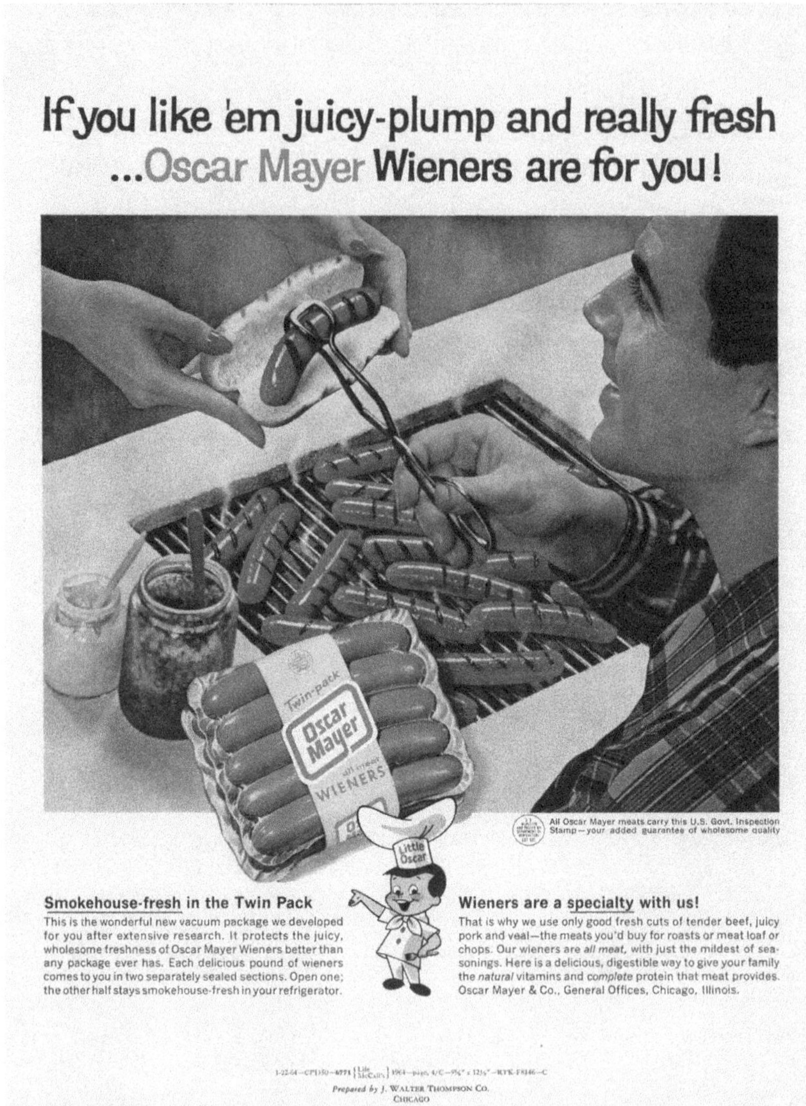

Figure 57. "Oscar Mayer Wieners" advertisement, 1964. J. Walter Thompson Company Archives, Domestic Advertisements, box OM1, John W. Hartman Center for Sales, Advertising & Marketing History, David M. Rubenstein Rare Book and Manuscript Library, Duke University.

neural pathways dedicated to sexual desire, as well as hunger, in order to do that, so much the better for the company's bottom line.

Putting aside the Freudian subtext of its backyard barbecue images, however, there was overall a certain innocence about its brand identity, which Oscar Mayer was able to use to advantage as advertising diversified, both in its appeal and in its depictions. For much of the twentieth century, its ads both envisioned the ideal meat-eating family as a white family and targeted primarily white consumers. Over the course of the 1960s, however, Oscar Mayer's ads became more likely to feature non-white individuals and non-white families.[52] In a 1962 ad depicting a Black family, two children, possibly twins, watch eagerly as their mother prepares hot dogs (fig. 58). The little boy eats a hot dog while his sister, with her arm around his shoulder, waits for hers. The ad resembles other Oscar Mayer productions in being relatively quiet and semantically unremarkable overall. Employing a retouched photograph, not a drawn illustration, as its primary image, the ad depicts the mother in typical fashion, as just a pair of manicured hands and an aproned skirt. A sense of safety, nostalgia, and homeliness suffuses the ad. Like other Oscar Mayer ads, it reinforces a number of basic values: male privilege (the boy eats first); the mother as domestic anchor for the nuclear family; the simplicity of childhood pleasures as the best pleasures; and the presence of abundant, inexpensive meat as a signifier of comfort. Nothing much has changed here, relative to Oscar Mayer's other ads, except the family's race.

In a mock-up of the ad, prepared by the J. Walter Thompson Co., the name of the intended media outlet is noted at the bottom: the *Philadelphia Tribune*, one of the nation's oldest Black-owned newspapers, long serving the Black community.[53] In placing the ad, then, Oscar Mayer was intending to appeal to and expand its consumer base, continuing to invoke the model of the female-anchored domestic sphere. The ad evokes, and attempts to tap into, the potential power of Black women's consumerism; at the end of the day, its goal was to generate profit. This shift in representation reads in some ways as a sea change from the days of Armour's happy servant, offering consumers a slice of rigid racial hierarchy along with their ham. But competition for market share drove the shift, at its core.[54] No matter what Mrs. Consumer's family looked like, no matter who stood or sat around her table, what really mattered to Oscar Mayer, and to the meat industry as a whole, was that *their* products should be at the center of it.[55]

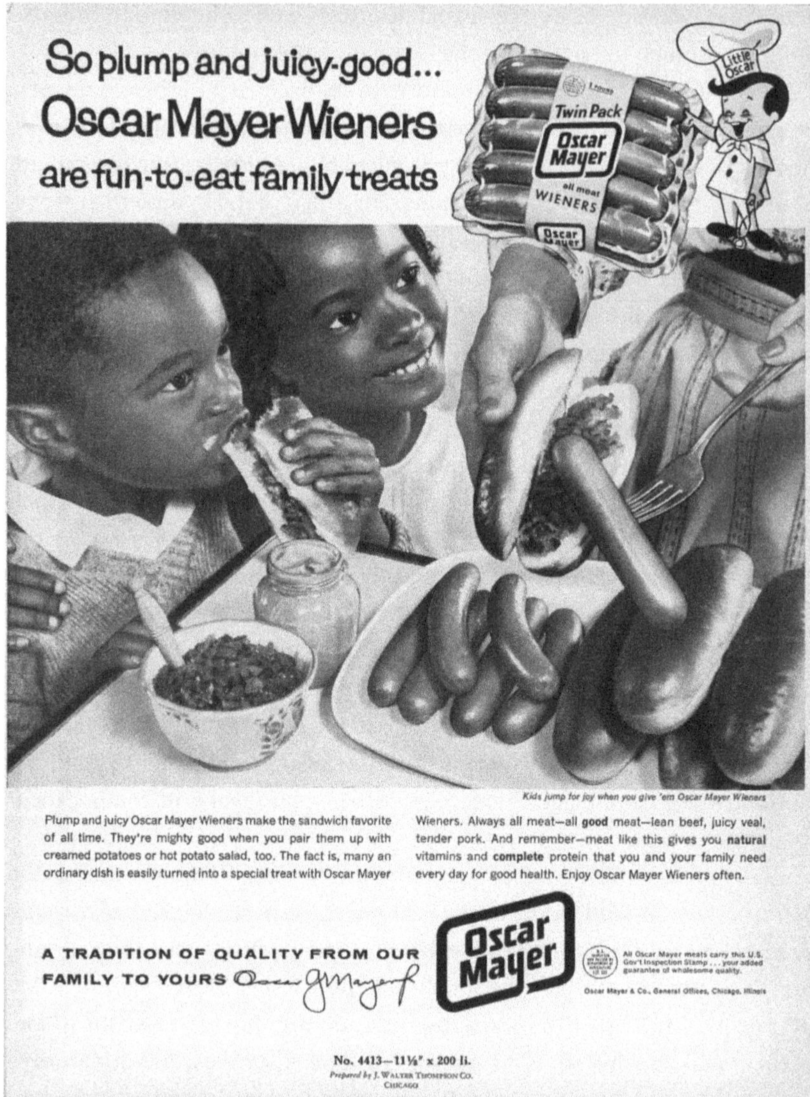

Figure 58. "Oscar Mayer Wieners" advertisement, 1963. J. Walter Thompson Company Archives, Domestic Advertisements, box OM1, John W. Hartman Center for Sales, Advertising & Marketing History, David M. Rubenstein Rare Book and Manuscript Library, Duke University.

Purity, Harmony, and Identity Converge: Home Economics at Swift

Gendered performances sat at the heart of many food companies' sales and marketing strategies. The Rath company's fiftieth-anniversary celebration of the early 1940s leaned heavily on performances of masculinity, conducted for an audience of women consumers. Oscar Mayer's advertising program in the mid-twentieth century depended in large part on embodying, enacting, and in the process, further imbricating hierarchies of gender within American families. Perhaps the most obvious manifestations of this phenomenon were the home economics divisions established at many food companies, including meat industry companies, from the early twentieth century. Always female, home economists were a group of workers defined by their gender and, more specifically, by ongoing performances of femininity, a function central to their role and to their value from food companies' perspectives.[56] Indeed, we have already seen home economists featured in several advertisements, including those of Armour and Oscar Mayer; and in the prior chapter we saw home economists featured within the pages of some of the photograph albums designed and commissioned by companies for internal circulation.

In the largest food companies, home economics departments might count dozens of employees. These women workers had broad imperatives —testing new products, developing recipes, communicating with customers, answering product queries—but companies essentially viewed them as proxies within the organization for the women who bought, consumed, and served their companies' products.[57] Certain food companies publicized the activities of their home economists more heavily than others, depending on these women's research and, especially, their performative activities to build consumer relations and increase market share. By far the most famous example, the character of "Betty Crocker" was developed by the Washburn-Crosby Company (later General Mills). Brought to life initially to respond to women's queries about the company's products, "Betty Crocker" ultimately became the name of an entire division of the company, employing a small army of women: recipe developers and testers, writers, and the performers who embodied her on radio and television.[58] As conceived by Marjorie Child Husted, the director of the company's Home Service Department, Betty Crocker's primary role was to inspire confidence in American homemakers, to help them feel valued and important. "Under [Husted's] direction, Betty Crocker became a figure of dignity who treated homemakers with respect."[59]

Within the meat industry, the closest parallel to Betty Crocker was "Martha Logan," the name assigned to the Home Economics Division of Swift & Company and to all the women who worked there. One of the Big Four, Swift had expanded rapidly in the late nineteenth century to become a dominant power in the Chicago meat business. Its founder, Gustavus Swift, pioneered the use of distributive technologies such as refrigerated train cars to convert his business from one based in the (relatively small-scale) trade of cattle to a behemoth and founding organization in the international trade of meat.[60] Keys to Swift's success lay in its founder's "farsighted" understanding of the industry and his "care for minute detail."[61] After Gustavus Swift's death, the company sustained the founder's shrewd foresight as it sought new avenues to increase its distributive influence and brand recognition. This included, after 1918, its Home Economics Division.

At first, the division was—like that of Washburn-Crosby/General Mills—modest in size and hardly a major aspect of the company's publicity program. Although Swift had begun using the name "Martha Logan" by 1933, Swift's home economists featured relatively rarely in its advertisements in the first decades of the division's existence. Unlike, say, Hormel or Oscar Meyer, both of which advertised their canned or processed products more heavily than their fresh meats, Swift branded itself from the 1920s as a purveyor of both convenience products and fresh steaks and roasts. Swift's canned product Prem competed with Hormel's SPAM in the budget processed-meat category. A 1943 ad for Prem incorporates as its primary image a color photograph of a plate of the product alongside vegetables (fig. 59). Over the top of this is superimposed a drawn image of an attractive young wife. Her hair pulled back into a pretty chignon, she points at a photograph of a can. The text below the image alludes to Second World War–era food rationing, noting how economical the product is. Although the ad itself is also economical and straightforward, even here a variety of manipulations occur. The food appears in a photograph, but the woman is a drawing, even more two-dimensional than the rest of the promotion around her. "Tonight we'll have a hot Prem dinner!" she chirps, implying it will be a treat, a come-up rather than a come-down from whatever effort she usually makes.

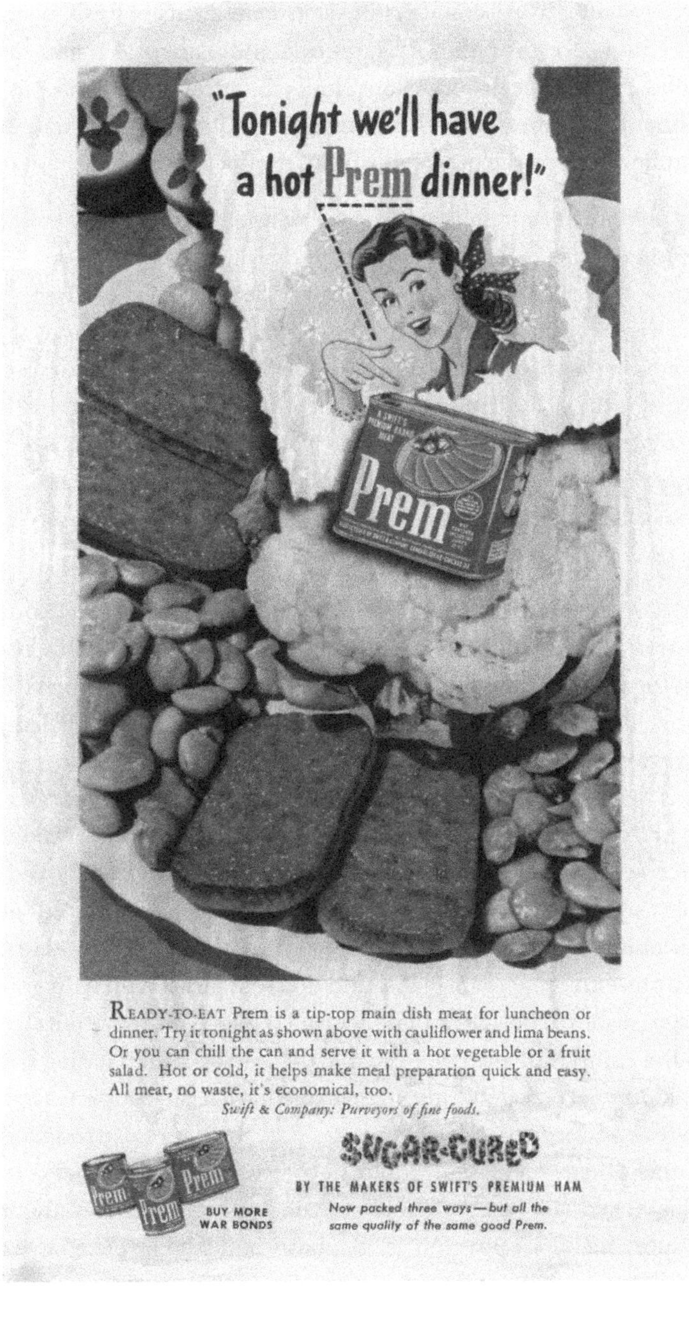

Figure 59.
"Swift's Prem" advertisement, 1943. J. Walter Thompson Company Archives, Domestic Advertisements, box SF1, John W. Hartman Center for Sales, Advertising & Marketing History, David M. Rubenstein Rare Book and Manuscript Library, Duke University. Used with permission of Swift Meats/JBS.

Many other Swift advertisements, however, touted cuts of fresh meat, such as veal cutlets, for everyday consumption and hams and roasts for Sundays, holidays, and special occasions. Indeed, in the 1920s Swift pioneered the branding of meat directly on the product itself, circumventing the long-standing issue in the meat industry of product anonymity:

> It was not until 1927 that fresh meats were identified. Swift was . . . one of the pioneers in this. Without identification there could be little promotion or advertising of such products, but with the identification perfected, fresh meats also were brought into the advertising program. Swift and Company . . . pioneered in the advertising and promotion of fresh meats to the consumer. We were the first to think enough of our products to place our brand names on them so that housewives could purchase steaks and roasts exactly as they do coffee, bread, canned goods, and other branded items. We have felt that by making this easier method of purchase possible we are encouraging greater consumption of beef, lamb, and veal.[62]

One 1945 advertisement for a veal roast featured as the largest image not a tasty preparation of the cut but an illustration, built from a heavily retouched photograph, of the raw product (fig. 60). "How to get fine veal every time. . . . Just look for a Swift brand: Swift's Premium, Swift's Select, or Swift's Arrow right on the meat!"[63]

These ads for Swift's roasts and hams focused solely on the product, depicting it in close-up, emphasizing quality and tantalizing preparation. Some, like the veal ad, included an image of the cut raw as well as cooked, at least in part so an inexperienced cook might know what to look for at the butcher counter; but many, particularly for easily identified products such as ham, presented the roast plated and dressed, gussied up with garnishes and side dishes. The ads implied that such lovely presentations lay well within the abilities of any cook capable of viewing and imitating the picture itself. Minimal text alongside might explain what the consumer was seeing: one 1945 ad for Christmas ham notes that the ham is surrounded by, among other things, "cinnamon apples decorated with cream cheese to look like ornaments" (fig. 61). Other ad copy includes superlatives about the roast's "supremacy," its position as "the best" and "the favorite," even through end-of-wartime shortages.[64]

Swift's Home Economics Division played a key role in producing these publicity materials. Employees of the division, which fell within the broader

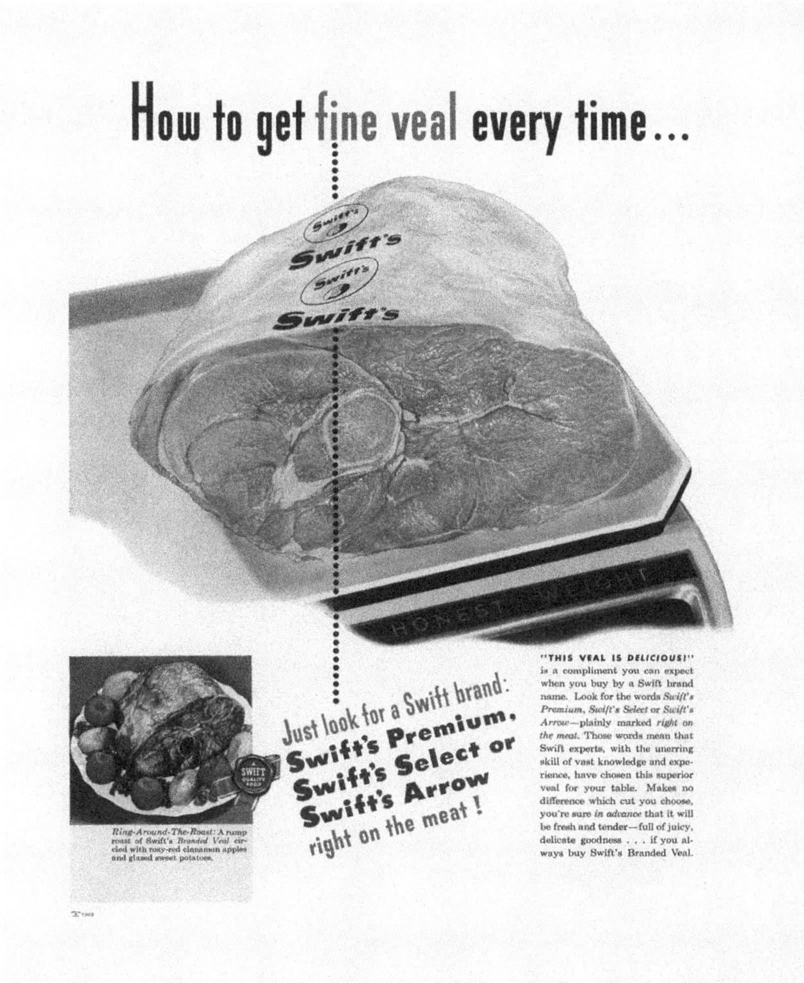

Figure 60. "Swift's Veal" advertisement, 1945. J. Walter Thompson Company Archives, Domestic Advertisements, box SF1, John W. Hartman Center for Sales, Advertising & Marketing History, David M. Rubenstein Rare Book and Manuscript Library, Duke University. Used with permission of Swift Meats/JBS.

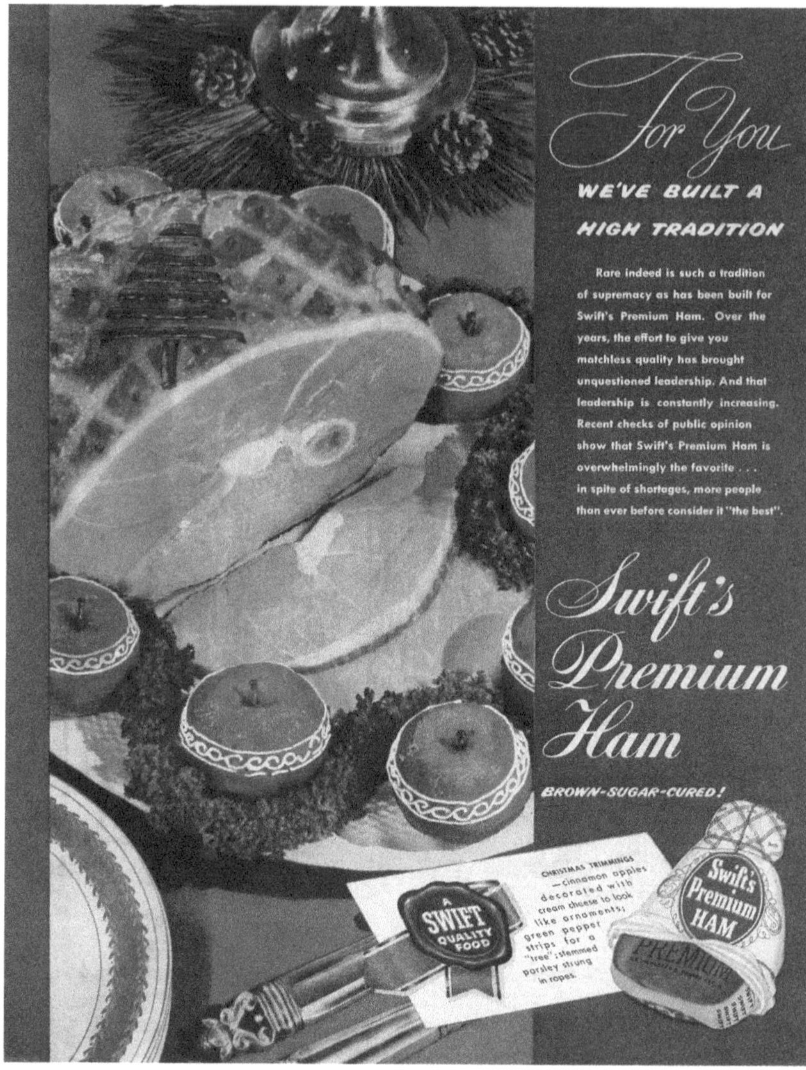

Figure 61. "Swift's Premium Ham" advertisement, 1945. J. Walter Thompson Company Archives, Domestic Advertisements, box SF1, John W. Hartman Center for Sales, Advertising & Marketing History, David M. Rubenstein Rare Book and Manuscript Library, Duke University. Used with permission of Swift Meats/JBS.

imperatives of the company's Research and Development Division, both developed the recipes and preparations shown in the ads and directed the photography. Not until after the end of the Second World War, however, did the home economists themselves move to the fore, coming to occupy an increasingly visible role in Swift's advertising and overall public presence. Ultimately, Swift's "Martha Logans" became, like Betty Crocker for General Mills, among the company's most visible, most valuable properties at mid-century.

Languishing a bit after its 1918 establishment, Swift's Home Economics Division expanded rapidly in the later 1930s. Prior to the mid-1930s, the division had consisted of just one person at a time, but in 1935 it moved out of the Marketing Division and into the company's Research and Development Division, at which time a number of new employees were hired in rapid succession.[65] These included a new director, Beth Bailey McLean, who joined the company in 1937, taking over operations for the division and further professionalizing and extending its activities.[66] Also during the 1930s, the division began using the name "Martha Logan" as the designation both for the unit and for all the home economists working within it.[67] Over the course of the 1940s and into the 1950s the division grew, reflecting the importance placed on home economists' activities by many food companies. At its largest, in the late 1940s and early 1950s, it employed upward of thirty-five people, the largest home economics division at any American food company.[68] The division contracted and became more streamlined in the later 1950s, employing around fourteen home economists under McLean's successor, Jeanne Paris.[69]

Swift's home economists created and tested recipes for the company's products, the results of which appeared as ad copy, on the sides and backs of the company's packaging, and in numerous cookbooks released by the company. They conducted taste testing in cooperation with the company's research scientists, carefully tracking and compiling the results—an activity Swift regarded as one of their more valuable contributions to its operations, both for its actual value in product testing and for the appearance of data-driven rigor it lent to the company's product development processes. The division also created educational materials for distribution to home economics teachers.[70]

In addition, the Martha Logan Division managed food photography for the company, including images made for advertisements and packaging.

According to one promotional booklet, "The Martha Logan in charge of this work must combine an artist's sense of color, texture, and composition with a practical home economist's skill."[71] The division's imperative expanded in later years to encompass films about food, as well. Some Martha Logans interfaced with the media, providing information to food editors and making guest appearances on radio and television, while others stayed behind the scenes, specializing in technical development and researching and collecting data for recipes, products, and cooking methods.[72]

A core group of Swift's home economists worked at the home office and facilities in Chicago, but in the 1940s to 1950s, at the division's height, there were also home economists associated with other locations across the US and in Canada. Swift had home economics facilities in Toronto, Los Angeles, Kansas City, Omaha, and several other North American cities, most of them attached to or associated with Swift's satellite meatpacking plants. (When the division contracted in the later 1950s, it jettisoned all but the Chicago and Toronto facilities.) In all of these, home economists held cooking schools in purpose-built test kitchens, each facing a seating area holding eighty to one hundred seats.[73] One or two Martha Logans performed in each of the kitchens, demonstrating products and recipes. There were also traveling Martha Logans who held demonstrations in theaters, auditoriums, and other venues.[74]

The Martha Logan Division's demonstration kitchens functioned as performative spaces. They promoted and demonstrated recipes not only for Swift's meat products but also—even primarily—for its baking-related byproducts, the lard and various shortenings that the company found so profitable.[75] The company also viewed these events as a kind of reconnaissance on its behalf: by interfacing with consumers, the Martha Logans could gain feedback and engage in a kind of ongoing product testing, "keep[ing] our finger on Mrs. America's pulse."[76] Publicity photographs and snapshots in Swift's Chicago archives show that the company conceived this "Mrs. America" as multiracial and multilingual. Audiences at some of its cooking shows and other events included non-white women, and one Martha Logan made regular appearances on Spanish-language radio.[77] But all the Martha Logans themselves, regardless of ethnic background or linguistic heritage, were white women.[78]

The company claimed that the Martha Logans had "a two-fold purpose" in their work: "We not only represent Swift & Company when talking to

Mrs. America; we also represent Mrs. American consumer . . . when working with the Swift & Company product departments."[79] But the critical function of the division was a form of soft selling, particularly effective because it looked like public service, not publicity. The company emphasized the pedigrees and training of its home economists, but the light of Martha Logan's rationalist approach to homemaking was designed to reflect onto the company as a whole. This distinction is critical to understanding the importance of Martha Logan's, Betty Crocker's, and other such large home economics divisions' contributions to their overarching companies: the purpose of corporate home economics divisions was not only to interface with the public but *to be seen interfacing with the public*, to engage with consumers and *to be seen engaging with consumers*. This self-reflexivity characterized many of the Martha Logan Division's activities.

Martha Logan flourished at roughly the same time as Betty Crocker, and she played a similar role, too, conceived by Swift as a figure of benevolent authority. Writing about Betty Crocker, Laura Shapiro quotes an anonymous advertising executive's 1957 assertion that such imaginary women should be "between the ages of 32 and 40, attractive, but not completely so, mature but youthful looking, competent yet warm, understanding but not sentimental, interested in the consumer but not involved with her."[80] The illustrators who imagined Martha Logan seem to have taken such directives to heart.

Over the course of her existence, "Martha Logan" had three distinct looks, updated periodically to reflect the feminine style of the era. She first appeared in 1946 as a smiling, attractive, competent-looking woman with a matronly air—closer to forty than thirty-two. Her hairstyle and white lace collar closely echo those worn by McLean, the division's director, in a 1944 headshot (fig. 62); but McLean's distinctive features and intelligent expression give way, in the drawn illustration, to a smoother visage, a bit like a homespun Donna Reed—not the girl next door, exactly, but certainly a lady who once had been (fig. 63).[81] Whereas McLean, as home economist and division director, was a professional woman in a leadership role, Martha Logan appears as a homemaker, just like the putative consumer—not a person with an office or title but a neighborhood mentor, older and more experienced with matters of kitchen, shopping, and home.

In 1957, the same year *Leave It to Beaver* debuted on television, Martha Logan was reborn as a June Cleaver–ish figure in pearls and a smart tailored

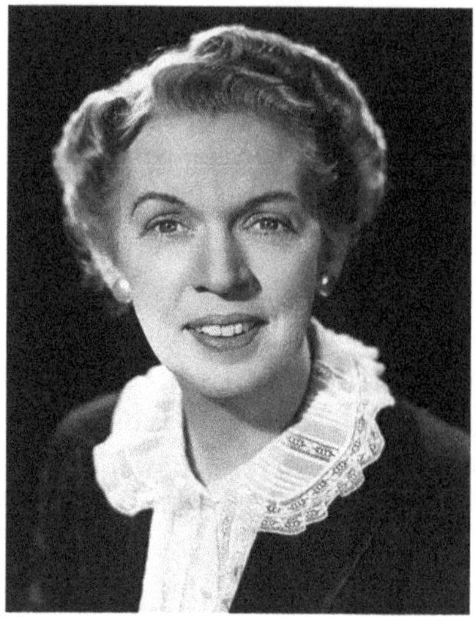

Figure 62. Headshot of Beth Bailey McLean, Director, Martha Logan Home Economics Division, Swift & Company, 1944. Martha Logan Photograph Collection, collection G1985:473, Chicago History Museum. Used with permission of Swift Meats/JBS.

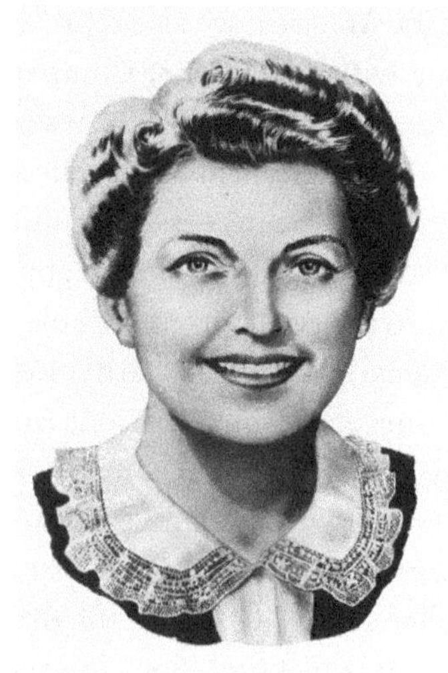

Figure 63. "Martha Logan" illustration, 1946. Martha Logan Photograph Collection, collection G1985:473, Chicago History Museum. Used with permission of Swift Meats/JBS.

jacket: still an ideal woman, still old enough to have a kindly authority, but perhaps a little worldlier and more cosmopolitan than the 1946 model (fig. 64).[82] Martha Logan's final makeover in 1970, just after the 1969 debut of *The Brady Bunch*, recast her as a perky Carol Brady–like woman with a short bouffant haircut, neckerchief, and dimples (fig. 65). Hipper and younger looking than the previous iterations, the last Martha Logan has the air of a busy, involved suburban mom. Introducing the new image at a press conference, Swift's president noted that "the new portrait is intended to symbolize the modern character of Swift as keen, capable, and 'with it' in the food industry."[83]

A raft of flesh-and-blood "Martha Logans" brought this vision of ideal wife and motherhood to life. At Swift, as at many other companies, the Home Economics Division fell within the purview of the Research and Development Division. The home economists did a great deal of recipe

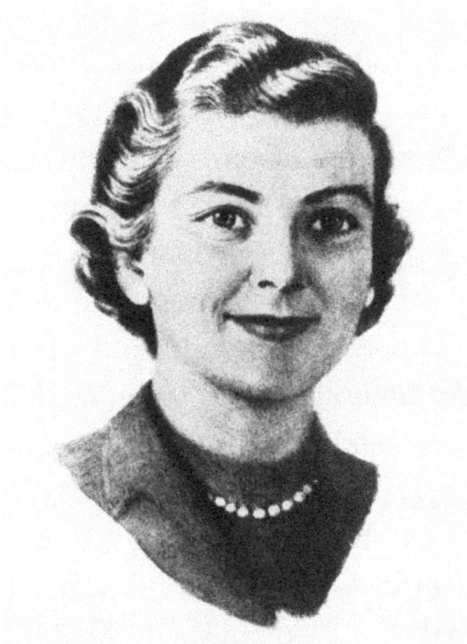

Figure 64. "Martha Logan" illustration, 1957. Martha Logan Photograph Collection, collection G1985:473, Chicago History Museum. Used with permission of Swift Meats/JBS.

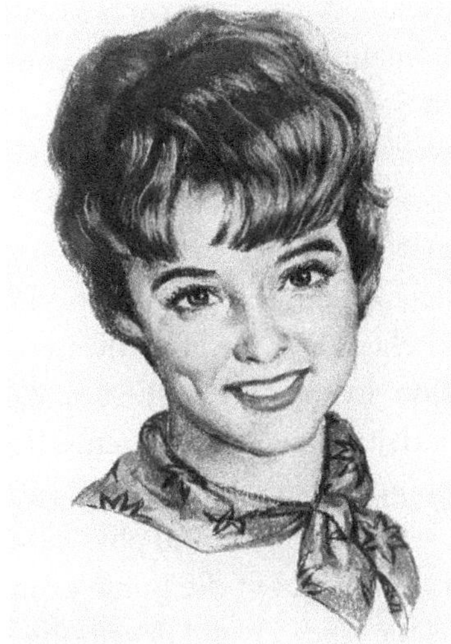

Figure 65. "Martha Logan" illustration, 1970. Martha Logan Photograph Collection, collection G1985:473, Chicago History Museum. Used with permission of Swift Meats/JBS.

development, product feedback, and taste testing. But the company quickly began distinguishing between this functional research and development work and a type of work that might be termed *performative* R&D. An undated promotional booklet, produced in the later 1940s or early 1950s, noted that

> the main advantage of the size of the Martha Logan staff is that it allows a home economist to specialize. . . . In general, there are two types of work, each requiring special training, ability, and aptitudes: technological activities and consumer service. Once a Martha Logan is hired she goes through a thorough training period in which she becomes familiar with Swift & Company operations, is given an opportunity to learn the various types of work, and is guided into the work for which she is best suited.[84]

For those home economists guided into "consumer service," interfacing with the public became a crucial aspect of their job. While this may have included activities such as taking calls and responding to letters from homemakers querying the company about products or processes, records in Swift's archives show that secretarial staff—women who, though they worked for the division, were not officially Martha Logans—actually performed many such activities.[85] The consumer-facing Martha Logans primarily engaged in more visible types of consumer service, making public appearances, running performances in the company's demonstration kitchens and traveling theater shows, and appearing on radio and television. Even some of the home economists counseled into "technological activities" concerned themselves with image making and publicity, as food photography fell under this rubric.[86]

Visual images and physical appearances, both of the Home Economics Division and of the home economists themselves, mattered just as much as the research and development work that they accomplished. All of the Martha Logans had professional headshots, and many appeared in the numerous publicity stills, promotional images, and photographs made to appear in a variety of sources: the public press, Swift's internal newsletters and magazines, media like cookbooks produced by the company, and advertising. For example, a home economist named Marie Kapranos appears in a 1951 headshot collected in a scrapbook covering the activities at Martha Logan's home facilities in Chicago.[87] The headshot, unsurprisingly, shows her coiffed and wearing dark lipstick. But she appears similarly made up for other images that show her in action. One photo shows her in the act of cutting a cake, her white uniform pristine, its Martha Logan logo clearly visible on the chest (fig. 66). Steadying the cake dish, her left hand curls as delicately as the hand of a lady in a Renaissance painting. Next to her stand stacks of paper plates and cups and, just visible at the right side of the frame, a piece of paper with a pencil atop it. The composition encapsulates her work as a cook and baker as well as her duties as collector and compiler of data beneficial to the corporation. But her pose, makeup, and composed air also bespeak her role as public symbol, her focus not only on "technological activities" but on how those activities will be received and perceived by the company's consumer base.

Another image shows Kapranos alongside a second home economist, identified in a label as Helen Stamm (fig. 67). The two women work together to prepare and photograph a tabletop tableau. Perched on the edge of the table, Stamm supports a platter containing a whole baked ham from

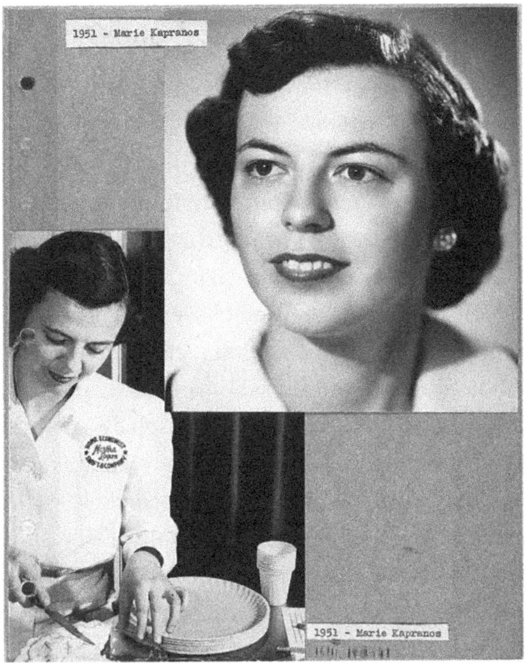

Figure 66.
Marie Kapranos, Home Economist, 1951. Martha Logan Photograph Collection, collection G1985:473, Chicago History Museum. Used with permission of Swift Meats/JBS.

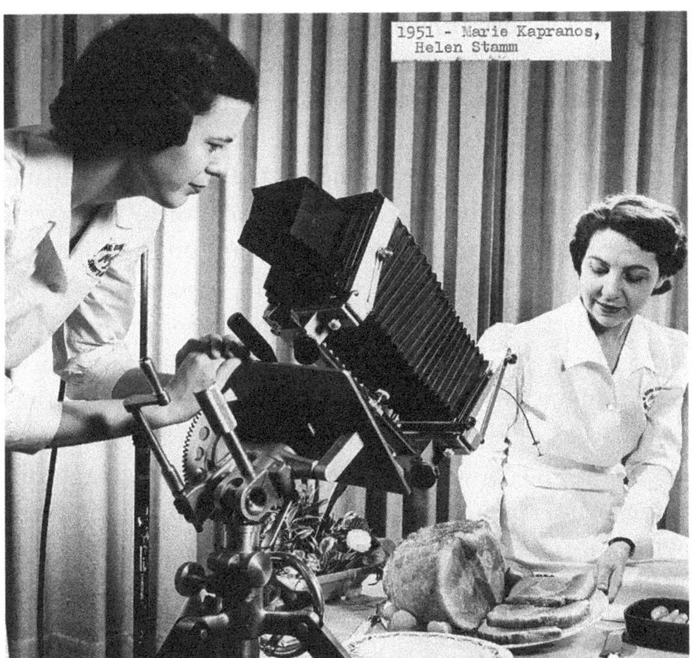

Figure 67.
Marie Kapranos and Helen Stamm, Home Economists, 1951. Martha Logan Photograph Collection, collection G1985:473, Chicago History Museum. Used with permission of Swift Meats/JBS.

which three meaty slices have been cut. Around the platter are other accoutrements of a spring dinner, including side dishes, a basket of flowers, and empty plates ready to receive the meal. Kapranos stands to the side of the table, next to a view camera on a tripod, looking into the viewfinder to check the framing of the shot. The image implies a lengthy photography session. The image offers tantalizing evidence (pun intended) that some Martha Logans operated cameras as well as performing for them. Although company literature implied a split between consumer-facing Martha Logans and those trained for "technological activities," Kapranos seems to have both performed for the camera and controlled it. In this picture, the two women set up a food tableau to be photographed, but the whole of the scene—the tabletop, the view camera, and Stamm and Kapranos themselves—constitutes a tableau of its own, set up and made ready to be presented to this second camera. The home economists work, but they also perform their work.

Such images helped Martha Logan become a more visible presence in Swift's publicity materials in the years just after the Second World War. Although the Home Economics Division had existed at Swift since 1918, and had carried the name "Martha Logan" since the 1930s, the activities of the division began to feature heavily in the company's advertising only in 1946. One remarkable advertisement from that year encapsulates this new visibility (fig. 68). A double-page spread, created for distribution as a centerpiece advertisement, the ad carries a full page of text facing a large picture.[88] In the image, a uniformed Martha Logan appears from the back, carefully arranging sprigs of parsley on a platter of cold cuts and accompaniments. A male photographer with a large camera stands nearby, along with a male technician controlling an assortment of lights and other equipment, while two more men, executives, observe the entire process. "Picture of a woman having the last word at Swift & Company," reads the header in elegant script. "And why shouldn't she!" the text below continues. "Swift quality foods go to the woman in the home. How logical, then, that another woman—Martha Logan, Swift's Home Economist—should have the last word in recommending how best to prepare, cook, and serve them." Next to this text is an inset photograph of the platter of cold cuts—the very image, presumably, that the home economist sets up in the larger illustration. The remainder of the ad discusses the jobs Martha Logan performs at Swift, describing her product and recipe testing, her taste testing and data collection, and her collaboration with the company's food scientists.

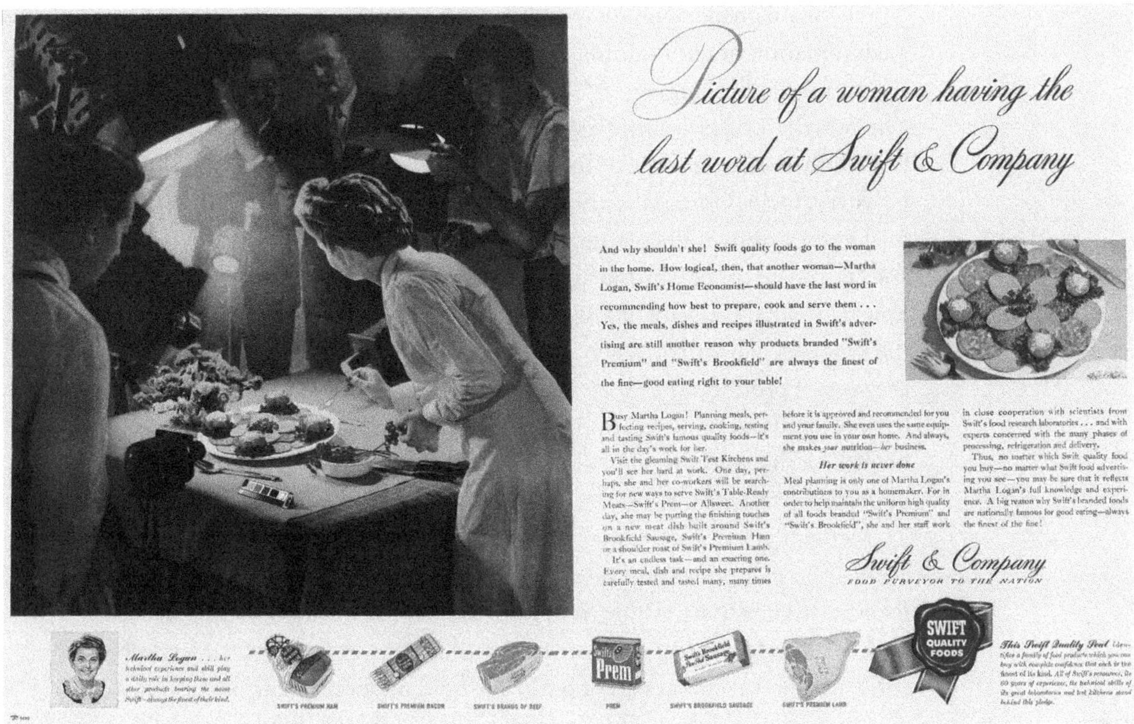

Figure 68. "Swift & Company" advertisement, 1946. J. Walter Thompson Company Archives, Domestic Advertisements, box SF1, John W. Hartman Center for Sales, Advertising & Marketing History, David M. Rubenstein Rare Book and Manuscript Library, Duke University. Used with permission of Swift Meats/JBS.

The company's foregrounding of the Martha Logan Division in the ad speaks to the importance Swift was coming to place on home economists as a key point of contact for consumer relations. From simply answering questions for the company, the division came to *embody* the company, as far as the public was concerned. It is no coincidence that Swift first gave Martha Logan a visage of her own in this same year, 1946. That illustration appears here at the bottom of the ad. Small by comparison with the large central image, the face of Martha Logan assumes outsize importance in the overall context of the ad, because hers is the only face we see. While the home economist shown in the illustration faces away from us, through the image of Martha Logan we can identify with her nonetheless.

The main image operates in two registers simultaneously: created from a combination of photographic and illustrative techniques, it speaks both the photographic language of the real and the language of the drawn ideal. The Martha Logan home economists not only prepared and styled food but handled at least some of the food photography themselves. In this ad, however, male photographers and technicians surround the home economist as she arranges the platter: she may get "the last word," as the headline asserts, but no other words of importance. In the background, the two male executives observe her work, whispering to one another. Their bodies block the flow of space into the deep background, and they control, by extension, the flow of product: they manage all aspects of its creation, from slaughter to development, up to the point when it comes into the home economist's hands. As she applies the final touches, both she and the product remain under their watchful eyes.

McLean, the department's first director, held undergraduate and graduate degrees in home economics and came to Swift from a faculty position at Iowa State University. After twenty years she retired from the company and took another faculty position at Oregon State University. McLean's successor, Paris, had a similar career path: she held undergraduate and graduate degrees in the field, had held a university position in West Virginia, and then entered the professional world, directing the Home Economics Division at Kroger before coming to Swift.[89] Their education and career experience are representative of Swift's impulse during that period—one shared with many other food companies—to systematize and professionalize its home economics activities. As the Martha Logan Division grew during the later 1930s and 1940s, the educational backgrounds of its home economists became a point of pride in Swift's promotional materials: "Each staff member must be a graduate with a major in home economics from an accredited university. Some members hold advanced degrees."[90] The company valued both scientific rigor and the prestige that rigor brought.

The career of home economist was, in itself, demanding and public. Many of Swift's home economists traveled widely, performing for audiences and cameras. But the career required, simultaneously, that they retain, maintain, and even work actively to reinforce appearances of domesticity. The tension present in the very title of the career path—"home economics," simultaneously domestic and scientific—inhered in many aspects of the job, as well. McLean addressed this balancing act directly in a book she wrote in 1953 (and expanded, with Paris as her coauthor, for a second edition, in 1962) titled *The Young Woman in Business.* From the first page, the book notes how women's roles in business were defined, in fundamental ways, by their gender:

> Most women in top positions . . . realize that women have different viewpoints and different innate abilities from men. They are not feminists; rather they develop an objective viewpoint and consider the interests and welfare of people in general, not just women. The successful woman is proud she is a woman and therefore accepts the responsibilities of being womanly with all the best traditional meaning of the word. She does not militantly compete with men. Rather, this woman understands that working *with* men adds a challenge and an inspiration which can sharpen her understanding of current problems, develop her ability to weigh facts objectively, and establish her reputation as a real co-worker.[91]

The text conjures a kind of bogeywoman here: an imaginary militant feminist, simultaneously hypercompetitive and filled with loathing at her own womanliness. The successful young woman will construct and conduct herself always with this anti-ideal in mind, hoping to become "a real co-worker" thereby.[92]

After this admonition, the book goes on to counsel young women about a variety of aspects of business. Much of the advice is practical and even instrumental: formats and suggested wording for business letters of various types, guidelines for preparing a report, reminders about appropriate tipping (including, in some instances, precise amounts). But some aspects of the book, particularly the 1962 edition, speak to the authors' own sophisticated understanding of image and performance in the context of corporate publicity. In a chapter titled "Good Publicity Is an Asset," for instance, they advise the aspiring businesswoman to consider the use and importance of films in making public presentations; and they counsel her on how to combine image and text for effective, impactful public presentations.[93] The book's illustrations, humorous cartoons by Harry Walsh, reinforce the overall message by showing bad examples or worst-case scenarios, to drive home the wisdom of taking the text's good advice.

The book's first page, with its admonition to avoid becoming a feminist monster, counsels the reader to aim for an impossible, almost oxymoronic state, a self-effacing ambition. But thereafter, both the 1953 and 1962 editions of the book reveal themselves to be fairly progressive documents, genuinely interested in providing concrete advice to women in business. The work described by the books is not solely or even primarily secretarial, for one thing. McLean and Paris imagine women workers who do not just passively facilitate but actively advance the overall goals of their organizations, in parallel to their male colleagues: they attend and conduct meetings, they collect data and prepare reports, they speak publicly.[94] They behave, in short, much like the authors and like other members of the Martha Logan team: a group of women whose work, defined in part by their gender, nonetheless required that they inhabit a corporate and industrial world and make active, ongoing contributions to the development and furtherance of their organization's goals.

Within Swift, however, gendered hierarchies and working patterns come through clearly in the company's own internal documentation of its research and development activities. One document from the late 1950s or early 1960s, apparently a manuscript for a shareholder's report or other internally circulating publication, describes the function of the Home Economics Division:

> The stay-at-home staff cooperate with the research men in developing new products and improvements in all established products. They also furnish . . . editorial copy, recipes, photographs, films, and recordings (for a variety of purposes). The Martha Logans handle direct mail and telephone requests for recipes and questions on cooking and baking. In all their work, the stay-at-home staff give the viewpoint of the women customers of the company.[95]

Although the reference to "stay-at-home" staff probably refers to those members of the Martha Logan Division who stayed in Chicago (or Toronto), as opposed to those who traveled regularly, in the context of the document the phrase is unclear. In any case, it is a telling choice of words, conjuring an image of someone who literally works from home—not from a combination test kitchen and office, as the division's workers actually did; and certainly not a laboratory, like the "research men." The memo implies that while the "research men" work on product development, the Martha Logans engage, at least in part, in glorified phone chitchat. The phrase "stay-at-home" implies homeliness, groundedness, even confinement: a worker who channels the limitedness of the company's consumer base

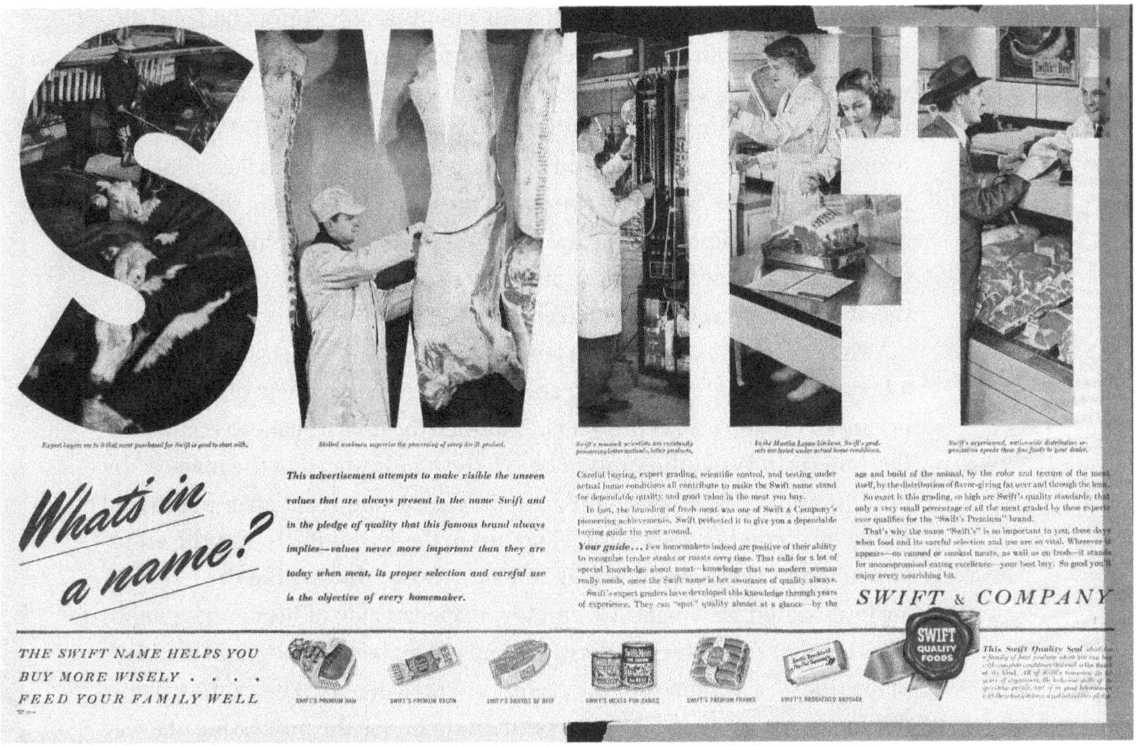

Figure 69. "Swift & Company" advertisement, 1947. J. Walter Thompson Company Archives, Domestic Advertisements, box SF1, John W. Hartman Center for Sales, Advertising & Marketing History, David M. Rubenstein Rare Book and Manuscript Library, Duke University. Used with permission of Swift Meats/JBS.

through her own limited experience, whose job consists of going nowhere in particular, laboring inside the domestic bubble.

In a variety of circulating materials, Swift lauded the Martha Logan Division's activities as critical to its overall mission. But women's roles were, as this document makes obvious, confined to one specific area, not distributed throughout the company. This contained role becomes quite clear in a 1947 advertisement. Part of the same postwar publicity push that produced the 1946 ad (see fig. 68), this full-spread, double-page promotional attempts to cement Swift's identity in the consumer's mind by supersizing and activating the very letters that spell out the company's name (fig. 69). In the ad, each letter in the name is created from a photograph. Retouched and illuminated with applied color, the photographs impart a brilliant, hyperreal vision of the company's activities. Beneath each letter sits a caption, while below these, the main ad copy stretches across the spread. "What's in a name?" reads the copy. "This advertisement attempts to make visible the unseen values that are always present in the name Swift and in the pledge of quality that this famous brand always implies—values never more important than they are today when meat, its proper selection and careful use is [*sic*] the objective of every homemaker."[96]

The company's values begin within the letter *S*, in which a man sits atop a horse in a corral filled with cattle—an acknowledgment of the animal origins of the company's products. Although the image evokes a scene from the American West, the caption below the *S* indicates that the man on the horse is an "expert buyer," not a cowboy, ensuring that "meat purchased for Swift is good to start with." The image depicts animals, but the ad refers to them, from the outset, as "meat." Next to this, within the letter *W*, a white-coated packinghouse employee opens a side of beef at the twelfth rib, a cut that allows inspectors to see and judge the meatiness and marbling of the ribeye, a key index of the carcass's overall quality. The caption below this image asserts that "Skilled workmen supervise the processing of every Swift product"—a typically vague, corporatizing reference to meat making, in which slaughter and disassembly become "processing," and the animal carcass becomes generic raw material, generic "product."

Within the subsequent letter, the *I*, a research scientist stands before an instrument in a laboratory. "Swift's scientists are constantly pioneering better methods, better products," the caption reads. Next to that, within the letter *F*, two of the company's home economists, both wearing pristine white uniforms, work in a test kitchen. The one in the background opens an oven door, while the other in the foreground attends to a raw rib roast on a rack. Both hands poised above the roast, she inserts a probe into the meat. Her notebooks and pencils, sitting atop the gleaming stainless steel tabletop, bespeak her rationalist objectives, her service to the company and, by extension, to the homemaker. Her fingernails and the watch on her wrist gleam, as well—the latter connoting her dedication to precision and timeliness, core values of her job; the former, to her care of her personal appearance, a core value of her womanhood. "In the Martha Logan kitchens, Swift's products are tested under actual home conditions," reads the caption. Again, the test kitchen is presented not as the workplace it was but as a kind of idealized, utopian cooking space, a "superhome": filled with all the utensils and gadgets one might need; replete with desirable and expensive cuts of meat; free from bothersome pets, children, and dirt (so, really, nothing at all like "actual home conditions"!).

Finally, at the end of the company's name, the image within the letter *T* shows a man in a suit standing at a butcher counter. At first glance, this image seems to disrupt the standard depiction of the consumer, who was nearly always shown as a woman. But the caption offers clarity: "Swift's experienced, nationwide distribution organization speeds these fine foods to your dealer." This man is a company salesman, not a consumer. Notebook in hand, he takes an order from the butcher for Swift products. Overall, the images provide a compressed, linear vision of the path of Swift's products: from field to butcher, outdoor to indoor, animal to human, farm to factory to store, impure to pure in five easy steps. Throughout, the ad emphasizes the know-how of the company's male employees: the "expert" cattle buyer, "skilled" packinghouse worker, "pioneering" scientist, and "experienced" salesman. Only the home economists inside the letter *F* receive no such amplifying adjective.

The main ad copy sustains, reinforces, and extends these gendered divisions. "Few homemakers indeed are positive of their ability to recognize tender steaks or roasts *every* time. That calls for a lot of special knowledge about meat—knowledge that no modern woman really needs, since the Swift name is her assurance of quality always." If the role of women within the company is simultaneously central and circumscribed, the company's vision of its women consumers is even more so. The ad copy addresses women as the possessors of their families' purchasing power, but it simultaneously strips them of their agency. The woman consumer, the ad copy implies, hardly even needs to think—the company will do it for her.

Selling Meat, Selling Sex

Considering these images and many others created by and for the Martha Logan Division, one recognizes Swift's sustained investment in visual imagery and visual control. The visual and performative aspects of home economics, in particular, functioned as a key means by which Swift promoted their products and, more broadly, contributed to changing patterns of meat consumption in the US. Of course, Swift was not alone in selling an image of femininity to its woman consumers alongside its actual products. Advertising firms and the companies they served made robust contributions to the construction of gender norms in the twentieth century. But as Carol Adams demonstrates in *The Pornography of Meat*, even within such a gendered field, women and meat have a special relationship. Images of women with meat, and women *as* meat, have been pervasive in Western visual culture. The phallic nature of hot dogs notwithstanding, the female body has most often been the one consumed. By equating women with animals and with meat, emphasizing women's animality, their sub-humanity, advertisers sell sex. By equating meat with a woman's body, emphasizing the product's desirability, its appeal to masculinity, advertisers sell meat.[97]

Swift's archives at the Chicago History Museum incorporate numerous photographs of Martha Logan activities, including displays at trade shows and conventions as well as dinners and other events that appear to have been more industry focused. Among other events, the Martha Logan Division sponsored beauty pageants and modeling contests. A photograph made in 1948 at a meeting focused on the company's "Swift'ning" shortening (a key by-product toward which the company dedicated immense promotional energy in the postwar period) shows Martha Logan director McLean standing on a stage next to a group of pageant entrants, each carrying a large tray of foodstuffs yoked around her neck. Beside and below the stage, two male company executives flank the group, one caught with cigarette in hand. McLean looks down on them solicitously. Martha Logan, itself a division comprising professional women, here hosts a display of an alternate femininity, in which women's bodies become consumables just like the foods they carry. Another pageant photograph shows a group of attractively dressed young women holding trays of food. Standing before a huge display of Martha Logan activities, beneath a banner reading "Nutrition is our business," the pageant contestants temporarily supplant the home economists, taking implicit responsibility for their activities. Domesticity becomes, in this image, not the rationalistic practice of white-uniformed professionals but an activity best conducted in heels and a strapless dress.[98]

At least one Martha Logan–sponsored stage display literally conflated women with meat and other comestibles. A photo of a promotional performance depicts a group of models dressed as various Swift products (fig. 70), their uniformly smooth legs emerging from huge costumes resembling vari-

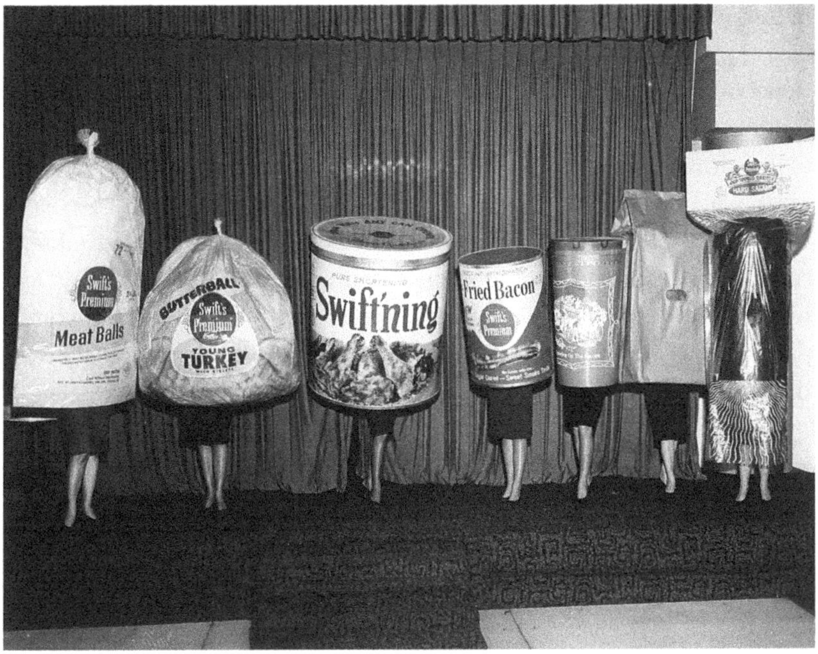

Figure 70. Models performing as Swift meat products, n.d. Martha Logan Photograph Collection, collection G1985:473, Chicago History Museum. Used with permission of Swift Meats/JBS.

ous food items: bagged frozen meatballs, a Butterball turkey, huge cans of Swift'ning and bacon, a salami. Their skirts fall at or just above the knee, relatively short for the time period; and they stand with models' stances, one leg extended a bit in front of the other, heeled shoes elongating and enhancing their calves.[99] Part of a larger performance, possibly one of the bespoke "industrial" musicals common in the mid-twentieth century, the still image neatly equates carnality and carnivorousness.[100] Women not only test, prepare, and serve meat but actually *become* it.

Engaged in the hard sell, advertisements informed women not only about meat products but about themselves: whom they ought to be, what they ought to aspire to, which products to cook and how to present them. Engaged in the soft sell, materials like cookbooks, anniversary books, and promotional events insinuated companies' names and products into women's consciousness, connecting meat and meat products with a certain kind of lifestyle, a certain kind of femininity. As we saw in an earlier chapter, materials and images designed to circulate within the meat industry overwhelmingly depicted meat production as a male activity, with women appearing only in circumscribed roles. In the public eye, by contrast, the industry represented the activities of meat purchasing and preparation as almost exclusively white and female. Performing these functions in promotional materials and ads, home economists essentially modeled behavior for the company's consumer base. Women's purchasing power was critical to the process, enacted over the course of the twentieth century, whereby meat consumption was normalized and regularized as a central aspect of American foodways. Twentieth-century women bought, prepared, and served meat, making it ubiquitous; promotional imagery helped companies to demonstrate and thus to propagate these behavioral patterns. As women purchased and prepared meat, however, they also became it: a transubstantiation of flesh into flesh, facilitated by promotional imagery.

CHAPTER FOUR

Picturing Meatpacking Work

THE MEAT INDUSTRY depends on a huge labor force to accomplish the constant processing of creature into meat. At any scale, meat production pivots on taking life from a creature who has possessed it. In industrial-scale packinghouses, it pivots on doing so as often, as quickly, and as smoothly as possible. Speedy processing hinges on the routinization of the labor of slaughter. In large-scale meat processing facilities, slaughter is compartmentalized into discrete acts, each performed by a single worker or a group of workers over and over. Each job is a cut, a twist, a particular motion with the hands, a specific movement of the body. Prior chapters of this book have focused on visions of meat production produced and promoted by the industry, whether to itself or to the public. Through bland language and corporate or mechanistic self-representations, the meat industry has encouraged its own workforces, too, to perceive the packinghouse as a factory like any other. Unsurprisingly, however, meatpacking workers' own representations of their work often differ from industry representations.

This chapter considers how representations of meatpacking workers created by industry compare with images created by, or on behalf of, workers. First, it examines photographs produced by industry to highlight what work looks like from a top-down perspective. Then it shifts to consider visions of meatpacking work from the inside, from the worker's perspective. The chapter considers worker photography more broadly; and it discusses the shape and preservation of union and labor archives and other materials. Workers may create, and have created, images related to their jobs and their activities—particularly if they were part of a labor union, and particularly if these were collective, union activities—for a wide variety of reasons. These are, really, the same reasons anyone would make photographs, from snapshots to more formal images: to commemorate, document, and remember; to record; to construct personal and collective histories.[1] Subject matter of workers' photographs includes everything from events and fundraisers at the local, to basketball games, to labor strikes and strike actions, to the work of slaughter and meatpacking itself. And photographs are not only made but collected, circulated, exchanged: histories created and accumulated, written and compiled.

This chapter also considers what happens when meatpacking workers take their cameras to work. What happens when workers image the packinghouse line, the packinghouse interior? What happens when they breach this space with their cameras—when, in addition to doing their work, they begin to *present* their work for consideration? What does it look like to see and to show oneself working? What does it look like to see and to show oneself working at the job of slaughter, specifically? What does slaughter look like to slaughterers and their coworkers? We are cautioned, aphoristically, never to ask how the sausage gets made; but what does sausage making look like to the people who make it?

Acting Like a Worker: Industry Visions of Meatpacking Labor

Historically, images *by* workers have proven far less common, in American industry, than depictions and representations *of* workers, the majority created or sponsored by companies and industry organizations. The field of scientific management, prominent in the early to mid-twentieth century, proves a case in point. The founder of the field, Frederick Taylor, argued that improvements in industrial labor could be made by analyzing workers' movements in order to eliminate unnecessary and time-consuming inefficiencies. Managers, often carrying stopwatches, observed and cataloged workers' motions, locating the most efficient workers and training others to emulate them. Brown notes that photography and other forms of image making played prominent roles in scientific management: "Photographic representation has been central . . . to the ways in which corporate managers have sought to secure the consent of workers, managers, and consumers in the unevenly successful project of rationalizing American capitalism."[2]

Following Taylor and building on Taylorist efficiency studies, industrial engineers Frank and Lillian Gilbreth incorporated photography and film extensively in their motion-efficiency studies of workers in many American industries. The Gilbreths contended that the camera was a neutral observer of reality, on the side neither of management nor of labor.[3] However, as Richard Lindstrom notes, the vision presented by the Gilbreths always became an institutional one.[4] Following the Gilbreths, an "aesthetics of efficiency, standardization, simplification, and economy" came to dominate in visual cultures of industry.[5] In images created or commissioned by and for the meat industry, workers virtually always remain nameless, defined entirely by their function(s) with regard to the conversion of animal into meat. The slaughtering processes depicted in ANCO's 1923 catalog, for instance, depended on images of workers to activate photographs of the company's machinery and to show how the machines worked. But such images were not intended to render labor itself visible. Virtually the opposite is true: workers operated ANCO's machinery in pictures in order to render the *machines* visible and legible, to demonstrate how the machines not only operate within but facilitate and dominate the human-animal-machine triangulation that characterizes packinghouse space (see, e.g., fig. 14; fig. 16). Human beings in the ANCO catalogs labor, essentially, to erase themselves, to minimize the value of their own work, or to fill in the space between machines. Not only do workers appear like automata in the catalog, all the same in their white uniforms; they appear to exist in order to facilitate the work of the machines, rather than the other way around. They serve as placeholders, occupying and signifying those spaces for which ANCO has not yet invented an appropriate device.

Exceptions to worker anonymity in industry-generated photographs do occur, once in a while; but even in these rare exceptions, the images exist to serve the ends of industry, not to represent the worker's perspective. A set of photographs created for Morrell & Co. in 1918, during the First World War, profiles a group of women workers hired as substitutes for men who had gone to serve in the military.[6] Well-made and carefully retouched, the pictures were clearly posed: in all the images the women stand statically, sometimes gazing at the camera or pantomiming jobs for the lens (fig. 71). Employment of women in meatpacking was, in itself, nothing new in 1918. Women had worked in meatpacking plants, especially in jobs like sausage making and canning, since the nineteenth century.[7] This group of workers, however, had taken jobs that would typically have been occupied by

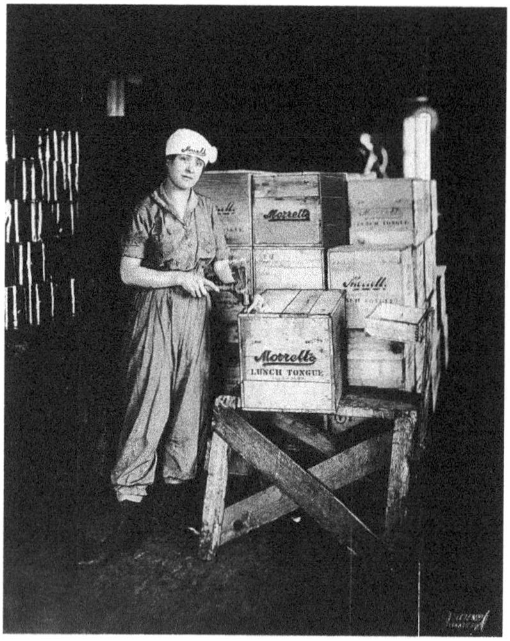

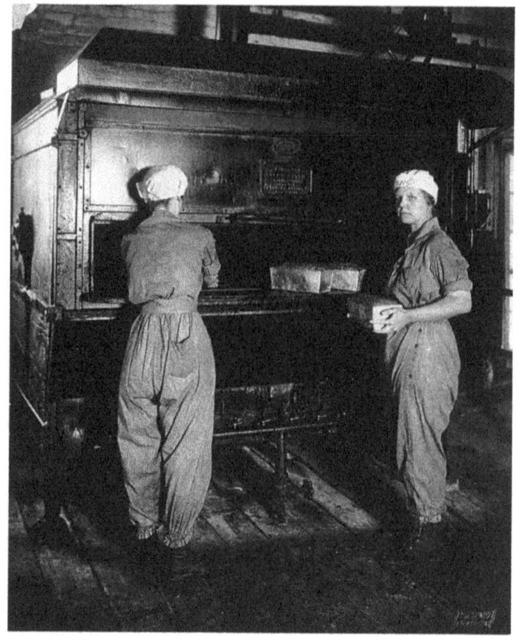

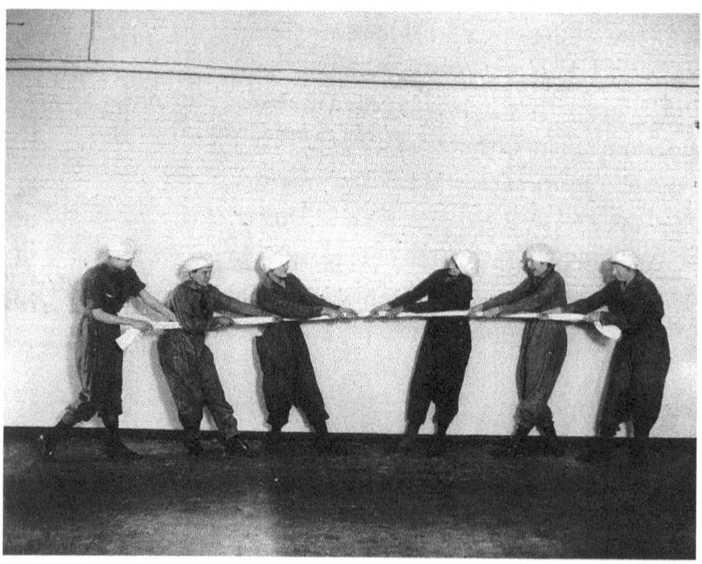

Figures 71 (top left), Woman nailing packing crates, ca. 1918; 72 (top right), Women workers baking, ca. 1918; and 73 (left), Women workers playing tug-of-war, ca. 1918. Morrell Meat Packing Company Collection, collection MSC 0100, Special Collections and Archives, University of Iowa Libraries.

men in peacetime, and the company had issued specially designed, semi-masculine uniforms for them rather than allow them to wear their own skirts or dresses, as women meatpacking workers typically would.[8]

As such, the photographs place more emphasis on the women's coveralls and their masculine work boots than on the work they have been hired to perform. One group photo depicts the crew of women workers in two rows, those in the top row standing and facing the camera, while those in the bottom row each prop one foot on a chair, pulling at their bootlaces. Another image shows two women removing loaf pans from an industrial oven (a nonthreatening, almost domestic activity), emphasizing the femininity of their custom-designed coveralls (fig. 72). With their flattering high-belted waists, buttoned flaps, and ruffles at the ankle, the outfits are masculine in basic form (i.e., they have pants) but feminine in detail. They seem designed to be photographed. The novel sight of women in pants and boots matters more, in these images, than the gravity of their contribution to plant operation. Indeed, many of these photographs highlight the women workers' *lack* of gravitas: one picture even shows them posing at playing tug-of-war (fig. 73).

A newspaper article, published in April 1918 in the local *Ottumwa Courier*, incorporates two of the pictures, a group portrait and the image of a woman sealing packing crates with a hammer and nails (fig. 74). The newspaper, too, emphasizes the women workers' clothing and appearance more than their jobs. Under the headline "Women Don Khaki Overalls as Butchers to Replace Husbands and Brothers Gone to War," subheadings read, "Girls Help Solve Labor Shortage in Packing Industry," "Morrell's Find Gentler Sex Able to Do Many Tasks That Only Men Were Assigned Before; Are Used All Over Plant," and "Find New Garb Safe and Sane."[9] Although the text in the body of the article acknowledges that the meat industry had employed women long prior to the First World War, the headlines make it seem as though these coverall-clad "girls" had only just come to work at the plant. Nearly one-half of the text focuses on the benefits—and the attractiveness—of their coveralls.

The lightheartedness of the pictures and the allure of the coverall-clad women together convey a sense of reassurance. The pictures imply that women's increased presence and importance to packinghouse operations will be temporary and hardly disruptive to their role as nurturers and homemakers; and the article text, too, asserts that the women will "keep home

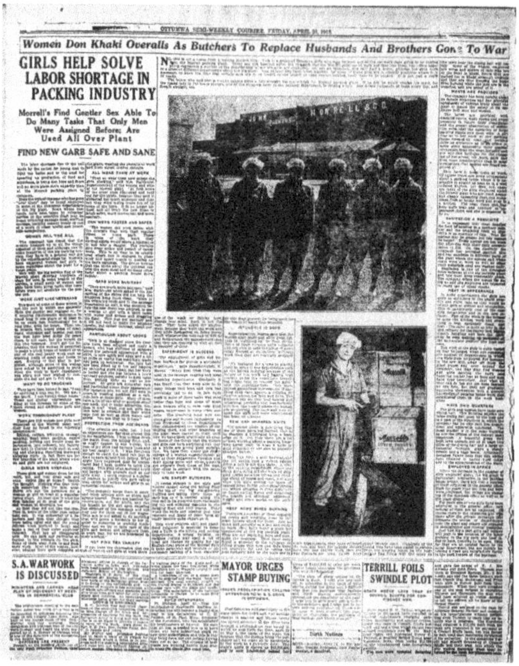
OTTUMWA SEMI-WEEKLY COURIER, FRIDAY, APRIL 26, 1918

Women Don Khaki Overalls As Butchers To Replace Husbands And Brothers Gone To War

GIRLS HELP SOLVE LABOR SHORTAGE IN PACKING INDUSTRY

Morrell's Find Gentler Sex Able To Do Many Tasks That Only Men Were Assigned Before; Are Used All Over Plant

FIND NEW GARB SAFE AND SANE

S. A. WAR WORK IS DISCUSSED

MAYOR URGES STAMP BUYING

TERRILL FOILS SWINDLE PLOT

Figure 74. "Girls Help Solve Labor Shortage," 7.

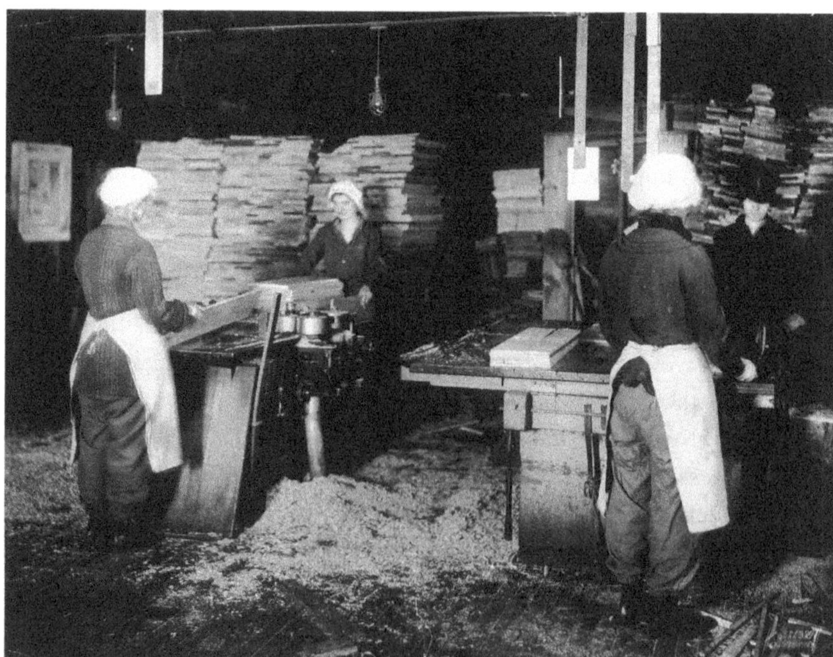

Figure 75. Women planing boards, ca. 1918. Morrell Meat Packing Company Collection, collection MSC 0100, Special Collections and Archives, University of Iowa Libraries.

fires burning" and that they "are not replacing boys and men, quite the contrary . . . by taking the jobs formerly held by the male sex in some departments, they have released the men and boys for heavier . . . work that the war demands for man power have made inroads upon."[10]

Morrell & Co. commissioned the photographs to emphasize the novelty of women's work—and women's trousers—in the packinghouse. But the news coverage and many of the photographs themselves, despite their contrivances, manage to convey that a woman worker is nonetheless a worker. The newspaper article notes that women workers both enjoyed the increased variation in their job assignments and performed well in a wide variety of roles, including skilled butchery. In this way it evinces something similar to the accidental assertions of worker personality that Lindstrom finds in the Gilbreths' time-study images.

Some of these photographs, too, permit a degree of worker expression to peek through. One picture in the group shows women workers preparing boards for packing crates (fig. 75). This picture, unlike most of the others, offers little visual commentary on their gender: their faces are too blurred to afford a means of evaluating their appearances, and their aprons and heavy sweaters obscure the details of their clothing. They are workers here, not women workers. Most notably, one woman in this photograph seems to have attempted a deliberate subversion of not only the gender assumptions that the photo series was designed to highlight but class hierarchies as well. Standing in the background at the right margin, she wears on her head not a white cap like the other women's but what appears to be a top hat: the headgear not only of a man but of a middle- or upper-class man, a businessman. Below it, her dark coat almost resembles a man's suit coat. Called upon to participate in one kind of gendered masquerade for the camera, the woman seems to have pivoted, taking her performance in a different direction and donning the uniform of the businessman.[11] In doing so, she calls into question the entire enterprise of the photo series and the gender and class assumptions it was commissioned to reinforce.[12]

Mining the Union Archive: Meatpacking Work, Unionism, and Preservation

The vision of workers vis-à-vis the meat industry—vis-à-vis their own work—proves considerably more heterodox than the vision presented by the industry itself: unsurprisingly more individual, more open to personal expression, to irreverence, to humor, to trenchant social commentary—sometimes to all of these things at once. Workers' visions of the meat industry often run counter, too, to the vision of the industry presented *by* the industry. Industry presents a vision of itself, its companies, its activities, and its products as unified and streamlined, modern, shiny, and clean. In industry-created media, workers often appear either as dehumanized functionaries—present to perform work but wholly lacking in agency—or as satisfied contributors to the capitalist endeavor. Workers themselves, on the other hand, are much more likely to admit and to represent the difficulties of meat production: filth, extreme temperatures, pain and injuries, repetition and boredom. They tend to acknowledge more frequently and more frankly the messy and difficult nature of meatpacking work, and the violence that underlies it.

In any industrial packinghouse, events occur in a certain order and tasks are broken down into constituent movements, as few as necessary to complete the job. Death comes first, but even it is compartmentalized into stunning and bleeding. For all its constancy and ubiquity, the passage of death is difficult to perceive in its entirety. Fragmentation is the rule. Only a few workers see any animal go from alive to dead, and even fewer participate directly—a situation that has been true for much of meatpacking's industrialized history, and that remains so still.[13] Trauma to the animal causes trauma, and thus value loss, to the meat, so even jobs directly involved with slaughter must feel, to the worker, as nonviolent and routine as possible.

Euphemism, too, is pervasive. The industry calls what happens on the cold side, or cut floor, of the plant, after the halved carcass comes out of the cooler, "fabrication."[14] The work incorporates a great deal of repetitive motion: knife cuts and twists, feeding substances into machines. Indeed, even on the hot side, or what the industry calls the "harvest department," worker training emphasizes removal and repetition.[15] But the scrim of propriety is sometimes thin: many workers and even managers persist in referring to the hot side of the plant not as the "harvest floor" but as the "kill floor."[16]

Compartmentalization of labor began very early in the meat industry, by the mid-nineteenth century. "As soon as mass production was used to turn the live animal into salable meat, sharp and thorough division into single operations became necessary, as in all mechanization," writes Giedion.[17] Work in American industrial-scale packinghouses has, historically, varied widely: from skilled to unskilled work; from relatively clean jobs to filthy, smelly, gut-wrenching jobs; from hot and humid environs to chilly, even frigid environs. Pay has varied, too, concomitant with skill level, among other considerations. The skilled mechanics who kept packinghouse machinery in working order often received good pay for doing a vitally important "clean" job.[18] Skilled "dirty" jobs could pay well, too: in the early to mid-twentieth century, the highest-paid employees included kill-floor workers who halved carcasses.[19] But less skilled workers throughout the plants, including some on the hot and messy kill floors, received low pay, even for jobs with a high rate of injury.[20] Offal departments, which received the internal organs and other by-products of slaughter, often housed semi-skilled workers who performed jobs among the dirtiest, smelliest, and lowest paid in any facility.[21]

As the work has varied, so have the workers assigned to it. White men historically had the best and highest-paid jobs, including skilled work on the kill floor, electrical and mechanical work, and, of course, supervisory positions.[22] Certain other jobs, such as the highly mechanized and visible work in bacon-slicing and bacon-packaging rooms, went to women who were selected for their physical appearances, a factor that included their race. At mid-century, "in sliced bacon, a particularly desirable and clean department, firms routinely employed young, native-born white women dressed in immaculate work clothing to impress visitors to this 'show' area."[23] Black women, by contrast, got the worst jobs: processing offal and washing sausage casings.[24]

> During the 1920s, black women came to dominate the casings and offal departments in Chicago's packing plants, partly because the work was considered particularly distasteful and partly because black women were considered more reliable employees; East European women usually quit work when they married and had children, whereas black women stayed in the workforce.[25]

Black men, too, tended to work in more dangerous positions requiring less skill: shackling live animals, beheading cattle, and scalding and dehairing hogs.[26] Black workers, male and female, tended to work in jobs near the beginning of the production line, or on the periphery, with offal and by-products; Warren notes that in the first decades of the twentieth century, many packers explicitly "prohibited blacks from touching finished meat products."[27] Such racial divisions within the industry in the early to mid-twentieth century would become a major source of workplace grievance. The degree to which meatpacking companies exploited racial divisions, and the effectiveness of responses to this divisiveness by organized labor, often determined the extent to which labor unions in the meat industry succeeded or failed at achieving their goals.[28]

All of these forces have shaped labor and the labor movement in the industry. "At best, working conditions were disagreeable, dangerous, arduous, and, in most jobs, monotonous. People working under these conditions may be expected to seek ways to improve their work environments."[29] Not that unionization efforts proceeded smoothly. In the early twentieth century, kill-floor workers and curing workers received hourly wages, but in many departments pay was allotted per piece, with significant incentives for higher production. Such conditions encouraged overwork and competition among workers, and they reduced workers' drive to join and support unionization efforts.[30] But as scientific management, efficiency, and profiteering rose within meatpacking as in many other industries, workers' desire for a strong collective bargaining position increased, and unions gained power and membership over the course of the 1920s and 1930s.[31] By the mid-1930s, meatpacking unions had evolved strategically, along with the industry itself:

> Producing a perishable commodity at very low profit margins, management incessantly struggled to keep labor costs at a minimum and exercise a maximum degree of control of the pace of work. . . . The intensely subdivided work process, dispersed into dozens of physically separate rooms, encouraged local unions to adopt a decentralized departmental structure so as to reach into all areas of a packinghouse.[32]

To have sufficient power to sway management, unions had to have a presence in all of the major meatpacking firms; and union locals in cities across the Midwest had to coordinate their efforts. While unions had to organize throughout packing plants, the kill floor—which stood at the beginning of the disassembly line and thus set the pace for the rest of the plant—became a key node in labor-management relations in most packing plants.[33]

Various unions served the industry, sometimes competing with one another. At mid-century, the United Packinghouse Workers of America (UPWA; a later iteration of the Packinghouse Workers' Organizing Committee, or PWOC) gained traction after it made racial integration and support for civil rights key components of its platform. Beyond a general concern for social equality, the UPWA recognized that in meatpacking especially, worker rights were synonymous with civil rights, since the meatpacking workforce tended to be racially and culturally diverse, especially in large cities like Chicago. The UPWA thus saw racial inclusivity as key to solidarity and, by extension, to a strong bargaining position for labor—particularly vis-à-vis organizational structures that would have preferred a divided and internally fractious workforce. Attempting to unify all workers, the UPWA contended that the interests of one group of workers must become the concern of all.[34]

Preserving materials and individual narratives becomes central to preserving collective stories and collective histories: a key way that labor movements control their own narratives and combat the erasures, both inadvertent and deliberate, that occur when the only available stories are those told by government and industry. Strong counternarratives depend on preservation of memory and on documents, images, and artifacts that may serve to prompt, anchor, ground, and represent memory in visible or tactile form.[35] As historian Alessandro Portelli points out, the attraction of official documents, such as judicial records, governmental reports, and even media reports, often lies in their precision, their air of certainty with regard to dates and other details—even when reports are compiled, as many are, from eyewitness accounts.[36] But this air of certainty belies their partiality. Dimensional understanding of historical events relies on the collection and preservation of narratives, documents, images, and other materials from varied and multiple sources, particularly when the interests of those sources are opposed to or at least not represented by the dominant forces of industry, government, and mainstream media.[37]

Archival preservation constitutes an interest that industry and labor have in common. But there are fundamental differences between the archive of a company or industry interest group, on the one hand, and that of a labor union, on the other. Archives of companies and industry organizations tend to be linear and authoritative, taking a magisterial view of meat manufacturing processes and presenting unified visions of company operations and products. It is in the industry's favor and interest to have and to present a unified vision of its operations, a unified sense of itself and its products. Industry has the resources to produce booklets, pamphlets, annual reports, and commemorative publications. Industry has advertising at its disposal.

The unified vision rises to the top; internal fracturing and disorganization, where it occurs, is tamped down and suppressed. The archive functions as "an apparatus of rationalization and social management."[38] Creating and organizing knowledge, it becomes a representation in itself.[39]

Labor unions, on the other hand, are fundamentally multivocal, democratic organizations, and their archives reflect this. While materials preserved in labor archives include pamphlets, brochures, posters, books, and periodicals as well as documents and communications—the same sorts of materials incorporated in industry archives—labor archives may also incorporate material submitted by union members or collected by union locals (which often vary widely in character, both from one another and from any national organization of which they might be a part).[40] Pictorial contributions to union archives come from all sorts of sources: some may be made by professional photographers hired by the union or by accomplished amateurs within the union; but members' casual snapshots may be included, too, as well as images from an array of other sources.[41] Materials might be collected piecemeal or amassed a little at a time, over time: a few photographs here, a snapshot there, a box of materials, a photo album. Perhaps these things are collected by and for the use of an entire group of people, kept or circulated at a union local; or perhaps they are in the possession of an individual and arrive through small-scale donation.

Labor archives may be very deliberately ordered, in service of an organization's endeavor to preserve and to relate its own histories. But there is a sense, in labor collections, that more people are being given more space to speak, to contribute their individual histories to the broader organizational history. The heterodoxy of labor archives reflects the nature of labor organizations themselves: flexible and collectivist, answerable not only to the demands of regional or national organizational leadership but to the needs of local chapters in local communities, experiencing localized conditions. Unions exist to serve and promote their members' interests and overall well-being—a collectivism reflected in the materials they gather and preserve. If a company, or the meat industry overall, has corporate personhood, members of the union have *actual* personhood, a sense that is preserved in union archives as representations of their organizations.

The archival process itself—the process of gathering, sorting, cataloging, and storing—is a sense-making process, a process of classification and arrangement. The making of meaning in the archive does not necessarily conform to the meaning that an individual or group would have made for themselves. What historian Rick Halpern calls "the occasionally visible hand of the archives" may well impose its own networks of meaning and value on collected documents, imagery, and other materials.[42] Archives of one organization may be spread over multiple institutions, too. Such diffusion of materials proves a persistent issue in archival research: archives may be widely dispersed into what Ricardo Punzalan has termed "archival diasporas."[43]

Meatpacking labor's archival diaspora includes both archived images and images that have circulated in advertisements, journals, and other publications. Larger archives, such as that of the Amalgamated Meat Cutters and Butcher Workmen (AMCBW)—which includes archives of organizations absorbed into the AMCBW, the UPWA among them—incorporate many thousands of photographs that were created, gathered, or collected in numerous ways.[44] Unions commissioned professional photographers to make some of them: pictures of union events, celebrations and conferences, and day-to-day operations both of the union itself and, sometimes, of the businesses where its members worked. Some professional photographers might have worked for labor unions just as for any other client; while others, as historian Larry Peterson discusses, specialized in—and often sympathized with—organized labor.[45]

The UPWA collected large numbers of photographs for its house periodical, the *Packinghouse Worker*, gathering and reproducing images of a wide variety of subjects off the press wires. Where a community newspaper sent its reporters and photographers to cover a meat industry event —anything from the opening of a new facility, to the novelty of an escaped animal, to a major strike action or other source of conflict—unions might request and compile copies of these materials, too. Meat industry unions also acquired photographs from government bodies such as the United States Department of Agriculture (USDA).[46] In addition, union members contributed personal photographs, with subjects running the gamut from snapshots of work-related activities to strikes and other conflicts. Consideration not only of union-collected and union-circulated photographs but of the nature of such photographs, the source from which they might have been acquired, and the way(s) in which they might have been deployed, used, and reused proves crucial to understanding how photography served labor causes.

Worker and Worker-Sponsored Photography in the Meat Industry

Across industries, spaces, and times, workers and worker organizations have long understood the value of visual representation just as their employers did. Internationally, impulses to a more democratic, inclusive photographic vision caught on rapidly in the early to mid-twentieth century. Leah Ollmann describes how leftist movements and leftist presses in Germany, for instance, came to depend on proletarian and worker photographers for a vision of the working classes that was created specifically *by* the working classes.[47] Workers' photography clubs in Germany often educated photographers both in the fundaments of the medium and in political rhetoric simultaneously.[48] Similar conditions arose throughout Europe.[49]

Despite having far fewer resources to marshal than industry, many American labor unions, too, endeavored to create representations of themselves and their work. Self-representation, for workers, mattered not only for public relations but for internal relations as well. As Peterson notes, "A crucial part of labor organizing involves breaking through the ideological influence of capitalist media over workers," by offering labor-produced media as an alternative.[50] Carol Quirke demonstrates that a dramatic increase in circulation and reproduction of photographs not only of workers but by workers occurred at the same time that photography itself was becoming more ubiquitous and visual culture more saturated in the US during the 1920s and 1930s, phenomena driven in part by the ascendancy of picture journalism. Depression-era impulses to unionism, strengthening simultaneously, both benefited and were hindered by press representations of workers, as journalism represented the worker sometimes as heroic, sometimes as subversive.[51] Quirke shows how the impulse to self-representation arose as unions and workers saw the power of images to affect both public and internal perception of unions, unionism, and workers. If industry media and mainstream journalism represented worker interests negatively or dismissively, labor could counter that narrative with its own, showing workers' valuable contributions to American culture and the importance of their concerns to overarching American ideals of openness and industriousness.

As such, unions worked actively to produce their own newspapers and illustrate them liberally with imagery, much of it created in-house or commissioned by and for the union. Steelworkers' unions had the *Steel Labor* newspaper, founded in 1936 (the same year as *Life* magazine); an influential distributive workers' union in New York had *New Voices*. Both publications were more likely than mainstream media to represent, for instance, the racial and gender diversity of unions and to emphasize the strength and solidarity of striking workers, rather than showing them as violent, disgruntled, or aggrieved, as the mainstream press might have done.[52]

In both Europe and the US, however, workers' photography has only intermittently been photography *of work*. While Ollmann highlights a few German photographers who created images of their coworkers in industry, many worker photographers made images in and of their streets and neighborhoods, not of their jobs or even of industrial work more broadly.[53] This absence of images of work itself was not from lack of interest in the subject: as early as 1926, the leftist journal *Der Arbeiter-Fotograf* wrote that

> of all the great sphere of topics that need to be covered . . . one [in particular] represents in our opinion the most important issue for worker photographers to address: picture reports on their places of work. . . . The entire revolutionary working class desires and demands that photographers capture the abuses of the capitalist system, . . . the insufficient safety regulations and the resulting accidents, and submit them to the sharpest criticism.[54]

Rudolf Stumberger notes that opportunities for such photographs were not often forthcoming, however. Since "factories were non-public areas . . . where political activity and, as a rule, photographic shooting were not allowed . . . worker photographers were therefore usually barred from using the factory or the workplace as a photographic subject."[55] Even within the USSR, where the distinction between capital and labor had been obscured and relations between workers and overseers were at least theoretically more egalitarian, Erika Wolf writes that worker photography of industrial workplaces proved difficult to achieve in practice.[56] Among worker photographers, then, documentation of the job, while *on* the job, has proven quite rare: suspicion of the worker-controlled camera has simply run too high, among capitalist and communist authorities alike, for opportunities to present themselves with regularity. Sanctioned, professional photographers and journalists might document working conditions; workers themselves most often could not (and cannot).

In American industry just as in European, workers have simply had far less time, opportunity, and resources available to photograph their own work. The best-known American parallel to European interwar worker photography organizations, the Photo League in New York, consisted in large part of professional or serious amateur documentary photographers and concentrated its efforts primarily on producing social documentary imagery of American streets, neighborhoods, and living conditions, not workplaces.[57] And despite their interest in creating visual representations of worker interests, photographers working on behalf of American organized labor have tended to concentrate their efforts on photographing union activities—not work itself. Quirke shows how, through circulation of images by volunteer member photographers in its *New Voices* periodical, the New York warehouse workers' union Local 65 "document[ed] union life and antiracism activism."[58] However, "*New Voices* photographers rarely showed life outside the union. Only during the [Second World] war years did employers allow *New Voices* photo shoots to show workers on the job."[59] Again, imagery of work itself—and of those who perform it, shown *as* they perform it—has tended in the US to be controlled by, and to flow out from, nodes of capital. As they control the workplace, so have employers controlled the very image of work itself.

Exclusion by employers provides only part of the reason why the workplace appears relatively rarely in labor-created imagery, however. Peterson shows that the goal of union and labor presses was, at least in part, to subvert the dominant position of the workplace itself—and, crucially, of the corporate interests that ran the workplace—in workers' lives, questioning in the process the economic and social order that corporate-created imagery implied:

> Corporations had long tried to convince workers that capitalist mass production was a rational, scientific form of industrial organization in which each class of employee fulfilled its function as part of a cooperative team and was fairly compensated for its contribution. This ideology was presented pictorially [by corporations] in photo essays of the stages of production. Workers' position—defined by the layout of architecture and machinery, designed by managers, and directed by foremen—appeared as the necessary consequence of technical and economic logic, not as a social relation of subordination and control.[60]

In avoiding direct depiction of the workplace and focusing on labor and union organizational activities instead, labor media "fill[ed] the company's visual sequence with a new content," and in so doing "subverted the pretense of scientific objectivity and made the authority of managers and foremen appear as mutable social relations."[61]

Meatpacking unions followed suit. In pictures made by, for, and on behalf of meatpacking labor organizations, the work itself comes up far less frequently than the working-class community more broadly. Social life, union life, and family life appear far more frequently and prominently than work life. The dearth of photos of meatpacking work by workers may be due to employers' reluctance to allow cameras in the workplace; but meatpacking unions also desired to foreground themselves as cultural institutions and forces. They may have grown within the meat industry, but they were not defined by it.

Hiring photographers to image union activities was a common practice, particularly in urban areas where a large professional photographic workforce made such outsourcing possible. Peterson notes how *Steel Labor*, a publication based in Chicago, took advantage of "the rapid expansion of commercial photography in the 1920s and 1930s [that had] spawned a large number of small photographic studios [in the city] . . . centrally located in and around the Loop."[62] Some advertised themselves as "union photographers," and others worked—and sympathized—readily with the goals of organized labor.[63] Meatpacking unions, too, took advantage of such resources. The credit stamps of freelancers Mickey Pallas and Jo Banks, among others, appear quite often on photos made for meatpacking labor organizations in Chicago.

Beginning immediately after the Second World War, the UPWA and other organizations hired Pallas with some regularity. Pallas was not solely a union photographer: his clients included large corporations as well as labor organizations.[64] But as a former factory worker and Congress of Industrial Organizations (CIO) labor organizer himself, Pallas understood and was sympathetic to the needs and goals of the labor movement. Moreover, his early involvement as an ally of equality and antidiscrimination movements also fed into his photographic career: he freelanced for *Ebony* and other magazines addressing primarily audiences of color, and he supported and maintained friendships with prominent Black entertainers, musicians, and other public personalities.[65] His active promotion of equal rights, both with his photographs and in his life, may have led the UPWA to hire him preferentially, since his interests and sympathies dovetailed with their own.

Subject matter in Pallas's photographs for meatpacking unions varies as widely as union activities themselves: he photographed meetings large and small, union events, and other newsworthy items, his pictures subsequently appearing in various meatpacking labor publications and industry periodicals. He also took pictures of meatpacking work itself, documenting interiors and labor processes in multiple plants, traveling to facilities throughout the Midwest to photograph their operations. Just as the work of a freelance photographer like Korling incorporates the point of view of the company or industry organization that hired him, Pallas's work reflects his own professional commitments. His photographs highlight workers themselves, often just as much as the tasks they performed. The photos represent if not the worker's point of view in itself, then something adjacent to it: images showing both the job and the person who does the job. Pallas treated workers as human beings, not simply as fleshly implements of mass manufacturing.

Contact sheets in the AMCBW/UPWA archives at the Wisconsin Historical Society show that among Pallas's assignments for meatpacking labor were groups of images created at the Tobin Packing Company in Fort Dodge, Iowa, and the Bookey Packing Company in Des Moines, Iowa, both independent slaughtering and packing operations. Dated to around 1947, the series created at Tobin shows primarily pork packing, while the images made at Bookey—undated, but from the same general time period—depict beef processing.[66] Though clear, authoritative, and matter-of-fact in presenting animal slaughter, Pallas's images make no attempt to offer a step-by-step depiction of all stages of processing. Pallas positioned himself, instead, in places throughout both plants where he could give a sense of how human beings operated in the spaces and performed the work assigned to them. Pallas offers far less sense of human-animal-machine triangulation in his images than what one typically finds in industry-produced publications such as ANCO's catalogs. Rather, Pallas conveys a sense of the realities of packinghouse work—the physicality of the work, the muscle memory that develops through repetitive tasks, the solitude inherent in some jobs and the interpersonal interactions that came along with others.

One Tobin contact sheet (fig. 76, *left*) incorporates images of diverse pork processing jobs, from sticking and bleeding carcasses to trimming of heads to deskwork with pen and paper. Even among just the twelve frames on this single contact sheet, Pallas's dedication to depicting the worker,

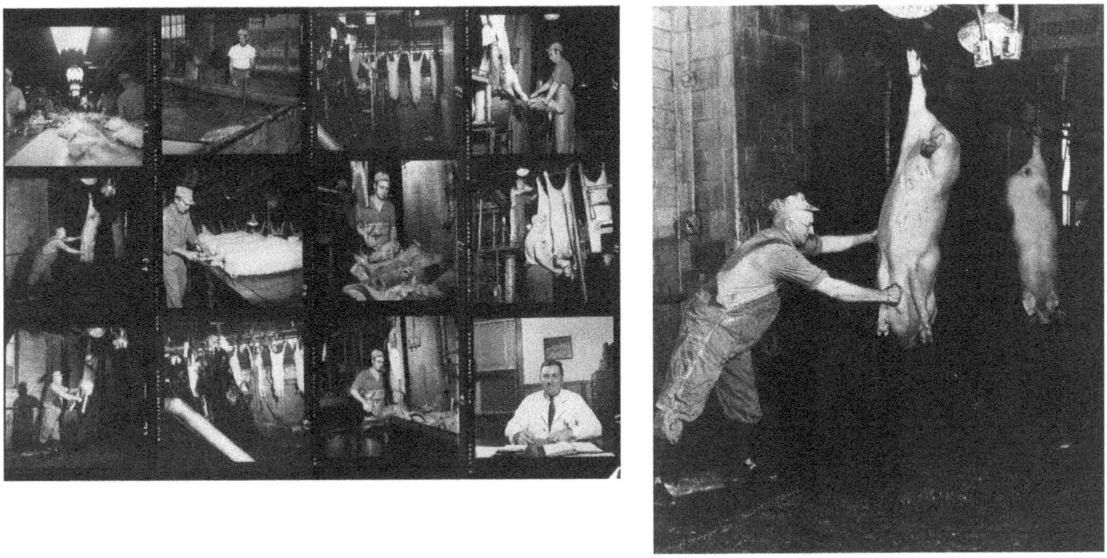

Figure 76. Mickey Pallas, Contact sheet and single image made at the Tobin Packing Company, Fort Dodge, IA, ca. 1947. Archives of the Amalgamated Meat Cutters and Butcher Workmen, collection M79-517, Archives Division, Wisconsin Historical Society. © Center for Creative Photography, The University of Arizona Foundation.

as much as or more than the work itself, becomes clear. While one or two images on the sheet seem to aim for a formalist aesthetic—like the image in the top left corner of the contact sheet, a nearly symmetrical picture of people laboring along a conveyor belt—the imperative to highlight the worker more often prevails. Pallas's most effective pictures give a sense both of the worker's experience and of the nature of the work itself. Two images on this contact sheet, at the center of the left edge and, below it, in the bottom left corner, show the sticker at work. In the top image of the pair—which Pallas and his clients regarded as successful enough to enlarge and print alone (fig. 76, *right*)—the sticker reaches out over the floor to catch and grip a hanging animal, pulling it toward him as he drives the knife into its neck.[67] Leaning almost precariously, he stretches his back and arms, his entire body involved in the work of slaughter. The posture would be awkward, ungainly for anyone, painful to repeat more than a few times. How many times has this worker, past middle age, performed this task, and at what cost to his body? The image makes clear both the physicality of the work and its constancy: another pig, nearly identical to this one, hangs behind it, referencing the endlessness of the hanging chain and the disassembly line. In the image below this one, the same worker bleeds a smaller, multicolored pig. Bringing the animal closer to him, he appears to operate more comfortably and less precariously than in the other picture, weight more properly centered over his feet. The pair of images together makes clear the variation in animal shapes, sizes, and colors; and a concomitant variation in the way the worker moves in encountering each one. The sticker employs his own accumulated experience and skill in the job; but the pictures make clear, at the same time, the unrelenting stresses and dangers the work presents to his body.

Another image, just to the right of center of the same contact sheet, shows a different worker trimming pigs' heads (fig. 76, *left*). It is one of a pair of images depicting the same subject, the other just below it. In both photographs, Pallas has centered the worker, showing him standing over a conveyor belt filled with heads. In the lower and less successful image, the material on the conveyor belt is partially obscured and the man's environs seem anonymously industrial. The upper image, however, shows clearly both the worker himself and the conveyor belt filled with pigs' heads. Still nearly intact at this stage, the pigs' heads retain their animal expressiveness, even without their bodies. Eyes glassy and lifeless, ears and snouts ajumble, they look in every direction: one appears to gaze at the ceiling, its erect ears hanging over the edge of the conveyor belt; while another, on its side, stares with something like shock directly into the camera's lens. Looming above the conveyor, the worker grips one head to trim it. Chin tucked to his chest, he looks down at the head with a matter-of-fact expression, concentrating on his task. Pallas successfully captures in this image the nature of much packinghouse work: a job both like many other factory jobs in the constancy and consistency of its routine, and profoundly unlike any other job in centering animal bodies, flesh, and substances at its core.

In the images created at the Bookey Packing Company, Pallas again manages to convey in just a few pictures a sense of how the routineness of cattle slaughter coexists with a demanding physicality and a profound existential estrangement. There is an impressiveness to the size of cattle that sets them apart from pigs and sheep. Cattle slaughter takes on an almost epic quality: mechanized slaughter, in its absolute rationalism, manipu-

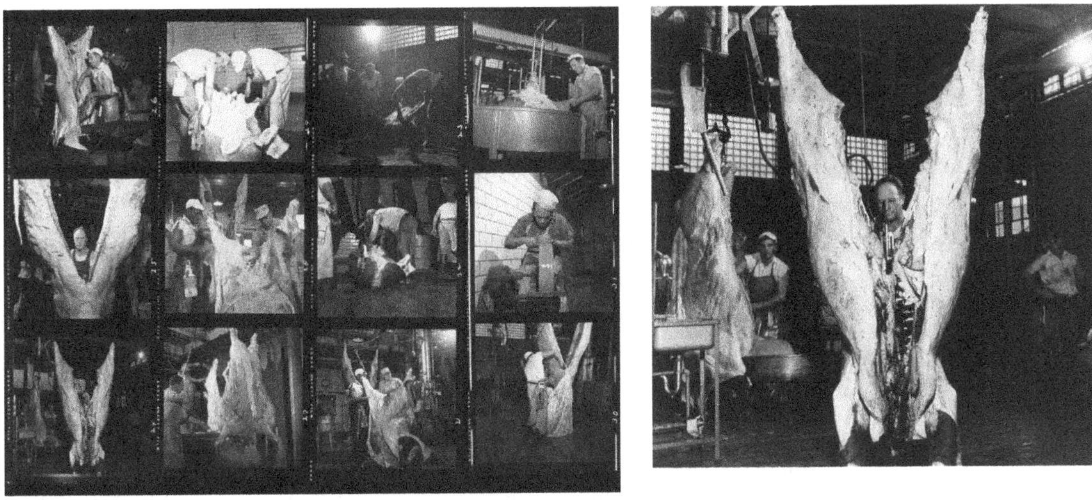

Figure 77. Mickey Pallas, Contact sheet and single image made at the Bookey Packing Company, Des Moines, IA, ca. 1947–1950. Archives of the Amalgamated Meat Cutters and Butcher Workmen, collection M79-517, Archives Division, Wisconsin Historical Society. © Center for Creative Photography, The University of Arizona Foundation.

lates and deconstructs even these huge bodies, converting them, too, into consumable flesh.

Due to the sheer size of the animals, the work of cattle slaughter demands cooperation. In one contact sheet, workers in image after image orbit around cattle carcasses (fig. 77, *left*). Many operate in pairs and teams, bending over prone carcasses to open and skin them, or hoisting them to remove hide and organs. A few teams comprise both white and non-white men, and Pallas photographs these multiracial interactions matter-of-factly, presenting interracial interaction not as a focus of the series but as inherent to the nature of meatpacking work. Indeed, such harmonious collaboration ensures speed and efficiency: Pallas's images foreground egalitarian interaction as crucial to both industrial and social progress.

Some meatpacking jobs, however, could not be shared. In Pallas's photographs from Tobin, this comes through in the pictures of the sticker, working alone to bleed pigs. In the Bookey contact sheet showing cattle slaughter, an impactful pair of images depict a single worker splitting a carcass. Located at the left center edge and lower left corner of the sheet, the images show the splitting of the animal from two vantage points: one from a remove, the other closer up. The massive carcass spans nearly the full distance between the hanging rails and the floor. Stripped of the hide, the subcutaneous fat gleams white in the monochrome images. In the more distanced photo, the halves of the carcass rise symmetrically like a pair of wings around the splitter, metamorphosing him into butterfly or angel (fig. 77, *right*). The image bears some kinship with Rembrandt's paintings of cattle carcasses, gleaming bright in the darkened back rooms of butcher shops. There is an epic, otherworldly quality about this massive carcass and the man splitting it: together they transcend the base, bloody packinghouse space.

The other photograph of the splitter depicting him from closer up, frames the worker's upper body in a *V*, created by the bright white spans of the animal's leg flesh. Split partially in two, the creature's bulging back muscle becomes a gleaming breastplate over the worker's bib overalls, converting him into a kind of Athena Nike of the packinghouse. There is, too, something coital in the spread of the carcass's pale body, as the worker uses his knife to penetrate and occupy the creature's core. In the pair of images, Pallas captures something of the essence of the slaughterhouse and of packinghouse work as a whole: at once matter-of-fact and heroic, debased and sublime, visceral and rational. His success at photographing the work of meatpacking must have satisfied his clients, who hired him repeatedly; but he also transcended the assignments to produce images that were, at their most effective, something very like art.

Less is known about Jo Banks, another photographer who worked for the Chicago meatpacking unions from time to time. A Black photographer and jazz musician, Banks ran a photography studio on Chicago's south side, advertising as "Jo Banks Creative Photography," available for "portraits, photo documentaries on assignment, [and] commercial photography of all kinds."[68] Portraits and other images by Banks appear here and there in a range of repositories, but there is no single archive of his work. He had a difficult youth, marred by episodes of violent racism. He apparently learned to photograph while serving in the merchant marine just before the start of the Second World War. Moving to Chicago, he worked as a photographer and became involved in the jazz world, running a small club and working as a musician. In the 1960s, he and his wife moved to Copenhagen, Denmark,

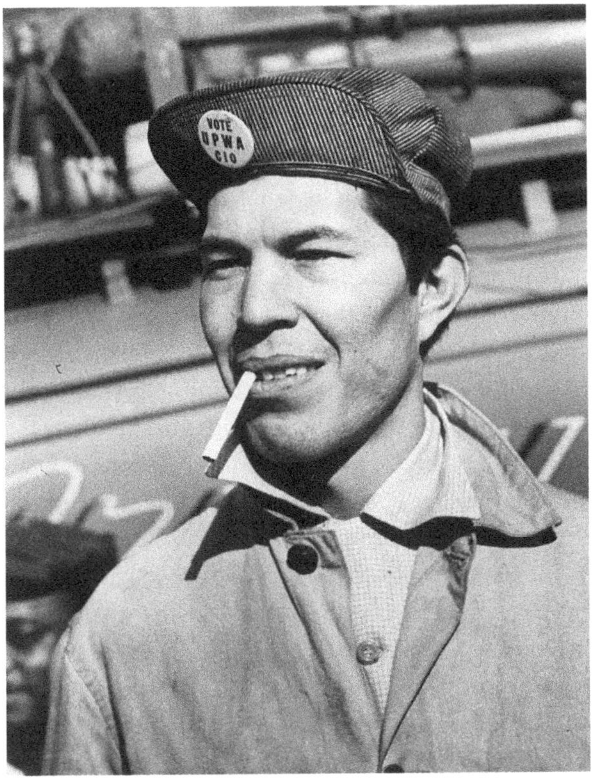

Figure 78.
Jo Banks, Portrait of a packinghouse worker, ca. 1942. Archives of the Amalgamated Meat Cutters and Butcher Workmen, collection M79-517, Archives Division, Wisconsin Historical Society.

where he continued to photograph while also playing music and running a jazz club out of his studio.[69]

Banks worked for the UPWA as early as 1942. Although his commissions for the UPWA probably fell under the rubric of what his advertisement called "photo documentaries on assignment," his experience as a portrait photographer, too, sometimes showed in his commissioned imagery. One picture of a UPWA union member, photographed outdoors at a rally or protest, shows him in a close-up one-quarter profile, squinting against the sun, a cigarette dangling unlit from his lips (fig. 78). Checked shirt buttoned to the neck under his jacket, the man wears an engineer's hat with an upturned brim, his union button pinned prominently on the visible underside. Though clearly not made under studio conditions, the photograph functions as a street portrait, the details of the subject's appearance and personage coming together to convey a sense of his fundamental individuality. He is not simply "a worker" or "a minority worker" in Banks's image, but a three-dimensional human being.

The AMCBW archives include one series of photographs created at Chicago's P. Brennan Packing Company in or around 1945 and credited to both Banks and Pallas.[70] The images focus on interracial, intergenerational, and intergender cooperation and collaboration in the meat industry. In virtually every image, two or more people of different races, genders, or generations work side by side or interact with one another. One of the images credited to Pallas shows a group of people, including an older white woman, a Black man of indeterminate age, and younger workers of various races, together around a worktable. On the back of this print, among other markings, a quote has been pasted. Reading "In the past these differences have been used to divide us and one group has been set against another," the quote is taken from the preamble to the constitution of the UPWA, first used by the organization in 1943.[71] Though labeled with the date of 1945 in pencil, the back of the image also contains a stamp dated 1958. Since the latter was the year that the UPWA merged with the AMCBW, and since the UPWA preamble—including those words—was absorbed into the preamble of the larger union subsequent to the merger, the images by Pallas and Banks were probably used to show the UPWA's ongoing concern with civil rights and racial equality and to contend that civil rights should continue to occupy a central place in the AMCBW's overall platform as key to maintaining worker solidarity.

One Banks image shows four people performing various jobs in front of a backdrop of hanging pig carcasses (fig. 79). Nearest the camera, a young Black man works at a trimming and washing station, a knife in his hand, while in the middle ground an older white man performs a similar job. Behind them two women work together around a deep tub, one smiling as she works. The image attempts to encapsulate, in a compressed and ambiguous space, both the variety of jobs on a packinghouse line and the variety of people who perform them. Another Banks photo shows two men, one Black and the other white, engaged in an exchange among huge boxes and barrels (fig. 80). The Black man, closer to the camera, stands above floor level on a forklift or pallet jack, looking down at the white man. One hand out in a relaxed manner, he appears to be issuing an instruction. The white man looks at him inquisitively, his own finger pointed toward

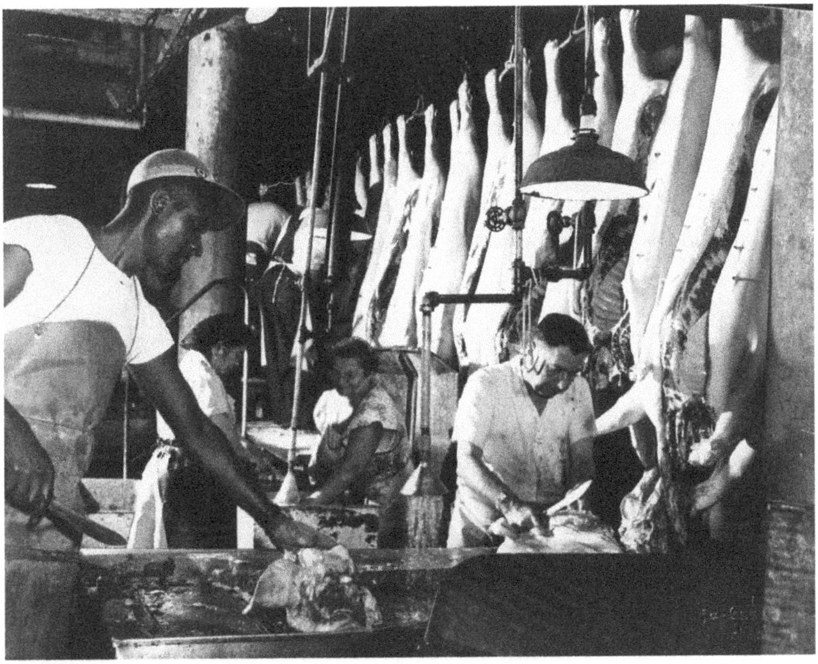

Figure 79.
Jo Banks, Interior of P. Brennan Packing Company, Chicago, 1945. Archives of the Amalgamated Meat Cutters and Butcher Workmen, collection M79-517, Archives Division, Wisconsin Historical Society.

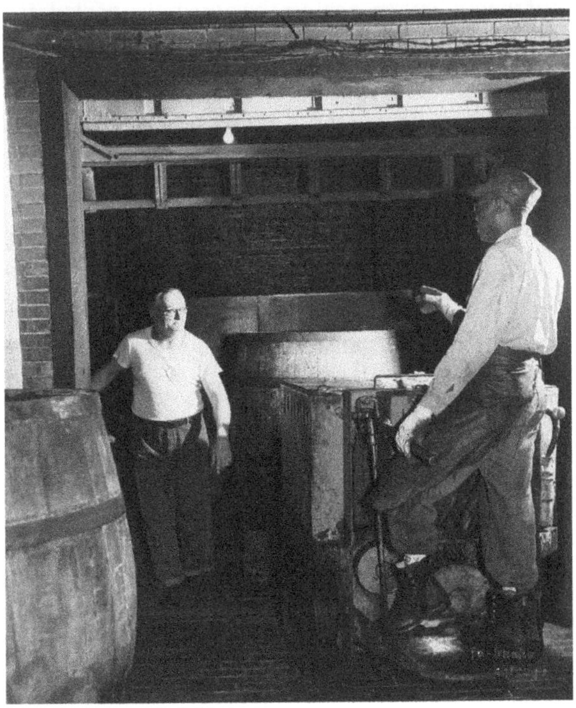

Figure 80.
Jo Banks, Interior of P. Brennan Packing Company, Chicago, 1945. Archives of the Amalgamated Meat Cutters and Butcher Workmen, collection M79-517, Archives Division, Wisconsin Historical Society.

the floor as though confirming. This picture in particular subverts typical power dynamics for the time, placing the Black man both in a position of primacy vis-à-vis the lens and in a position of actual authority within the packinghouse. Banks, who had himself overcome extraordinary odds to achieve success and authority as a photographer and musician, here recognizes and highlights, with his lens, an inversion of racial hierarchies within the space of the packinghouse.

Meatpacking Worker Snapshots: Pugilism and Play, Subversion and Sociability

In photographs made by workers themselves, packinghouse operations and pictures of the line appear far less often than photographs of other union activities. What both Quirke and Peterson note of steelworkers' and other unions' image making proves true for meatpacking as well: union activities generally constitute a large proportion of worker photography output, with many of these photographs focused on the union as a proactive and productive cultural and social institution. But unions existed, at base, to safeguard and defend the interests of their members against the possibility of exploitation or mistreatment by industry. When the threat of exploitation loomed, then, or at any point when tensions between labor and management became heightened, worker photography reflected these stresses. Strikes and strike actions, then, also constitute a major component of meatpacking workers' photography. Such images may serve to commemorate strike actions or to honor the participation of particularly active union members. Group photographs of striking workers in front of hastily built "strike shacks," for instance, serve as one way that union members documented and celebrated their own participation, in the process cementing their identities as part of the group.[72]

Photography could play more than a commemorative role in strikes, however. Management and labor alike used photography during labor strikes to identify and label: industry to identify striking workers for purposes of retaliation or even dismissal at the end of the strike period; labor to identify, label, and even harass strikebreakers. Both industry and labor deployed the camera itself as a means of intimidation—each side was not only photographing the other but *wanted the other to know* they were being photographed.

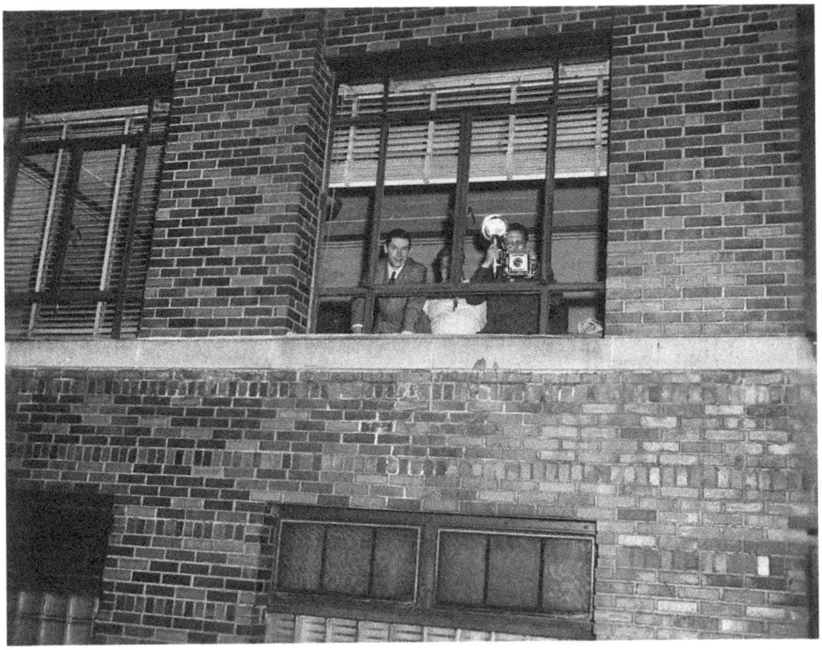

Figure 81. "James Newman, Rath public relations director, with photographer," ca. 1948. Archives of the Amalgamated Meat Cutters and Butcher Workmen, collection M79-517, Archives Division, Wisconsin Historical Society.

A number of such images were made during a particularly contentious 1948 strike in central Iowa.[73] One image from Waterloo, located in the AMCBW archives, shows three people looking out of the window of a building (fig. 81). One holds up a camera, while the two others look on amusedly. A note on the back reads, in part, "James Newman, Rath public relations director, with photographer." The Rath photographer hardly attempts to hide his camera, a standard press rig with a sizeable flashbulb. The flashes alone, projecting down from the building onto picketers below, would have announced the camera's presence. Perhaps this was the intention: at that distance the flash cannot have done much actual good as a means of illumination. Indeed, a second note on the back reads, "Head: 'Flash bulbs didn't scare Rath workers,'" recognizing—and dismissing—the presence of the Rath camera as a scare tactic. The camera that has made the picture we now examine, too, can hardly have been inconspicuous: the three Rath employees look down and directly at the lens, apparently aware they are being photographed from street level. This image thus encapsulates a kind of pictorial standoff. Seeing one another seeing, depicting one another in the act of depicting, labor and management involve the camera in their interaction just as they involve signs, leaflets, and other tools of disruption. This photograph of a photographer photographing functions as a nod of mutual recognition: one threat mirrors, and is mirrored in, the other.

The camera might also be deployed to commemorate ephemeral guerrilla-style strike actions. Within the first weeks of the 1948 strike in central Iowa, both of the two meatpacking companies involved, Rath and Wilson & Co., had begun hiring strikebreakers to keep their plants running. Strikebreaking workers drove or were driven through picket lines to get into the plant, and there was a great deal of intimidation on both sides. One image from a union local in Cedar Rapids shows a handmade sign put up by striking union members close to the substitute workers' entry door at Wilson (fig. 82). It reads, "You are walking through blood now." Having already run a gauntlet to get to the packinghouse door, the strikebreakers would be confronted by this sign. Whose blood would they be walking through? The sign draws an implied parallel, even encourages a certain identification, between humans and animals. Some workers would, of course, step in the blood of animals once inside the plant; but the sign implies that by crossing the picket line they are walking in the lifeblood of the regular workers, and potentially even shedding their own. The materiality of animal bodies becomes, here, part of the material of the strike itself: human blood may be shed just like animal blood, the sign suggests, if strikebreaking workers persist in crossing the picket lines. In the sign, the two Os of "blood" look like eyes: it promises surveillance, monitoring, the word building outward from the materiality of meatpacking work toward a totalizing threat.[74] But the sign, located on company property, likely did not hang for long before it was removed. The photograph thus memorializes the sign's presence, taking something impermanent and ephemeral and making it into something lasting, a hanger or container for memories of the strike.[75]

The sign engages in, and the image commemorates, a relatively rare acknowledgment that the nature of meatpacking work differed, in its fundamentals, from other factory work, inasmuch as it entailed dealing directly with animal bodies and animal substances, converting living creatures into meat and its many by-products. The overwhelming majority of worker photographs of meatpacking simply do not address the nature of the work itself. But when they do, the acknowledgment is often designed and intended as a confrontation: the blood and violence of meatpacking work is evoked to warn strikebreakers that the work they are taking on may shock

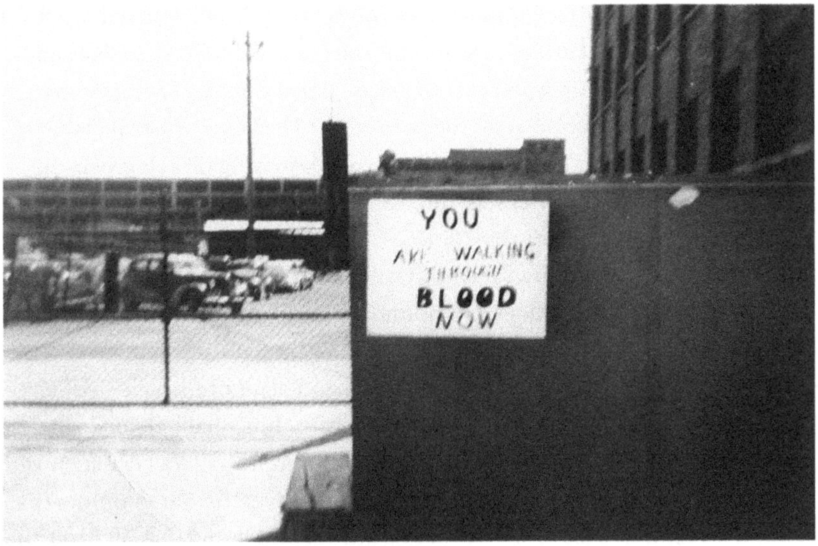

Figure 82.
"You are walking through blood now" sign, Wilson & Co., Cedar Rapids, IA, ca. 1948. State Historical Society of Iowa, Iowa City.

them; and possibly even as an implicit threat, a notice that the violence so routinely deployed against animals may also be deployed against human beings, should they behave in a manner unfair to organized labor.

In this way, the sign and the picture engage with a form of protest that, while not often photographed, did occur from time to time in meatpacking plants during periods of labor conflict: workers' acknowledgment and evocation of the violence of meatpacking activity as a threat against strikebreakers or even management. Recalling union activities in the 1940s, for instance, a Chicago worker named Sam Parks reports leading a work stoppage action among Black workers from the kill floor at Wilson's Chicago plant. Black workers, particularly women, at the plant—like most American plants at mid-century—were generally hired only for dirty, low-paying jobs. Parks and a group of his coworkers occupied the plant's main office one day, in an attempt to convince the company to begin hiring Black women to work in coveted positions in the bacon-slicing room:

> Hog kill, beef kill, beef offal, fertilizer department—those were the black jobs in that plant. And I started a crusade. I sent the hog kill down to the company to grieve, to force them to hire some black women in sliced bacon. I led a bunch of black workers, beef kill and hog kill, with blood on 'em and

every other mother-fuckin' thing and went into the Wilson office and we sat all on top of the fuckin' desks. Scared the shit out of the superintendent and everybody else. Black workers with knives, blood dripping, sweat, scared them poor white women in that office to death! They were screaming; they figured a revolution had come! I said we're not moving till you give us an agreement. . . . The answer was, they would hire black women.[76]

In *Every Twelve Seconds*, Pachirat writes of the wall between the packing plant and the front offices as a boundary and a barrier: "This wall both demarcates and enables the volatile combinations of citizenship, race, class, and education that separate the industrialized slaughterhouse's zones of privilege from its zones of production."[77] By breaching that barrier, bringing blood into a space reserved for paper, Parks and his coworkers achieved their desired result. Of course, part of their power, as Parks points out, lay in the fact that they were workers from the kill floors, which lie at the start of the factory disassembly line: "No packing plant's worth a damn without the ability to process the meat."[78] In transposing taboo substances and taboo tools—blood and knives—into the pure, intellectualizing space of the office, the kill-floor workers stopped animal slaughter, at least for a brief time, and instead created an atmosphere redolent with the threat of human violence.[79]

The workers on strike in Iowa, unable to enter the plant during a strike without either trespassing or betraying the union, hung the handmade sign instead: a reminder to replacement workers that meatpacking work was not like other manufacturing work; and that livelihoods and lives hung in the balance if they proceeded to cross the picket line. In both Iowa and Chicago, workers evoked animal blood—whether in idea, via written word, or in actuality, via literal blood dripping from knives—as a material sign of their power. Representing the work of meatpacking, workers took into consideration the nature of the raw materials with which they worked—in this case, animal materials and animals themselves.

In workers' own rare photographs from inside the packinghouse, the visceral immediacy of the work often comes through in a way it does not in industry-sponsored representations. Workers' photographs of meat manufacturing tend to be less big picture, more close view, showing the job in a manner more fragmented and also, sometimes, bloodier—not as a form of protest but because blood and guts are a fact of the job. In depicting their own work, packinghouse workers often take into account their own experiences, their own perspectives, what it means to be a worker. Snapshots, as Catherine Zuromskis notes, tend to be "embroiled in the pleasures and politics of mutual social relations."[80] The rare images that show packinghouse workers at work often reveal their subjects not only as meatpackers but as people: smiling, looking at the camera, interacting with one another, participating in the act of pictorial representation, determining how they want to be seen.

A comparison serves to demonstrate the differences that arise when workers, rather than industry, depict the work of meatpacking. One page from Korling's *Waterloo Packer* includes an image at the top center of the page in which a worker splits hog carcasses with a cleaver (see fig. 32). Shown from the side and back, he wears a pristine white butcher's coat and white hat over coveralls and boots. Below, the caption reads, "The splitter's job is one for Dead-Eye Dick. A few deft blows along the spine and what was a hog becomes two sides of pork, but never stops moving." The image presents the work of slaughter as a clean process, and the caption draws attention to the (anonymous) worker's dexterity and skill. Like *Waterloo Packer* overall, the image supports a vision of meat making as clean, mechanistic, organized, and essentially nonviolent.

An image preserved in an album held by a union office in Fort Dodge, Iowa, represents a similar subject far differently (fig. 83). Made, like others in the album, at the Tobin Packing Company in the 1950s or 1960s, the snapshot shows two men beginning the process of opening pig carcasses.[81] The workers in the image seem to have felt comfortable with the photographer: one of them, piercing a carcass with a knife, has a wry grin on his face, while the other stands back and purses his lips in a smirk. They are not wearing the clean white coats of skilled butchers but, rather, the coveralls of factory workers. Bare-chested under their jumpsuits, they look like overgrown boys. The work of packing here seems less a dignified vocation performed by professionals and more just a hard job, done only as well as necessary.

The nature of the work, its physicality and viscerality, comes through clearly: the man with the knife plunges it in, his hand making a fist around the handle to facilitate the stabbing-and-slashing motion he deploys in opening the carcass. With the other hand he grips a leg tight for leverage. The snapshot lacks the slickness of Korling's photograph for Rath, and no caption sits alongside to provide narrative context. It does not go out of

Figure 83. Packinghouse workers at Tobin Packing Company, Fort Dodge, IA, ca. 1950s–1960s. State Historical Society of Iowa, Iowa City.

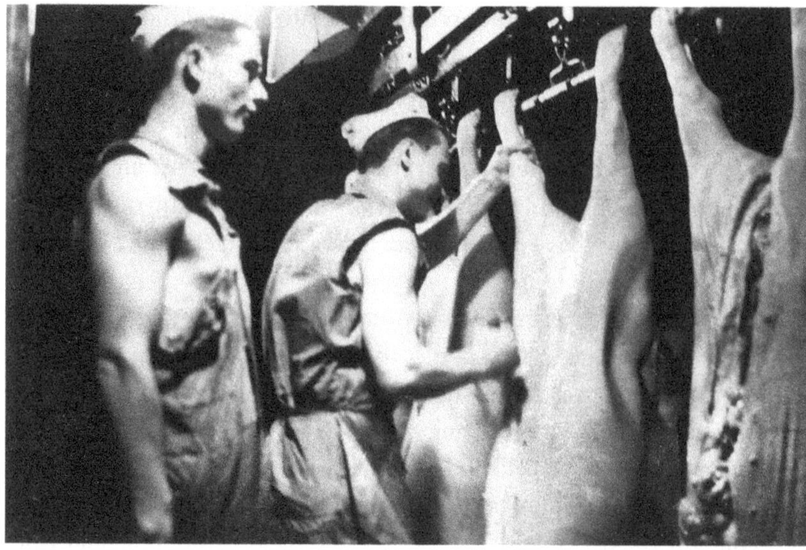

Figure 84. Packinghouse workers at Tobin Packing Company, Fort Dodge, IA, ca. 1950s–1960s. State Historical Society of Iowa, Iowa City.

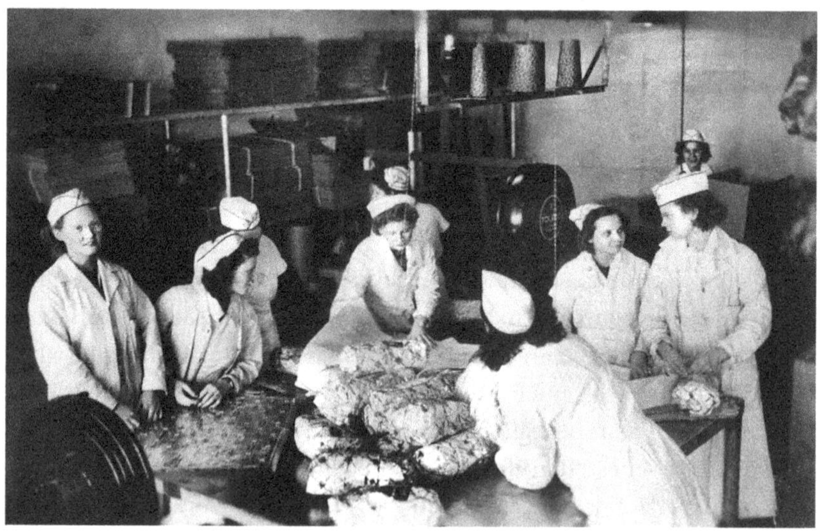

its way to dignify these men or their work. It does, however, individualize them. They appear to have personalities, to be actual human beings. The man in the Rath photograph is a blank: he has skill and dexterity, qualities that aid him in his work; but he possesses no qualities of his own, at least not as far as the image is concerned. He works for the company; but the men in the Tobin photo work for a living—and their livingness, their liveliness, comes through in the image.

Collectively, the pictures in the Tobin album provide a rare depiction of a packinghouse interior, and of the work that went on inside it, from a worker's point of view. Perhaps supervisory or management positions were understaffed that day; perhaps management agreed to turn a blind eye to the rare presence of a camera on the kill floor. The series of pictures hardly offers a comprehensive depiction of the plant's interior, certainly nothing comparing to the linear progression shown in industry documents like ANCO's catalogs. But the photographer clearly had access to and familiarity with the plant and with its employees. In picture after picture, groups of people are shown looking up from their jobs at the plant to smile and interact with the camera and the photographer. Some pictures show workers chatting or joshing with one another, leaning across their work surfaces, demonstrating and enacting the networks of interpersonal relationships that form out of, and constitute a key aspect of, their working lives (fig. 84).[82] The materiality of their work, the flesh they manipulate with their tools and their hands, holds little immediate significance—in these photographs, what stand out are human qualities and human relationships, far more than those of animals or machines. Divisions of labor by gender become clear: certain photographs show jobs, such as the linking of sausage or the trimming and wrapping of certain cuts, being performed by groups of women, while other jobs, including those related directly to slaughter and dressing, are done by men.

The pictures emphasize, too, the solitary nature of some jobs, as opposed to the collectivity of others. While most images in the series show groups of people working around conveyors or tables, an image of a sticker depicts him working unaccompanied, the bodies around him those of pigs, not humans (fig. 85). An evocative photograph, dramatically lit, the picture carries an unusual gravity. Holding a knife in his right hand quite near the camera, the sticker pushes a pig's body along the rail to its next station. Blood jets from the carcass's neck. Looking down at his work, he stoops

Figure 85. Packinghouse worker at Tobin Packing Company, Fort Dodge, IA, ca. 1950s–1960s. State Historical Society of Iowa, Iowa City.

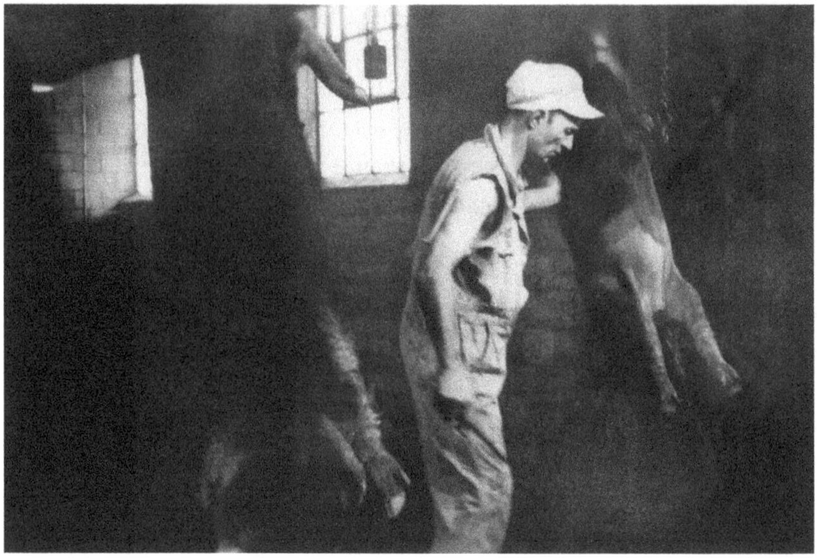

slightly to angle the trajectory of the blood away from himself as the carcass swings on its chain. The scale of the photograph is a bit confounding: near the left side of the frame, a second carcass hangs, so close to the camera that it appears nearly as large as the man himself. The distortion that generates this illusion, an artifact of the lens, becomes at the same time a serendipitous metaphor for the pervasive disorientations enacted by the disassembly line. Presumably next in line, this hanging body looms massive in its vulnerability: waiting to be bled, waiting to be reduced.

The most remarkable image of the set depicts a man, his coveralls stained with blood, wearing a pig's head over his own (fig. 86). Thick eyebrows have been drawn on the pig's face with a marker, and a real wooden pipe rests between its teeth. The head is huge, and it must have been heavy —the man holds it with both hands, one hand invisible behind its flapping jowl skin, the other with fingers curled into the flesh of the side of the face. Perhaps it received some other preparation before he put it on: it looks relatively clean, washed free of blood and offal; and the ears stand remarkably upright, perhaps propped or wired in place, their identification notches symmetrically aligned. As he wears the pig's head he assumes an identity other than his own: hybrid human-animal, a hilarious and frightening monstrosity.

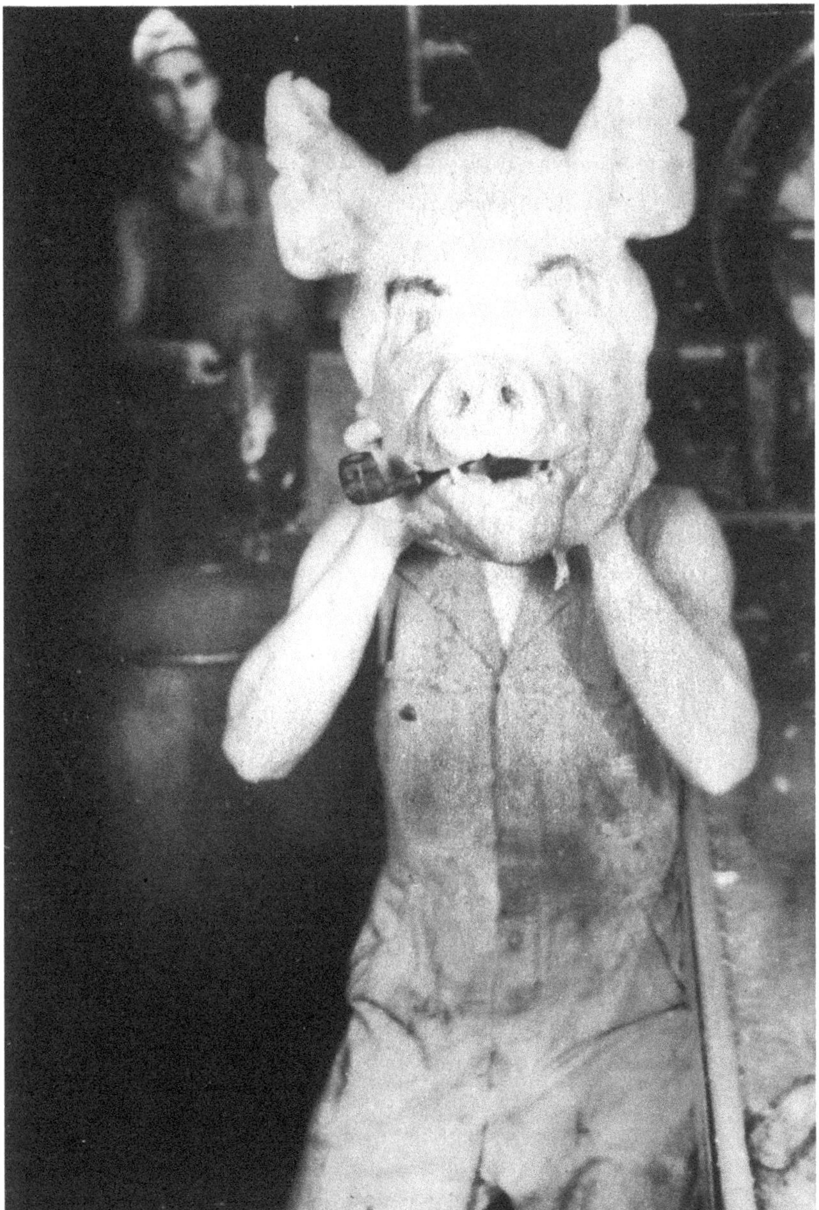

Figure 86. Packinghouse worker at Tobin Packing Company, Fort Dodge, IA, ca. 1950s–1960s. State Historical Society of Iowa, Iowa City.

Human-animal hybrids populate myth, art, and literature from earliest days, and across space and time: the ancient Greek Minotaur; the Egyptian sphinx; the elephant-headed Hindu god, Ganesha. Hybridity may connote royalty or divinity, as with Ganesha and other hybrid deities; or it may connote, as with the Minotaur, the potential for grisly chaos. Even human-pig transformations, specifically, have their own long record. Circe, sorceress of the *Odyssey*, turned men into pigs. In that account, human-pig hybridity carried connotations not of divinity (accursed or otherwise) but of abjection and servitude, even the potential for cannibalism: in turning Odysseus's sailors into pigs, Circe also turned them into dinner. Odysseus ultimately saved them from this fate; but their pigheadedness, while it lasted, proved a decidedly vulnerable, unexalted state of being.[83]

Donning a pig's head anywhere is rather like donning the head of a mule or other abject animal: one becomes reduced, silly. Like Bottom, the hapless, ass-headed actor of Shakespeare's *A Midsummer Night's Dream*, the pigheaded meatpacker assumes an identity simultaneously animal and human, but somehow less, not more, than the sum of its parts.[84] In the packinghouse, especially, pighood proves an extraordinarily vulnerable status: a pig's fate is sealed simply by virtue of its species. Unlike Odysseus's pigheaded sailors, no packinghouse pig ever escapes the dinner plate. In assuming porcine identity in a meat factory, then, the pigheaded worker assumes a state of absolute vulnerability, a state of wretchedness.

The slight alterations to the pig's visage, however, confer a pointedness to the worker's commentary. The pig sports bushy marker-drawn eyebrows and a pipe. These read a bit like three-dimensional equivalents of the mustache-and-goatee graffiti that schoolboys make in their textbooks: juvenile sneerings, carrying little weight. But, as Marcel Duchamp made clear with the mustachioed *Mona Lisa* in his work *L.H.O.O.Q.*, even the schoolboy's crude commentary may carry more meaning than meets the eye.[85] Inversions of social norms, rejections of whole systems of propriety and hierarchy, inhere in and radiate out from such simple gestures, as Peter Stallybrass and Allon White show:

> Inversion addresses the social classification of values, distinctions and judgements which underpin practical reason, and systematically inverts the relations of subject and object, agent and instrument, husband and wife, old and young, animal and human, master and slave. . . . Hybridization . . . produces

> new combinations and strange instabilities in a given semiotic system. It therefore generates the possibility of shifting *the very terms of the system itself*, by erasing and interrogating the relationships which constitute it.[86]

The pigheaded worker's bushy eyebrows and pipe carry the same Dada-esque rejections of dominant narratives, the same refusals of hierarchy, as Duchamp's altered *Mona Lisa*. In giving his pig's head the characteristics of a fat-cat, fat-pig capitalist, the worker identifies the animal with management or ownership, with capital.

The pig's head becomes a medium for commentary, and the action of wearing it represents the same kind of pointed play—play as critique—that characterizes Dada's carnivalesque mode: "brush[ing] the carnival at one extreme and the gallows at the other," to borrow E. P. Thompson's formulation.[87] In submerging himself beneath—within—the pig's head, the worker also asserts himself, enacts a form of resistance through joking and mockery. Just as much as through the formalized community-building activities seen in many workers' and labor photographs, workers declare themselves and their identities through failure to conform to expectations, failure to respect packinghouse hierarchies. This man wears a pig's head not—or not only—to take on the pig's identity but to project the pig's identity onto someone else entirely; to enact the role of the fat-pig capitalist by assuming his very face, in all its piggy fleshiness.

Again, one wonders at the occasion that permitted such carnivalesque inversion of normal hierarchies: absence of oversight, perhaps? A shop floor manager's vacation or sick day? But the size of the management staff at any industrial packinghouse was significant—it seems unlikely that the entire range of personnel responsible for enforcing packinghouse norms would simply disappear, and on a day when someone happened to have a camera. The more likely explanation would seem to be that the occasion for this particular antic was the presence of the camera itself. Given the range of packinghouse locations shown in the series of images, the camera traveled widely on the shop floor that day, from kill floor to trimming rooms, from hot side to cold side of the plant. The camera itself provided the excuse for levity, and the camera captured the results of its own presence.[88] While the image of the solitary sticker imparts a sense of gravitas to packinghouse activity, acknowledging and even highlighting the existential weight of the packinghouse's perpetual cycle of slaughter, in performing for the camera the pigheaded meatpacker brings the reality of packinghouse activity to the fore in a wholly different manner, the crude performance conveying the casual nature and massive scale of packinghouse butchery with a visual, visceral shock.

This worker was hardly the only packinghouse employee to identify with and even assume the identity of an animal on the line. Although photographs prove elusive, workers' descriptions of activities at other plants sometimes include accounts of similar tricks, playful but semantically loaded. Several workers employed in the mid-twentieth century at the Dubuque Packing Company in eastern Iowa, for instance, reminisced about wearing pigs' tails. "You made your own fun," one worker notes. "On the kill you would have pig tails that were cut off and get a piece of wire and wrap it around and hook it to the belt loop of your pants and walk around with a pig tail."[89] A second worker recalls that "there was this one guy . . . [who] hated his boss. He'd take the pins that they used to put tags on the carcasses and a tail and sneak up behind his boss in the cooler and pin a tail on the long frock so the boss was walking around all morning with a tail on."[90] Another worker reports that "we used to have one guy in the kill floor, every time we got a new employee, especially if they were a woman . . . he'd take a pig tail and make it look like a penis and pin it to his pants and walk around."[91]

The worker as pig, the piggish boss, the sexist pig, and the capitalist pig: in all of these instances, the human-pig hybrid, evoked with humor, becomes at the same time a charged mode of commentary on the human-animal dynamics of the slaughterhouse. Rejecting expectations and throwing off the implied narrative of the factory—one of constant forward movement, of flow (wholly without ebb)—workers joke to register their resistance and assert themselves *as selves*, as individuals.[92] "The historical record is not a simple one of neutral and inevitable technological change, but is also one of exploitation and of resistance to exploitation."[93] In resisting, workers enact "a caustic commentary on their not being pre-programmed robots, but humans with a mind of their own."[94] The routinization of slaughter, the fact that animal detritus is literally everywhere, is the point of the joke: in a place where pig heads and tails are easy to come by, assuming an animal identity proves no difficult task.

Routine here is disrupted, and violence evoked, through joking and kidding—anything to break up the monotony. "Humour is used by workers to enliven their work, but moreover, to assert their identity, to parody their managers, and to register their resistance."[95] But these violent disruptions are also deployed as a form of commentary on slaughterhouse hierarchies, the flow of money and power in the packinghouse. In the case of the pigtail penis, pighood functions as a medium for harassment, a way of *asserting* power. Carnivalesque modes of social inversion "often violently abuse . . . *weaker*, not stronger, social groups—women, ethnic and religious minorities, those who 'don't belong'—in a process of *displaced abjection*."[96] Whether for commentary on the power of capital or for assertion of masculine supremacy, the pig becomes less a creature than a medium—both in the artistic and expressive senses and in the spiritualist sense. Through the animal medium the worker gives vent to his spleen, channels his aggressions, and in the process calls attention to the packinghouse as a kind of perpetual bizarro world. "Only hybrid notions are appropriate to such a hybrid place."[97] Macabre deviations draw attention to the macabre nature of the norm.

In language and in image, the meat industry has encouraged both the public and its workforces to distance themselves from the violent realities of meat production. Despite the meat industry's best efforts, however, it has never succeeded at exerting *full* control over linguistic or imagistic representations of meat production. In picture and word, workers represent their own work in ways that are fundamentally different from industry representations. Workers are far more likely than management or industry to evoke the animal nature of their work, the fleshly nature. Representations of slaughter by workers tend to be franker, less mechanistic, and more fragmented. Unlike meat industry advertising, such snapshots are useless for promotional purposes. But they are legitimate representations of the realities of meat production. If workers are less inclined to give a sense of the big picture, they are far more cognizant of and inclined to acknowledge the smaller picture, more likely to represent and even to instrumentalize the violence, both potential and actual, in their work.

Posing with Pigs

Care, Control, Dispatch

ALIVE OR DEAD, animal bodies in meat production systems constitute the primary material that must be managed. Historically, photography has offered valuable assistance in this process. In depicting perfectly managed animal bodies, photographs also reproduce and extend the conditions under which such management is maintained. Meat industry photographs generally seek to diminish or erase the animality of animal bodies: even when shown as living creatures, animals' ultimate destiny as meat becomes their sole condition. Meat animals are meat first, animals second.

When creating, commissioning, or circulating photographs foregrounding animal bodies, alive or dead, meatpacking companies and industry organizations often elide animality, animal existences, and animal experiences. Animals are simultaneously everywhere and nowhere in industry media: depicted frequently but acknowledged rarely. Industry organizations have used pictures to convey a sense of living animals not as creatures but as raw materials. Images have assisted in the meat industry's ongoing endeavor to de-emphasize the animal nature of the animal, a key strategy in many self-presentation programs within the industry.

Meat animal bodies follow a trajectory: from farm to stockyard or feedlot to packinghouse. Depictions of meat animals follow suit, in a parallel arc of animal imaging. Farm animals sometimes appear in images as a collective, imbuing the entire trajectory of meatpacking processes with the folksy flavor of the family farm.[1] But by far the most common type of live animal image depicts individual stock animals, especially prizewinning or otherwise special animals, as paragons of their breed or even of their species. Such images run a sort of gamut, from casual and snapshot-like to formal, even hieratic. Animals in the images may stand alone or with a human handler. Photographs may be made on farms, at livestock shows, or at auctions.[2] But the overall purpose of such imagery is to laud the individual animal as a prime specimen, and by extension (and implication) to honor the skills of the breeder. Such breed-standard photographs constitute a very common, even ubiquitous, type of animal-focused imagery.

Stand-alone images of paragon animals predate photography. Certain painters have specialized in them: by the nineteenth century, portraits of high-value animals such as prized horses and dogs constituted a genre unto themselves. But paintings that focus on everyday domesticated and working animals are less common.[3] Instead, the genre of images depicting paragon farm animals developed primarily within print media, a workaday format appropriate to workaday subject matter. Over the course of the nineteenth century, breed-standard imagery coalesced as one aspect of the increasing professionalism and rationalization of farming: as manuals for the raising of animals proliferated and as breeders experimented with new animal breed lines and techniques, images of paragon animals helped novice farmers to understand better the characteristics of various animal breeds. They also, crucially, helped to propagate breeds, by cementing characteristics and codifying knowledge about breed standards.[4]

As photography became increasingly capable of imaging the world and photomechanical reproduction in books became cheaper and more ubiquitous, photography took over from printmaking the duty of imaging paragon animals, as it took over so many other duties. The form has proven remarkably stable: photographs of ideal animals from the present day look much like those of a century or more ago.[5] In breed-standard photos the photographer attempts to show the animal's conformation to the ideal for its particular breed. In conforming to the ideal, the animal demonstrates its proximity to perfection, as it is conceived and defined for the breed.[6] But the conceiving is a human conceiving; conformation imagery deindividualizes, and in some ways deanimalizes, the animal body. Its perfection inheres in its ability to perform its function, as that has been imagined and defined by humans.[7]

The side view of the animal is a classic, even ancient view: the animal stands in hieratic profile, all body parts clearly visible.[8] A four-legged animal should stand with all its legs visible, as though arrested in a stately walk. The photograph should show the line of the back and the angle formed by the meeting of the back and the hindquarters. Cattle would typically be photographed with head raised, clearly depicting the shape of the head and the line of the neck as well as the shape, presence, or absence of horns, dewlap, or other breed-specific characteristics. If the legs are properly staggered, the viewer ought to be able to see all four feet and judge the leg conformation, a key index of health and stability.[9] Profile conformation photographs exist for other species, as well, with species-specific adjustments; in all instances, a skilled viewer should be able to look at a well-made photograph and judge from it, with relative accuracy, the animal's conformation and, thereby, its value.[10] The quality of the picture matters a great deal, as farmers often depend on images when making decisions about breed stock.[11] Circulation of images translates directly to circulation of animal bodies.[12]

Images of prize breeder animals depict the individual animal as a paragon, valued not just for its individual characteristics in themselves but for how those characteristics hew to an ideal breed standard. But the standard, in itself, only genuinely exists as an idea and an ideal, against which all real individuals are measured. "The type is not a fixed thing, but a living, imaginative hope and memory."[13] Moreover, value often inheres not solely in the body of the animal itself but in its potential offspring. Breeders prize certain animals not, or not only, for possessing certain traits—lean meatiness, a tendency to fatten, disease resistance, physical attractiveness—but for the heritability of those traits, for the promise of a stronger and more valuable herd. A prize breeder animal promises fecundity, reproducibility, prosperity: the one for the many and for the very idea of muchness.

Images of prize meat animals—animals raised to be eaten, not bred—appear in some ways quite similar to those of prize breeder animals. But key differences arise in why such images were made and in how they might be used. Breeder animal photographs stand for the animal, or at least for its genetic material. Meat animal photographs, by contrast, do not as often need to serve as a form of visual currency. While a meat animal photograph might be used in a sale of the animal itself, there's less potential for a circulation of the image independent of the actual body of the animal.

Where a prize breeder animal would often be shown on a farm, images of prize meat animals are often created at livestock shows or right in the stockyards, the animal already absorbed into the body and space of the industry. One photograph created in 1918 at the stockyards in Sioux City, Iowa, for example, depicts a high-value steer accompanied by a detailed caption inscribed directly on the image: "Steer fed by Chas. Hurley, Ponca,

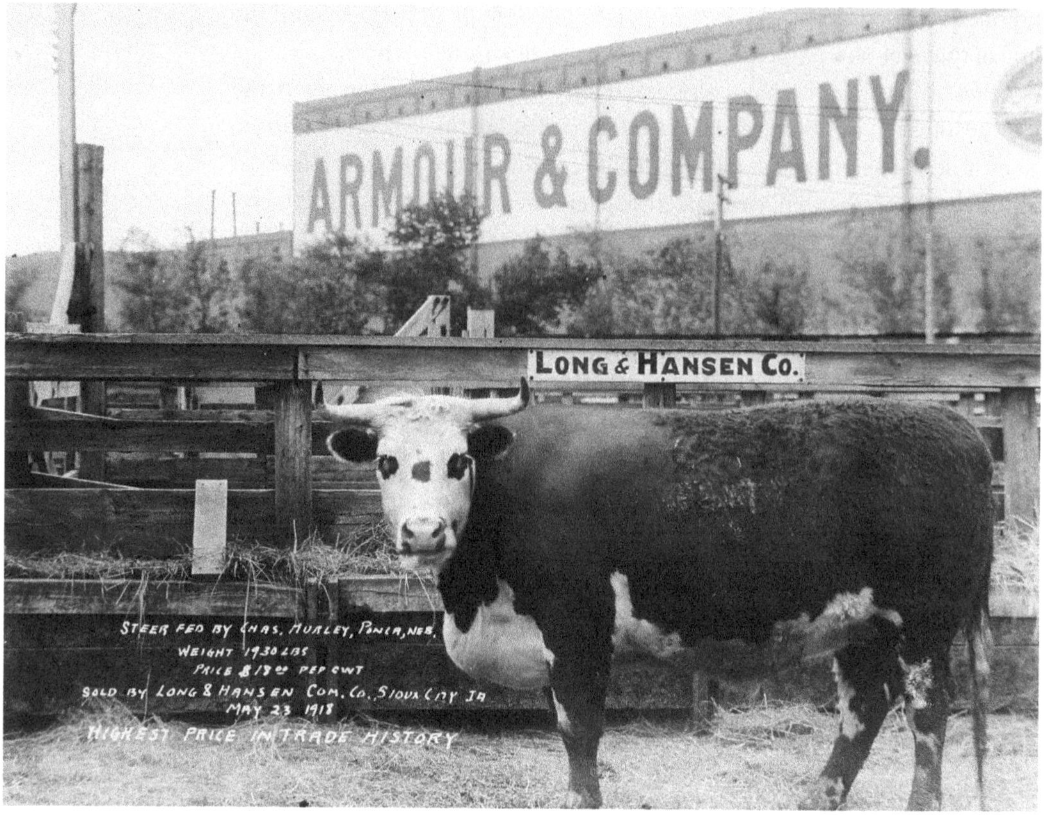

Figure 87. "Steer fed by Chas. Hurley [. . .]," 1918. Courtesy of the Sioux City Public Museum, Sioux City, IA.

Neb. Weight 1930 lbs., Price $18.00 per CWT. Sold by Long & Hansen Com. Co., Sioux City, IA, May 23, 1918. HIGHEST PRICE IN TRADE HISTORY" (fig. 87). A steer is a castrated male, an animal raised specifically and solely to become meat. The caption on this photograph places this steer within a network of overlapping human interests, including those of the farmer and the auctioneer. It assigns a specific dollar value to the animal's body, but at the same time it conveys remove or disinterest: the steer is not owned but only "fed by" the farmer, he was "sold by" Long and Hansen Com. Co., and it is not entirely clear what company paid such a high price for the animal (though the Armour sign, dominant in the background, may be intended to indicate that this company will process the steer). The caption makes clear the high value of the animal's meat, even as it remains unclear who will reap the profits: in this commodified presentation of the animal, the copious textual detail nonetheless permits capital to remain anonymous.[14]

The ability to raise a prize animal reflects on the human.[15] A significant subsection of prize animal images, then, depict the animal alongside the human farmer, breeder, or handler. Images like this would typically not be used for functional or commercial purposes—they would have less direct use or value than the sorts of conformation photos used, for instance, to advertise breeder animals. The purpose of such human-animal imagery is most often commemorative: the sort of image published in newspapers, collected in scrapbooks, or framed for the family prize cabinet or photo wall.[16]

Posing with pigs—or sheep or cattle—memorializes quite effectively the human effort that went into raising them. This is particularly true when the effort is that of a child or youth. Images of young people with prize animals commemorate the young farmers' victories at fairs and livestock exhibitions, but at the same time they show key points in the children's education, points at which they have achieved a certain degree of mastery over animal bodies and behavior.[17] Pictures of children with their animals often function as multifigure, multispecies human-animal portraits: a paragon animal, under the control of a child rapidly developing skilled command. In one image, created in the late 1940s at a central Iowa meat animal exhibition sponsored by the Rath Packing Company, a child about six or seven years old pats and guides a pig with his left hand, while simultaneously holding out his crook with his right, checking and controlling the direction of the animal's walk (fig. 88).[18] The image conveys in equal measure the boy's care of and his authority over a pig weighing four to five times as much as he does.

For some breeds, the paragon image proves relatively easy to create. Cattle tend toward a self-contained stillness, their phlegmatic nature making them better suited to photography. Not only their stolidity but their sheer size lends itself to dignified dual portraits, as a human can stand, comfortably and without crouching, alongside a full-grown animal. Portraits of humans and cattle often depict them standing together, in implied comfort and mutual regard: humans and the creatures under their care. Compliant sheep, too, prove relatively easy to corral and maneuver into position with a halter, although the smaller size of the animals often demands that their humans kneel alongside them.[19]

Pigs, however, can prove recalcitrant subjects, indifferent to the lens and resistant to posing. Images that show both pig and handler on the move

Figure 88. "Tommie Faint, Havelock, Iowa," Waterloo Meat Animal Show, 1949. Rath Packing Company Records, collection MS 562, Special Collections and University Archives, Iowa State University Library.

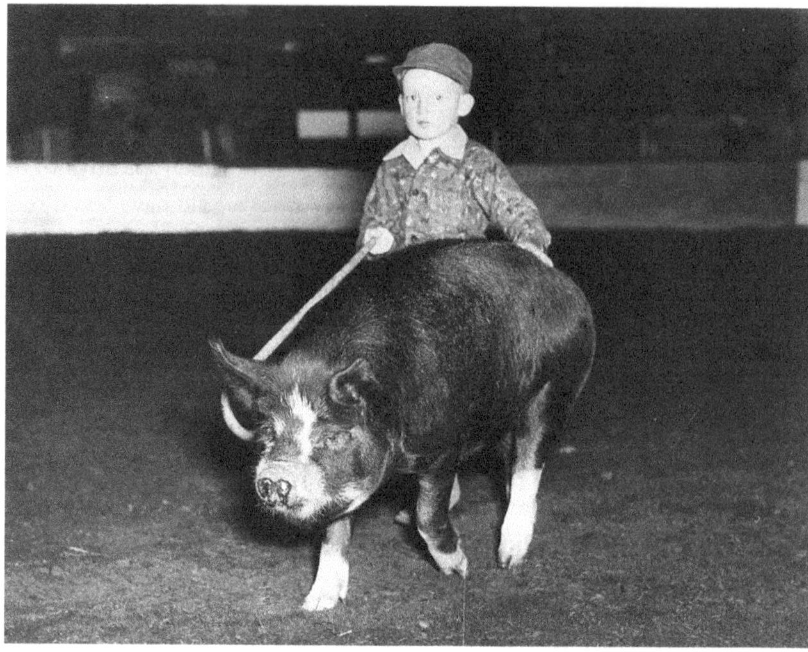

are one common solution to this problem (fig. 88). But to show a pig at rest requires external persuasion. The animals tend not to stand still on their own: given the choice, they root and forage, always moving. Food offerings, then, constitute a commonplace, even a characteristic, feature of many pig portraits (fig. 89). When piled on the ground, the food persuades the pig to pose with lowered head. When held by an unseen handler, the food persuades the pig to raise its head, the better to feature a specific head shape or ear conformation.[20] In all instances, photographing pigs requires acknowledging their animal nature, negotiating their fundamental pigginess. Even in meat animal shows, where the contestants are destined not for breeding but explicitly for consumption—their fates written into the very title of the event—pigs' agency, their specific ways of being, demand acknowledgment.

This piggy ontology makes frequent appearance in human-animal portraits of prizewinning pigs. At meat animal shows, photographs of prizewinning animals with their handlers are routinely created during show events; and in human-pig portraits in particular, the machinations required for the creation of the image—machinations that could often be concealed or sub-

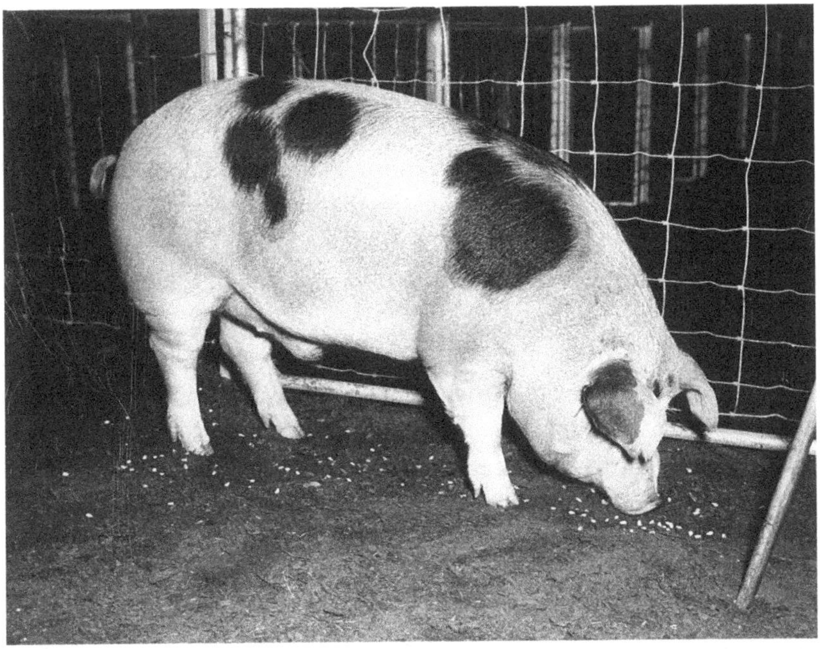

Figure 89.
Spotted pig with corn, Waterloo Meat Animal Show, 1947. Rath Packing Company Records, collection MS 562, Special Collections and University Archives, Iowa State University.

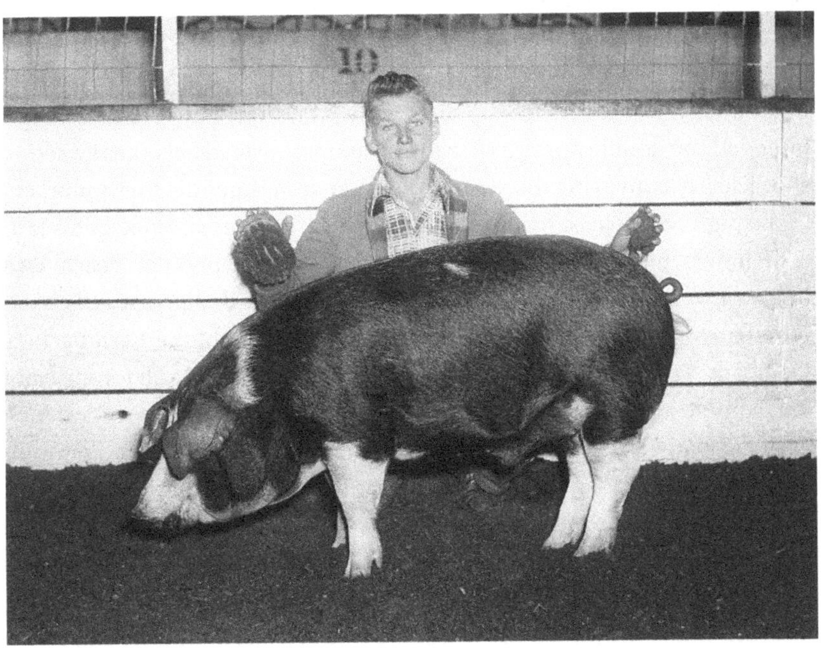

Figure 90.
Youth with pig, Waterloo Meat Animal Show, 1949. Rath Packing Company Records, collection MS 562, Special Collections and University Archives, Iowa State University.

limated in portraits of other animals—are often on full display. Sometimes, a pig might be persuaded to pose solely by the presence of food or a trusted handler. Sometimes perhaps a bit of care might help or even substitute for the food: one young teenager pauses from his grooming activities for the duration of a photographic exposure, smiling good-naturedly upward (fig. 90). Activities of care, necessary for the animal to have a show-worthy appearance, here constitute a point of human-animal exchange: through the bristle brush, the boy offers attention and pleasant scratches, which the pig pauses to accept.

Posing for the sake of posing, however, proves problematic for pigs. One feels the sheer weight of their indifference to the machinery of animal display. Shown at a 1947 Waterloo meat animal show, a second-place barrow stands not for his picture or at the request of his handler but for the pile of food under his nose (fig. 91). The information on the ribbon in the handler's hand, along with that given on the card attached to the photograph concerning his weight and his sale price, all constitute a data set based in his body. The words and numbers reflect his history and tell his fate beyond the moment of the photograph; but they are, at the moment of the making of the image, wholly external to him. The first-place barrow at the same 1947 event likewise ignores his handler and his prize (fig. 92). Shifting his bulk to move closer to his food, he alarms the handler. The boy looks down, holds out his hand, tries to stop the barrow's motion. The resulting image, intended as a double portrait of man and animal, instead becomes an action shot, clearly conveying the sense of a moment in time: the finest moment in the pig's life, at least from a human perspective; but also one of his last.

Banners, hierarchies, handlers' moments of glory: pigs don't care. Two images from the 1948 Waterloo Meat Animal Show depict a grand champion barrow with his handler and the then president of the Rath Packing Company, R. A. Rath. In one, the animal appears to stand still for the camera, submitting to the photograph while his handler and R. A. Rath look at the lens, the executive smiling slightly (fig. 93, *top*). The handler holds out a switch but does not touch the pig with it; its very presence appears sufficient to still the pig. The image conveys a firm, stable sense of human dominion over the animal: the handler as disciplinarian, the packing company executive as capital's representative, commanding the disassembly line into which the pig's body will shortly be absorbed. The second image (fig. 93, *bottom*), however, demonstrates the pig's general indifference to

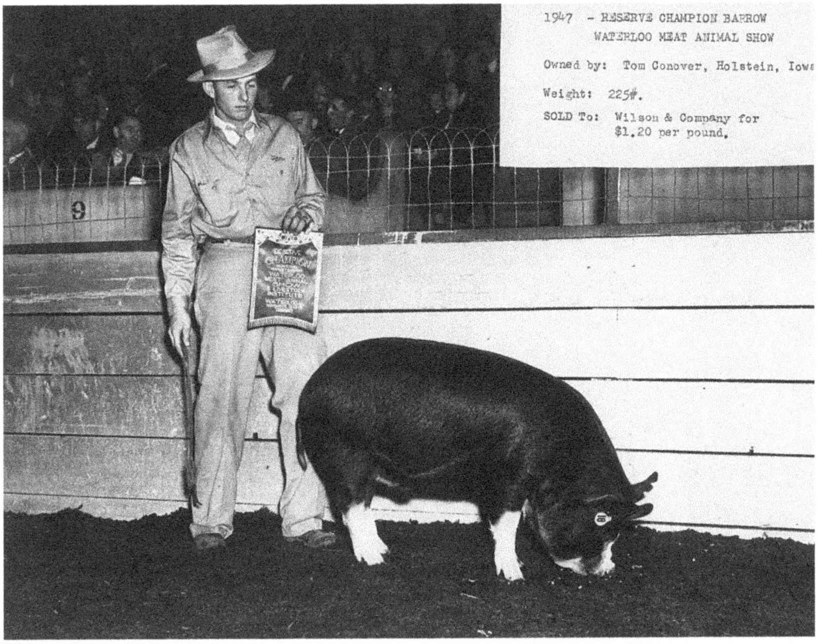

Figure 91.
"Reserve Champion Barrow, Waterloo Meat Animal Show," 1947. Rath Packing Company Records, collection MS 562, Special Collections and University Archives, Iowa State University Library.

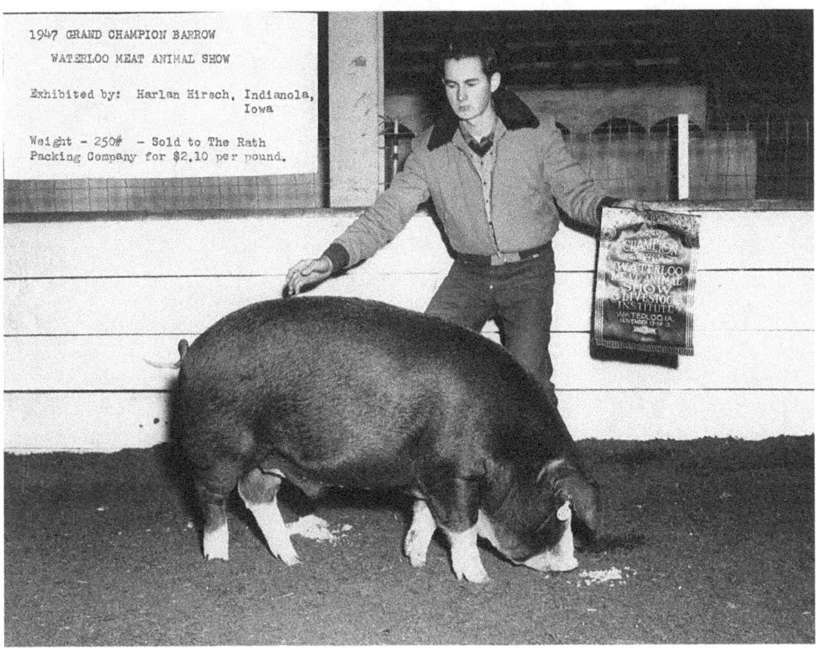

Figure 92.
"Grand Champion Barrow, Waterloo Meat Animal Show," 1947. Rath Packing Company Records, collection MS 562, Special Collections and University Archives, Iowa State University Library.

Figure 93.
"R. A. Rath presenting Champion banner to Bob Faint, 1948 Meat Animal Show," 1948. Rath Packing Company Records, collection MS 562, Special Collections and University Archives, Iowa State University Library.

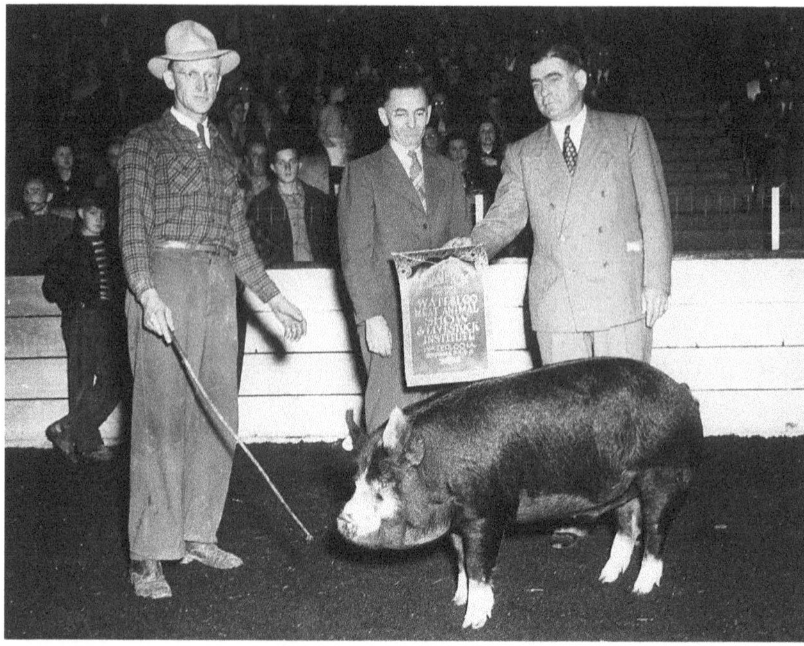

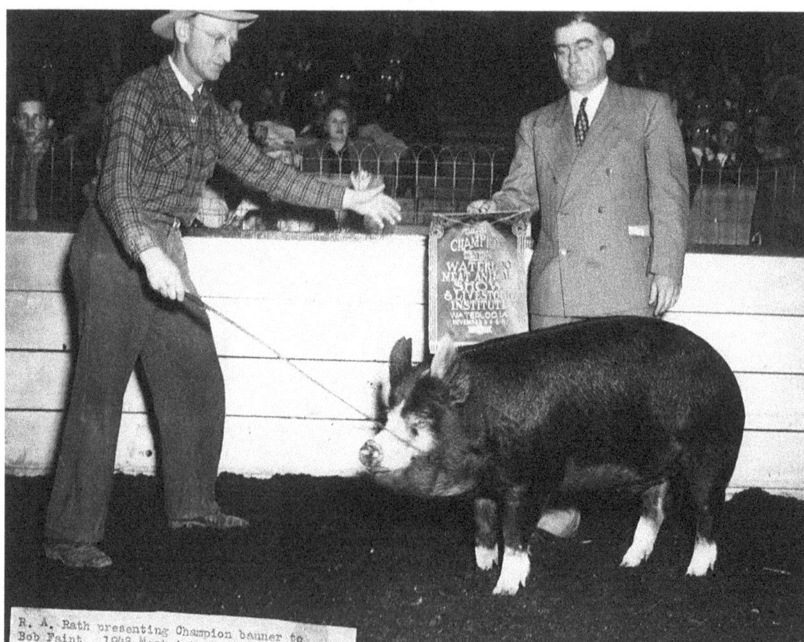

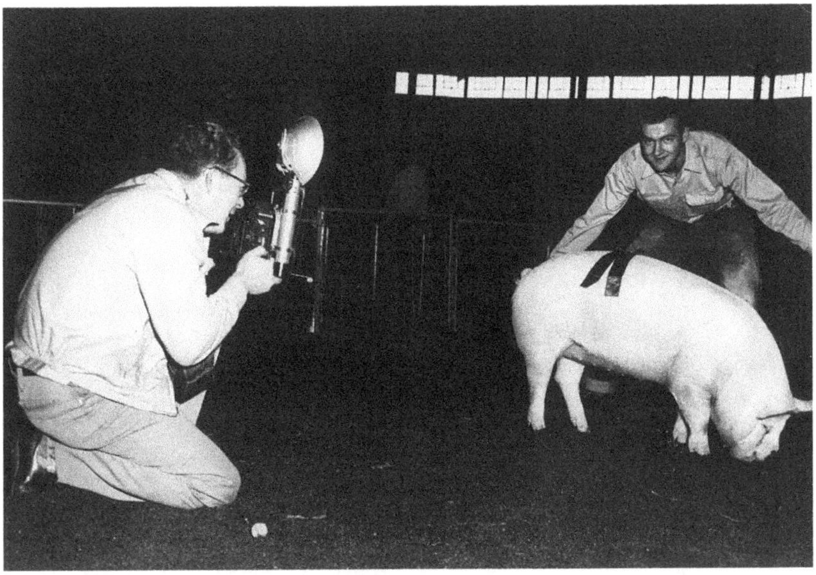

Figure 94. Photographer and handler with prize animal, Waterloo Meat Animal Show, ca. 1950s. Rath Packing Company Records, collection MS 562, Special Collections and University Archives, Iowa State University Library.

this hoopla, revealing the fictions of the first. Here, even as his handler receives the championship banner directly from the hand of R. A. Rath, the pig prepares to depart. His handler raises the switch in front of the pig's face to arrest the animal's progress, but the gesture appears inadequate. Disciplinary order breaks down, at least temporarily: the handler's face registers exasperation, while the executive looks at the pig with an inscrutable expression, as though truly seeing it for the first time.

Show rings are places designed to showcase human dominion over animals, human management of animal bodies. Photography of animal bodies constitutes a crucial part of the equation: photographing animal ideals allows them to be frozen, their characteristics captured and preserved even beyond the event of the animal's death. Photographs of prize animals, even casual snapshots, fuse with and contribute to the development of the idea and ideal of the perfect animal, the paragon against which all future contestants will compete. Yet even in the controlled environs of the show ring, photographing animals and human-animal relationships requires complex sets of negotiations: between humans and animals; among humans; between humans and lenses. In one image, a photographer kneels in the dirt and manure, his camera held up to his face (fig. 94). The animal subject of the image, a white gilt, stands with her nose to the ground, eating.[21] Her

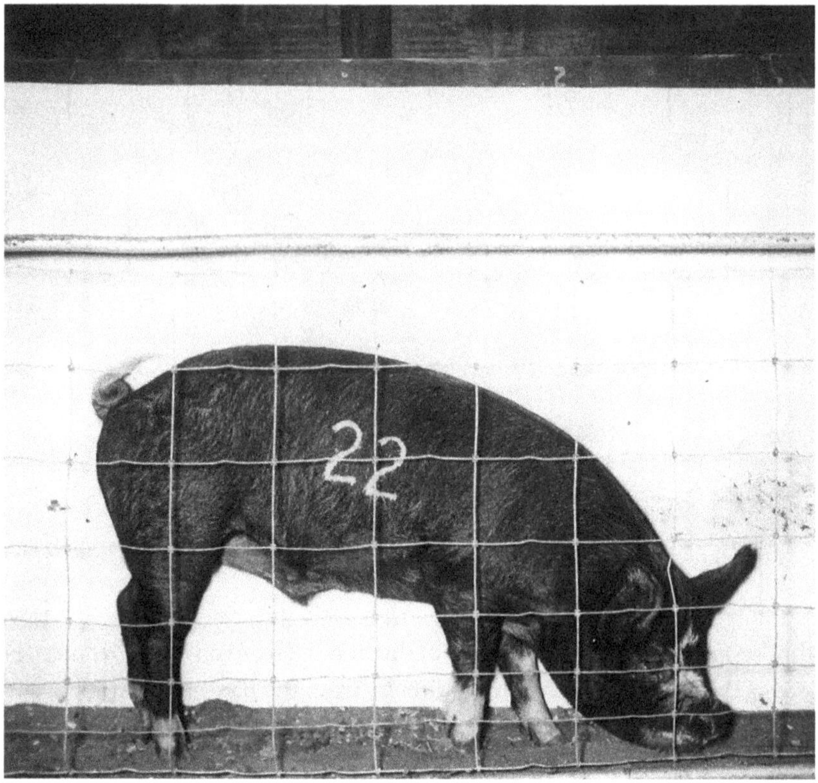

Figure 95. "Carcass contest barrow," Waterloo Meat Animal Show, 1951. Rath Packing Company Records, collection MS 562, Special Collections and University Archives, Iowa State University Library.

handler crouches awkwardly behind her, arms spread wide to catch and contain her for the duration of the shot. To free his hands, he has laid the prize ribbon directly across her back. Smiling, he looks at the wrong camera —not at that of the kneeling man in front of him but rather at the one making *our* picture. The pig, however, is again indifferent to everything but the food under her snout, placed there to entice her to participate in an event in which she would otherwise be completely uninterested.

Photographs solidify the sense of human command, human mastery over the meat animal. Whether instrumental, informational, or commemorative, they convey a sense that in achieving carcasshood, the meat animal reaches its own logical conclusion, the higher purpose for which it has existed all along. In an image from the 1951 Waterloo Meat Animal Show, a pig is depicted with a number painted on his side, rooting on the ground (fig. 95). Taken from quite close, the picture lacks the commemorative, companionship quality of human-animal meat show portraits; rather it appears dry and affectless, like a piece of evidence. The animal stands behind a flimsy fence, the purpose of which seems to be not containment but rather a kind of gridded measurement—between the fence and the light-colored backdrop, the animal's body can be measured, seen, and known. On the image's reverse, an inscription reads, "Carcass contest barrow." The Waterloo Meat Animal Show had incorporated a "carcass contest" in which a group of animals would be made available for evaluation and ranking by contestants. The animals would be slaughtered and their dressed weights, quality, and other relevant information taken. A winner would be declared based on the proximity of their evaluations to the animals' ultimate value and to expert livestock buyers' assessments. Designed to train young and novice farmers in estimating the worth of a living animal, the contest also functioned to make carcass evaluation a more scientific, measurable, and regular practice.[22] Photographing the carcasses constituted one more data point, a means of comparing carcasses across years and thus assisting in regularizing the sublimation of the animal body to values of mass production. Photography here enables, indeed creates, a form of knowledge not previously present or at least not measurable.

This image and others like it complement an existing knowledge base, a way of evaluating animals' bodies via expert judgment, a kind of close looking by trained eyes (fig. 96). Evidentiary photographs lend scientific credence to this body of experiential knowledge; but they also, simultaneously, signal its potential obsolescence: the passing of visual observation as a craft form in favor of hard science and hard evidence. Photography here augments and extends human judgment; but it also raises the specter of judgment's obsolescence, the human eye supplanted by the machine.

Humans use photography in their endeavor to create a more perfect meat animal and to commemorate points at which they believe themselves to have achieved some milestone in this unending progression toward the ever-receding goal of perfection, however they define it. The photograph commemorates this effort; the carcass testifies to it, becomes its very product. Indeed, the living animal itself becomes just a kind of vessel for the carcass, a carrier or temporary housing for the meat it contains. The purpose of its life is to carry its own body, to feed its own flesh, to live only as well and as much as necessary to become sustenance for others.

Industry and science meet in the more perfect carcass. Conferring

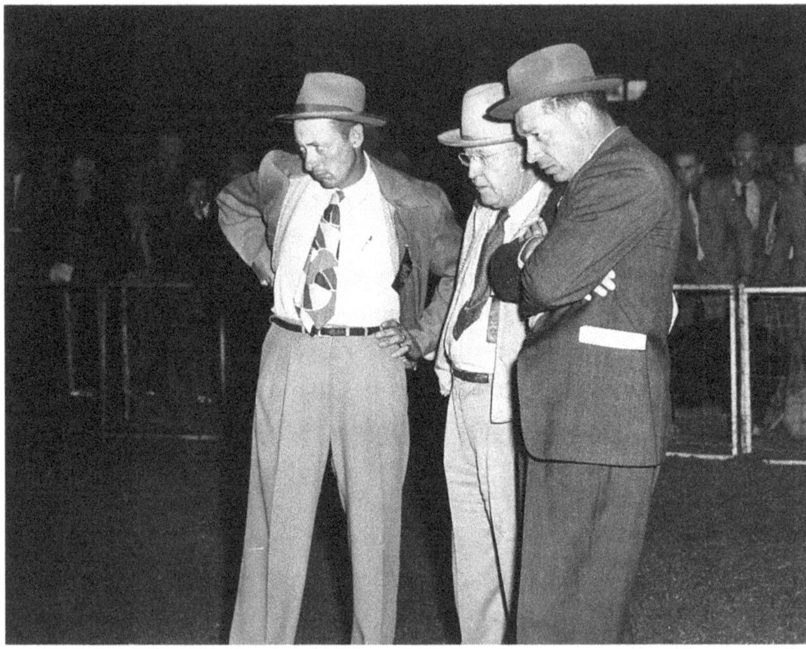

Figure 96. "Hog Judges—Leon Franz, E. L. Quaife, and Carroll Plager," Waterloo Meat Animal Show, 1953. Rath Packing Company Records, collection MS 562, Special Collections and University Archives, Iowa State University Library.

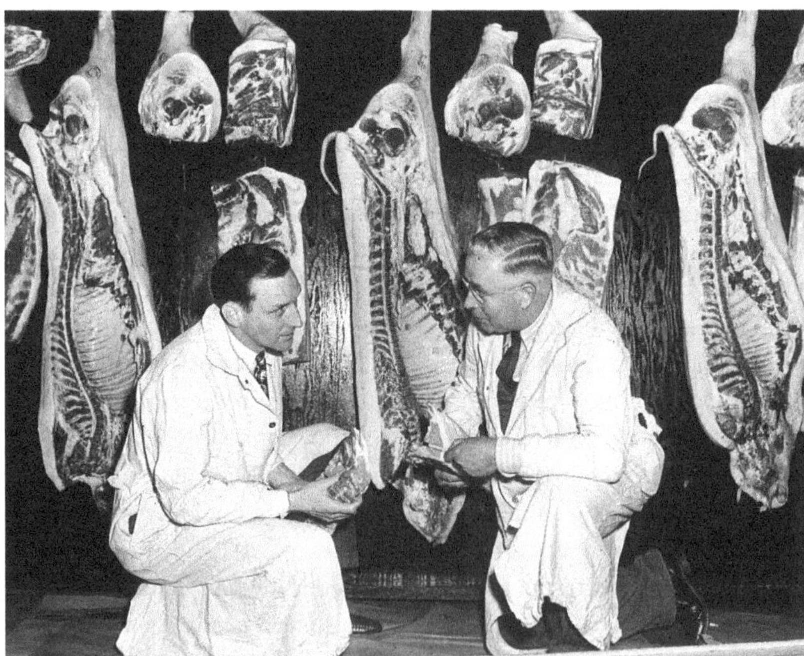

Figure 97. Louis Smith, Dubuque Packing Company supervisor, and C. W. McDonald, livestock specialist at Iowa State University, conferring before hanging carcasses, October 20, 1946. Des Moines Register and Tribune Archives, collection PH11438, Special Collections, State Historical Society of Iowa, Des Moines.

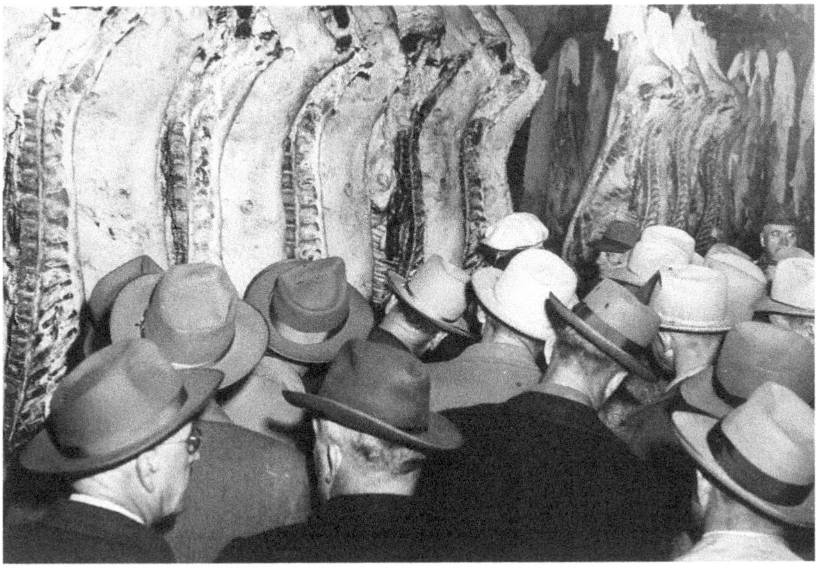

Figure 98.
Men inspecting hanging carcasses, ca. 1950. Rath Packing Company Records, collection MS 562, Special Collections and University Archives, Iowa State University Library.

before a display of cuts and carcasses in a 1946 press photograph, an industry representative and a university professor enact this meeting: exchanging knowledge, conveying expertise to one another, creating and drawing on a human knowledge base grounded in animal bodies (fig. 97). Trading information about the carcasses, they celebrate the accumulated experience and data that have yielded such bounty. The carcasses, in turn, become evidence: data points in themselves, elements in the edifice of knowledge built around, by, and through the animal body. Animal bodies themselves also become teaching tools: viewing carcasses, farmers learn better how to raise animals that will become yet more (and perhaps better) carcasses (fig. 98).

Representing these processes, these events, these occurrences, photographs elevate them and those who participate in them, down to and including the animals themselves. These images cast the state of carcasshood as a kind of achievement for the animal, who arrives there not through a process of reduction but as a culmination of life. In images circulated by the meat industry, meat production becomes a divine calling for all involved, human and nonhuman animals alike: farmers and meat producers *serve* meat animals by assisting them in realizing their potential. Photographs exist to bear witness to, and to perpetuate, this process of transubstantiation.

Notes

Introduction: Wonders of the Flesh

1. ANCO, *ANCO Machinery*. A copy is stored in the Trade Catalog Collection, Special Collections, Chicago Public Library.
2. The term *rückenfigur*, meaning "back figure," is often used to describe figures often, but not solely, in landscape paintings, who are shown facing away from the viewer and who are generally perceived as proxies for the viewer. In the works of German Romantic painter Caspar David Friedrich, the rückenfigur "functions as a placeholder we can imaginatively occupy, allowing us a virtual existence in the [image], and shaping our lines of sight within the spatial frame" (Haladyn, "Friedrich's 'Wanderer,'" 49). For a discussion of the ideological uses of the rückenfigur in landscape photography, see Nickel, *Francis Frith*, 137–41.
3. See Horowitz, *Putting Meat on the American Table*; Perren, *Taste, Trade and Technology*; Warren, *Great Packing Machine*; Cronon, *Nature's Metropolis*, esp. 207–59.
4. Sinclair, *The Jungle*. Publication of Sinclair's novel caused a massive outcry over food safety in the meat industry, drawing public and regulatory scrutiny to the industry and resulting in major changes to US food law. See Young, "Pig That Fell into the Privy," 467–80; Wade, "Problem with Classroom Use," 79–101.
5. See Armstrong, *Scenes in a Library*; Brunet, *Circulation*; Parr and Badger, *Photobook: A History*.
6. Certainly an impulse to realism is built into the fabric of photography as a technology. Among others, see Bazin, "Ontology of the Photographic Image," 237–44.
7. A voluminous literature addresses photography's relationship to truth. See, e.g., Sontag's influential essays in *On Photography*, which offer an incisive critique of the medium's capacity to manipulate.
8. See especially the work of Tagg, which considers deployments of photography within power structures: *Burden of Representation*; *Disciplinary Frame*.
9. At base, even the decision to photograph one thing over another represents the photographer's intervention in the scene being photographed. "Photographs bear witness to a human choice being exercised in a given situation.

A photograph is a result of the photographer's decision that it is worth recording that this particular event or this particular object has been seen. . . . A photograph is already a message about the event it records. . . . At its simplest, the message, decoded, means: *I have decided that seeing this is worth recording*" (italics in the original). Berger, "Understanding a Photograph," 291–94.

10. Barthes addresses how photographs and other images carry meaning in many of his studies. See, e.g., *Mythologies*; *Image, Music, Text*; *Camera Lucida*.
11. See, e.g., Hine, *Men at Work*; Lucic, *Charles Sheeler*.
12. Writing in 1927, Kracauer notes that photography "is a secretion of the capitalist mode of production. The same mere nature that appears in photography flourishes in the reality of the society produced by this capitalist mode of production." "Photography," 434.
13. See Sekula, *Fish Story*.
14. Sekula, *Fish Story*, 32.
15. See Lucic, "Charles Sheeler."
16. "Unlike most manufacturers, [a meatpacking company] does not bring together a group of raw materials in its meat operations and combine them into a manufactured product. On the contrary, a reverse process must be followed. The company purchases a live animal and 'disassembles' it into a great many products differing widely in character and marketability." Swift & Company, *Year Book*, 15. A copy is stored in the Chicago Stockyards Collection, Special Collections, University of Illinois–Chicago.
17. The journal titled *National Provisioner* is one of the meat industry's leading publications. Originating in the late nineteenth century and illustrated with photographs since the early twentieth, it is still published by the North American Meat Institute (originally the American Meat Institute), the industry's main trade organization. Accessed April 11, 2023, https://www.provisioneronline.com/.
18. Cronon, *Nature's Metropolis*, 208.
19. See Horowitz, *Putting Meat on the American Table*, 43–45.
20. See Warren, *Great Packing Machine*, 8–10; Anderson, *Capitalist Pigs*, 105–34.
21. See Cronon, *Nature's Metropolis*. Other early meatpacking centers included Omaha, Nebraska; Waterloo, Cedar Rapids, Ottumwa, and Sioux City, Iowa; and St. Joseph, Missouri. See Warren, *Great Packing Machine*, 8–13.
22. See Cronon, *Nature's Metropolis*.
23. See Wade, *Chicago's Pride*, 11–218.
24. Pacyga, *Slaughterhouse*, 45.
25. Giedion, *Mechanization Takes Command*, 218–19.
26. Weems examines how photographs of the extensive stockyard fencing grids, many created from an elevated viewpoint, were distributed as postcards

and used to instill "a sense of panoptic mastery" in their viewers, part of a program for envisioning the stockyards as a locus for animal management and meat production on a massive scale. Weems, "Scale, a Slaughterhouse View," 116.

27. Weems, "Scale, a Slaughterhouse View," 227–28.
28. See Humane Methods of Slaughter, sec. 1902, for directives about stunning. Animals slaughtered according to kosher law are handled differently. There are far fewer depictions of kosher slaughter than of conventional, however, and the topic is beyond the scope of this book.
29. "Reflection on dirt involves reflection on the relation of order to disorder, being to non-being, form to formlessness, life to death. Wherever ideas of dirt are highly structured, their analysis discloses a play upon such profound themes." Douglas, *Purity and Danger*, 7.
30. Federal Meat Inspection Act; Code of Federal Regulations, Title 9, Chapter III, Subchapter A, Part 310: Post-Mortem Inspection, https://www.ecfr.gov/current/title-9/chapter-III/subchapter-A/part-310.
31. Compartmentalization of labor persists after chilling, when the half-carcasses are broken down into cuts of meat and sent off to be packaged, but it is especially critical during the slaughter and dressing processes, when the dangers of contamination are highest. See Pachirat, *Every Twelve Seconds*, esp. 53–72.
32. For a thorough description of the many individual positions and jobs on the kill floor in an industrial-scale slaughterhouse, see Pachirat, *Every Twelve Seconds*, 257–70.
33. The *Lancet*, a British medical and public health journal, carried an extensive series of articles in the first decade of the twentieth century focusing on the spread of municipal abattoirs in European (and a few British) cities. Most were written by the journal's "special sanitary commissioner," a socialist journalist and sanitarian named Adolphe Smith. See, e.g., "Municipal Slaughter-House of Brussels"; "Paris Abattoir at La Villette"; "Basle Model Municipal Abattoir"; see also Morgan, " 'Those Truly Augean Stables.' "
34. Indeed, many British cities maintained this model into the twentieth century. British slaughter practices were slower to modernize than either European or American, with some cities adopting a European-style municipal model while others maintained systems of small, private butcheries serving localities and neighborhoods. See Ayling, *Public Abattoirs*.
35. Brantz, "Slaughter in the City," 22.
36. Vialles, *Animal to Edible*, 17–18, 23. Vialles notes that the word *abattoir*, used with reference to animal slaughter, has been euphemistic from the start.
37. Giedion, *Mechanization Takes Command*, 210.
38. See Brantz, "Slaughter in the City," esp. 149–200.
39. Giedion, *Mechanization Takes Command*, 211. La Villette alone has been the subject of a number of surrealist visual meditations on the existential.

Photographer Eli Lotar produced a surrealist-inflected series at the facility in the late 1920s, part of a commission by philosopher Georges Bataille. See Lusty, "Eli Lotar's Para-urban Visions," 87–103. French surrealist filmmaker Georges Franju's *Le sang des bêtes* (*Blood of the Beasts*) from 1949 incorporates documentary footage of a series of Paris slaughterhouses, depicting the slaughter of horses, cows, calves, and sheep. At once lyrical and confrontational, the film deploys a modernist idiom—unusual angles, animal's-eye views, and monochromatic rigor—to show animal slaughter and butchery as simultaneously horrifying and aesthetically pleasing. Franju demonstrates, in proper surrealist fashion, that reality is disorienting enough if one looks at it the right way. But the film, though a documentary, is generally read as a commentary on aspects of the human condition: a postwar meditation on mass slaughter recalling the Holocaust; a confrontation of the violence that lies at the heart of the everyday; a critique of capitalism's exploitative nature. For the film, see Franju, *Blood of the Beasts*; for an example of a typical reading of the film, see Williams, " 'Blood of the Beasts.' "

40. See Pacyga, *Slaughterhouse*; Cronon, *Nature's Metropolis*.
41. For discussion of meat industry involvement in the negotiations leading to the 1906 passage of federal legislation governing food production and meat inspection, see Young, *Pure Food*.
42. Sinclair, *The Jungle*.
43. In addition to his own observations and research, Sinclair drew on other accounts of the Chicago meat industry, particularly the extensive series of articles published by Adolphe Smith in the medical journal the *Lancet* in 1905. See Morgan, " 'Those Truly Augean Stables.' "
44. Pacyga, *Slaughterhouse*, 125–29.
45. "I aimed at the public's heart, and by accident I hit it in the stomach," Sinclair writes in late 1906. "What Life Means."
46. Wade, "Problem with Classroom Use."
47. See Executive Order No. 13917, "Delegating Authority." The language of the order was based on a draft by the American meat industry's leading trade group, the North American Meat Institute. See Grabell and Yeung, "Emails." The order came at the end of April 2020, after it became clear over the course of that month that COVID spread rapidly in meatpacking facilities and many were forced to close. Subsequent to the executive order, plant closures slowed dramatically, but the spread of the virus among workers did not. See McCarthy and Danley, "Map."
48. Data USA, "Butchers."
49. See, e.g., Artiga and Rae, "COVID-19 Outbreak."
50. For an example of an article illustrated with a plant exterior image, see Leys, "Coronavirus"; for one with a plant interior image—from Shutterstock—see Nosowitz, "Meat Production Has Rebounded." Stock images of cans,

machinery, animals, and piles of personal protective equipment (PPE) were also common in stories about the COVID-19 pandemic in the meat industry.

51. Scher, "Food Plant Workers."
52. See, e.g., Estes, "America's Meat Shortage," which is illustrated with three stock photographs from Getty Images: one of a cow, taken by Amilcar Orfali; one of a group of pigs, taken by Carsten Koall; and one of a group of chickens, taken by Dimas Ardian.
53. Rod Nickel, "Meat Plant Infections Rise."
54. Lewis, "Where Are the Photos."
55. Lewis, "Where Are the Photos."
56. See Dudziak, "Numbers."
57. Cartmill, *View to a Death*, 30.
58. Proto-Cuneiform Tablet with Seal Impressions [. . .], ca. 3100–2900 BCE, Sumerian, Uruk (likely), Mesopotamia, Metropolitan Museum of Art, New York (accession number 1988.433.1), https://www.metmuseum.org/art/collection/search/329081.
59. Ostracon with Pharaoh Spearing a Lion and a Royal Hymn on Its Back, ca. 1186–1070 BCE, Egyptian, Upper Egypt, Thebes, Valley of the Kings, Tomb of Tutankhamun (KV 62), debris near the entrance, Carnarvon/Carter excavations, 1920. Metropolitan Museum of Art, New York (accession number 26.7.1453), https://www.metmuseum.org/art/collection/search/544076.
60. The Acheloos Painter, Pointed Neck-Amphora (storage vessel): Herakles (Hercules) and the Stag in the Presence of Athena, ca. 510 BCE, Attica, Greece, Toledo Museum of Art, Ohio (accession number 1958.69a), http://emuseum.toledomuseum.org/objects/56424/pointed-neckamphora-storage-vessel-herakles-hercules-a.
61. See Cummins, *Hound and the Hawk*; Goldberg, "Louis the Pious."
62. See, e.g., Koslow, "Law and Order."
63. Alexander, "*Labeur* and *Paresse*," esp. 448.
64. See Ronan, "Capturing Cruelty," esp. 56–60.
65. See Jones, "Deer Drinking."
66. Model of a Slaughter House, ca. 1981–1975 BCE, Egyptian, Tomb of Meketre at Thebes, Middle Kingdom, Metropolitan Museum of Art, New York (accession number 20.3.10), https://www.metmuseum.org/art/collection/search/544257.
67. Pitcher (Oinochoe) with Butcher's Shop Scene, ca. 500–490 BCE, Greek, Museum of Fine Arts, Boston (accession number 99.527), https://collections.mfa.org/objects/153541/pitcher-oinochoe-with-butchers-shop-scene.
68. Hans Schäufelein, *A Farmer Slaughtering a Hog* [. . .], 1517, woodcut, Germany, Metropolitan Museum of Art, New York (accession number 59.507.157[1]), https://www.metmuseum.org/art/collection/search/670503.

69. "Boucher," in Diderot and d'Alembert, *Encyclopédie*, 2:350 (translation by the author).
70. Pieter Aertsen, *A Meat Stall with the Holy Family Giving Alms*, 1551, North Carolina Museum of Art, Raleigh, North Carolina (object number 93.2; accessed May 19, 2022), https://ncartmuseum.org/object/a-meat-stall-with-the-holy-family-giving-alms/.
71. Craig, "Pieter Aertsen," 2.
72. Craig, "Pieter Aertsen," 7.
73. For an overview of readings of *A Meat Stall* and Aertsen's work more broadly, see Moxey, "Interpreting Pieter Aertsen," 29–39.
74. Houghton, "This Was Tomorrow," 282.
75. Rembrandt, *Le Boeuf écorché* (*Slaughtered Ox*), 1655, oil on canvas, Louvre, Paris, France (inventory number M. I. 169; accessed May 19, 2022), https://collections.louvre.fr/en/ark:/53355/cl010065605.
76. See Honig, "Making Sense of Things."
77. See Baatsen, Blondé, and De Groot, "Kitchen."
78. Although outside the scope of this volume, the works of artist Sue Coe stand among the most influential contemporary explorations of large-scale slaughter. Her book *Dead Meat* incorporates prints of slaughterhouse scenes, some observational and others imagined, created over the course of the 1980s and early 1990s. Animal rights and existences have constituted a primary focus of Coe's art throughout her career; see her website at https://www.suecoe.com.
79. Garval, "Visions of Pork Production."
80. Garval, "Visions of Pork Production."
81. Weems, "Scale, a Slaughterhouse View," 111.
82. Weems, "Scale, a Slaughterhouse View," 118.
83. Berger, *Why Look at Animals?*; Baker, *Picturing the Beast*.
84. "It can look at me. It has its point of view regarding me. The point of view of the absolute other, and nothing will have ever done more to make me think through this absolute alterity of the neighbor than these moments when I see myself seen naked under the gaze of a cat." Derrida, "Animal That Therefore I Am," 380.
85. Brower, *Developing Animals*, xv.
86. Brower, *Developing Animals*, xvi–xvii.
87. People for the Ethical Treatment of Animals (PETA) has long relied on the persuasive power of images as a core aspect of its messaging. It uses images and videos of packinghouses and animal industrial sites to demonstrate the existence of poor conditions and mistreatment; and sympathetic images of animals to generate empathy and engender feelings of protectiveness. See, e.g., PETA, "Videos." PETA has also used images of women and female nudity to make animal rights activism look "sexy," a move that has drawn

criticism from scholars including Adams: "[PETA uses] the *gender that has been burdened with species to try to eliminate the oppression of species that have been burdened with gender*. If animals are burdened by gender, by gendered associations, by the oppression that is gender, then they can't be liberated through representations of gender oppression" (italics in the original). *Pornography of Meat*, 169.

88. The work of Canadian photojournalist Jo-Anne McArthur, for instance, functions as a wide-ranging and sustained critique of exploitative human-animal relationships across space and time, including in small- and large-scale meat production facilities worldwide. Accessed May 4, 2023, https://joannemcarthur.com/.
89. See Heinzerling, "Varieties and Limits of Transparency."
90. Drafted by the American Legislative Exchange Council (ALEC), the full text of the model legislation is available; see ALEC, "Animal and Ecological Terrorism."
91. Cultural, economic, and social histories include Horowitz, *Putting Meat on the American Table*; Cronon, *Nature's Metropolis*; Perren, *Taste, Trade and Technology*; Warren, *Great Packing Machine*. Historians focused on labor history include Barrett, *Work and Community*; Fehn, "Striking Women"; Halpern, *Down on the Killing Floor*; Horowitz, *"Unite and Fight!"*; Perry and Kegley, *Disintegration and Change*; Stromquist and Bergman, *Unionizing the Jungles*; Wade, *Chicago's Pride*; Warren, *Struggling with "Iowa's Pride."* Anthropological and sociological studies include Blanchette, *Porkopolis*; Fink, *Cutting into the Meatpacking Line*; Pachirat, *Every Twelve Seconds*; Stull and Broadway, *Slaughterhouse Blues*; Stull, Broadway, and Griffith, *Any Way You Cut It*. Adams has written numerous feminist critiques, including *Sexual Politics of Meat*; *Pornography of Meat*.
92. Organized labor, on the other hand, constituted a major concern at midcentury—what the meat industry both perceived and publicly declared a major threat to industry success. See especially Halpern, *Down on the Killing Floor*; Horowitz, *"Unite and Fight!"*
93. See Beers, *Prevention of Cruelty*. Beers describes in chapter 7 the passage of humane slaughter legislation in 1958. Even this reform, Beers notes, was not particularly limiting to the meat industry: "Humane slaughter was, in some ways, the quintessential moderate reform. Easing the suffering of food animals required no dietary sacrifices or inconveniences for consumers; the balance between human interests and animal needs could be easily achieved through regulation." *Prevention of Cruelty*, 159–60.
94. Cronin addresses the use of visual imagery, including but not limited to photography, for animal advocacy at the turn of the twentieth century in her book *Art for Animals*.
95. See Blanchette, *Porkopolis*.

Chapter One: Inside Views

1. Sekula, "Traffic in Photographs," 22.
2. Sekula, "Traffic in Photographs," 21.
3. See Klein, *No Logo*, esp. 8–26. Klein identifies branding as a recent phenomenon, but looking at the materials covered in this chapter and at other similar materials, it becomes clear that corporate identity building, for purposes of internal coherence and cohesiveness as well as for market-directed messaging, is a long-standing phenomenon and virtually inherent to capitalist enterprise.
4. Roland Marchand addresses corporate identity and power relationships in his essay "*Life* Comes to Corporate Headquarters" (in Doss, *Looking at "Life" Magazine*, 123–38), where he focuses on house-produced press organs for internal circulation, such as company newsletters and annual reports. In publications like these, Marchand finds, adoption of innovative, picture magazine–style layouts skyrocketed after initial publication of *Life* in late 1936. Company executives adopted such strategies in part as a way of connecting with their own workers, building a cohesive internal identity and thus possibly diminishing the influence of labor unions: "As labor unions increasingly flexed their muscle, corporate leaders sought to regain what they imagined nostalgically as close and trusted communications with their own working force. . . . Perhaps company publications should image their audience in the *Life* fashion as a broad and ostensibly classless amalgam, reachable through stories told with pictures." Doss, *Looking at "Life" Magazine*, 127.
5. Examples of corporate photo collections abound. A few have occasioned scholarly attention. Mildred Simpson, archivist of the collections of Atlantic Richfield Company (ARCO), a petroleum company, describes the extent of the collection she oversees in "Photographs in a Business Setting." Shipley analyzes works from another collection, now held in the Smithsonian National Museum of American History, in "Town That Photography Built."
6. Shipley, "Town That Photography Built," 89.
7. "Circulation is not only motion; it is also currency, and, at least to a certain extent, circularity, recirculation, or recurrence of signs and objects within a given cultural territory; reproduction, reuse, re-mediation, repurposing, and return of the same are components of circulation just as essential as physical transportation." Brunet, *Circulation*, 32.
8. Gray, "Managing Motivation," 111.
9. Gray, "Managing Motivation," 118.
10. Nye, *Image Worlds*, 3.
11. Foucault, *History of Sexuality*, 140–41 (italics added).
12. Tagg, *Disciplinary Frame*, xxxi.
13. Numerous studies consider biopower in relation to animal studies and

human-animal relationships. See, e.g., Holloway et al., "Biopower"; Wadiwel, *War against Animals*; Rosenberg, "Race Suicide among the Hogs."

14. Wadiwel, *War against Animals*, 15.
15. Management of herds and flocks via electronic sensing and surveillance is burgeoning—witness the rising use of video feeds and even facial-recognition software to manage flocks and herds. See Owen, "Here's Looking at You"; Grodkowski et al., "Automatic Herd Control Systems"; Bellet, "Reconfiguring the Senses."
16. Shukin, *Animal Capital*, 51 (italics in the original).
17. Shukin, *Animal Capital*, 20–21.
18. Mitchell, *Cloning Terror*, 112–36.
19. Mitchell, *Cloning Terror*, 112–36.
20. Mitchell, *Cloning Terror*, xvii.
21. "Invention has no great role" in the industrialization of meat production, Giedion notes; the real problem that had to be solved for meat to be produced on an industrial scale was, rather, that of distribution. *Mechanization Takes Command*, 224.
22. Giedion, *Mechanization Takes Command*, 230.
23. For a discussion of the mechanization of bread making, see Giedion, *Mechanization Takes Command*, 169–208.
24. Code of Federal Regulations, Title 9, Chapter III, Subchapter A, Part 309, https://www.ecfr.gov/current/title-9/chapter-III/subchapter-A/part-309.
25. Giedion, *Mechanization Takes Command*, 230–31.
26. For a useful overview of humane slaughter laws in the US, see Welty, "Humane Slaughter Laws."
27. To address the problem of willing movement, trained decoy Judas animals have been a common solution: pigs, cattle, or, most commonly, goats, released into a crowded pen and trained to walk calmly toward the packinghouse, up ramps and through chutes, leading the flock behind them. Judas animals do not have to be the same species as the target animal population: goats, themselves slaughtered only rarely in American packinghouses, have been used to lead animals of many species. The use of Judas animals constitutes an almost naturally determined design solution, essentially weaponizing animals' own proclivity to follow a leader.
28. Giedion, *Mechanization Takes Command*, 230–31.
29. See Pacyga, *Slaughterhouse*, 10–11.
30. See Weems, "Scale, a Slaughterhouse View," esp. 114.
31. Sinclair, *The Jungle*, 39–41.
32. Hurford, Hog hoist.
33. Farmstead Foods Collection, Series 8, box 11, Brucemore Historic Site, Cedar Rapids, IA.
34. "Technological improvements constantly altered the industry. . . . The packers devoted much time to innovate labor-saving equipment that would

make everything as regular and predictable as possible. Management first installed an endless chain to move the bodies by power around the killing floor, probably the first time any industry attempted to control by a power source the flow of raw product during any manufacturing process. With the endless chain and refrigeration, the plant necessary to power these innovations became one of the most important features of any packinghouse." Pacyga, *Slaughterhouse*, 107. For more on the Corliss engine and its role as a symbol of industrialization in nineteenth-century US, see Rosenberg and Trajtenberg, "General-Purpose Technology."

35. For information about recording and accounting practices in nineteenth-century American industry, see McGaw, "Accounting for Innovation."
36. *Harry Schultz Journals*, Chicago History Museum (accession number 2004.136.1).
37. See Rosenblum, *World History of Photography*, esp. 442–53, 630–38; Newhall, *History of Photography*, 154–55.
38. Warren, *Struggling with "Iowa's Pride,"* 13.
39. Morrell Meat Packing Company Collection, collection MSC 0100, General Series, Photographs and Drawings, box 30, Special Collections and Archives, University of Iowa Libraries (hereafter cited as Morrell Collection).
40. Giedion emphasizes this difference repeatedly in his discussion of mechanized slaughter. See *Mechanization Takes Command*, esp. 212–46.
41. See, e.g., Keefe, "Surveying Damage."
42. Morrell Collection, box 31.
43. See Horowitz, *"Unite and Fight!,"* esp. 18–19.
44. Pages in the album show scenes out of sequence, possibly because the album has been unbound and, in the process, reordered—in the album as currently stored, an image of the pork cut floor appears between the two pictures of the beef killing room.
45. See Federal Meat Inspection Act, Appropriations Enacted by Fifty-Ninth Congress, Session 2, Chapter 2907, 1907, esp. 1260–61.
46. Weems, "Scale, a Slaughterhouse View," 127–29.
47. The goal in many packinghouses, up to the middle of the twentieth century, was sixty animals per hour, although more could be processed if space and the size of the workforce permitted. See Horowitz, *Putting Meat on the American Table*, 39.
48. The pork images carry no dates but presumably were made at about the same time as the beef sequence in 1925. The first of these depicts the pork cut floor, where dozens of men pause in their work and turn to face the camera. After this initial image, most other pictures of the hog kill and cut lines do not show human workers, who must have been asked to leave just while the pictures were made. Though similar to the images in the beef kill sequence, these pictures carry no indication that the facilities for hog slaughter and processing are new. Other characteristics, however, remain

much the same as in the beef kill photographs. Repetition and volume, for instance, still connote abundance: myriad hog carcasses, dozens of bins of viscera, racks filled with hams, hundreds of slabs of bacon.

49. There are other series of images in the album, all probably added well after they were made. One series shows a flood or a series of floods that covered the packinghouse grounds in deep water. Sparsely populated and free from any sign of animal life, these pictures depict the grounds as an industrial site overwhelmed by a natural phenomenon. In this way, they constitute an interesting counterpoint to pictures made inside the slaughterhouses. Meat-packing facilities in river towns often depended on rivers as water sources and vital avenues for transport; but floods served as a reminder of the risks of such dependency. Whereas other photographs in the album document human mastery over nature, this series documents a rare interval when the opposite condition reigned and technocracy was immobilized as a result.
50. Warren, *Struggling with "Iowa's Pride,"* 30, 40–41.
51. Warren, *Struggling with "Iowa's Pride,"* 34–40.
52. "New Beef Building."
53. Warren, *Struggling with "Iowa's Pride,"* 42.
54. "New Beef Building," 3.
55. "New Beef Building," 3.
56. Such unintentional revelations appear repeatedly in issues of *Morrell Magazine*. The magazine regularly featured retirees and printed the obituaries of deceased former workers, touting their contributions to the company; but many of them worked for Morrell literally into their dotage. Honoring their service, the company reveals both its lack of any retirement scheme and the relentless, body-breaking nature of packinghouse work. One retiree, Frank Vivian, processed blood in the beef and hog kill departments for nearly forty years before "serv[ing] as a doorman until the time of his retirement" at age eighty; see "Four Veterans Retired." Another employee, John Lynch, described as "one of the real 'old-timers' at the Ottumwa plant," collapsed while still at his post; see "Death Takes Three Employees."
57. One auxiliary meat industry supplier, chemical company B. Heller and Company, was a prolific publisher of magazines, pamphlets, and other media. A manufacturer of supplies for the making of sausage and by-products, Heller had a primarily wholesale trade with a small inventory created for public sale. Heller's publications, also created for industry circulation, included cookbooks, a monthly publication titled *Scientific Meat Industry*, and a booklet titled *Heller's Secrets of Meat Curing and Sausage Making*, first issued in 1904, reissued almost yearly thereafter, and extensively rewritten in the late 1920s. All these publications included illustrations—some, engraved images made from, or with reference to, photographic originals; others, photographs printed in halftone. Heller's publications functioned as thinly disguised advertising. Although issues of

the *Scientific Meat Industry* alternated between advertisements and articles like any other periodical of the time, it carried advertisements solely for the company's own products, such as the Freeze-em-Pickle meat preservative, dried canned garlic for flavoring sausages, and antiseptic washing powder for sanitizing meatpacking facilities. But advertising was not confined to the explicitly commercial content: many of the articles in the journal likewise promoted Heller's products and those of its affiliates. See, e.g., B. Heller, "How to Produce a Fine Red Color," B. Heller and Company Collection, box 6, folder 5, Special Collections, University of Chicago Library (hereafter cited as B. Heller Collection).

Even where *Scientific Meat Industry* incorporated more substantive information, it seems to have served primarily to legitimate the promotional content by surrounding it with a cushion of ostensibly noncommercial information. For instance, the December 1905 issue also includes an extensive article, with engraved images, detailing how to slaughter various animals—cattle, sheep, and hogs—on a farm (which is to say in the traditional manner, one animal at a time). The article justifies itself as an instructional series for small-town butchers who are buying livestock directly from farms. It is true that the publication probably had some circulation among butchers: a section of classified advertisements at the back includes Help Wanted and Position Wanted ads as well as ads selling butchery equipment, indicating that readers regarded the journal as a resource connecting them with other members of their profession. But it is hard to imagine that such individuals would not already know how to slaughter an animal; and it is equally difficult to believe that anyone could learn, comprehensively, how to do something so complicated by reading the article, with its rather minimal line engravings. The article seems to exist, then, primarily as padding for the rest of the journal's promotional content. See B. Heller, "Killing on the Farm," B. Heller Collection, box 6, folder 5.

Heller's Secrets of Meat Curing and Sausage Making also promotes the company's products. First written in 1904, the publication was repeatedly reissued, with slight alterations in content, through 1928, then rewritten in 1929. Both before and after 1929, the publication constituted a brief instructional manual for best practices in the meatpacking industry, including sanitary practices, knife sharpening, treatment of animals prior to slaughter, and so on. Though seemingly exhaustive in the list of topics it covers, from "Hog Tongues, How to Cure" to "Sauer Kraut" to "Killing on the Farm," it offers only superficial information on each topic. An entry titled "Hoisting Hogs in a Large Packing House, with a Hog-Hoisting Machine," though accompanied by an image of a hog wheel, actually focuses primarily on how to avoid injury to hogs just before hoisting, giving more detail about proper tools for driving a hog and proper size of a killing pen than about the operation of the wheel. See B. Heller, *Secrets of Meat Curing*, 33–34, B. Heller Collection, box 6, folder 8. Again, it is difficult to imagine the target audience, butchers and packinghouse executives, needing to gain such knowledge from a book, particularly as the Heller publication leaves out a great deal of information. The booklet appears to have functioned, much like its periodical cousin, as a means of multiplying Heller's presence in the packinghouse, thus increasing its name recognition, its cachet as a company with expertise in the meat-packing field, and ultimately its appeal as a supplier.

58. ANCO, *ANCO Machinery*. Trade Catalog Collection, Special Collections, Chicago Public Library.
59. ANCO, *ANCO Machinery*, 6.
60. Retouching photographs for printing and reproduction was standard practice. Degrees of finesse tended to correspond to the quality of the practice. For portrait photography, retouching of both negatives and prints could be extensive, and skillful retouchers were regarded as essential workers in high-end portrait studios. Publications like the ANCO book also employed retouchers, and retouching was a routine part of the production process; but the demands of high-volume work, coupled with other stresses like turnover among production staff, led to variations in quality. Some retouchers might have formal art training; others might receive on-the-job training; and there were training manuals to allow for some degree of self-teaching. One such book advises readers that while its pages "do not pretend in the least to make either an artist or a retoucher of one who has had no experience of such work," a reader might at least learn some of the basics of the trade by using it as a reference. Johnson, *Art of Retouching*, v. See also Nadeau's *Encyclopedia*, which includes discussion of multiple retouching processes and methods of the time period.
61. See the start of the introduction for discussion of this image.
62. As, for example, in the lower caption in figure 16.
63. See Horowitz, *Putting Meat on the American Table*, esp. 67–68; Morgan, "*Waterloo Packer*: Selling Slaughter," 142.
64. In representing the sheep's death, the image implies the killable nature of all sheep. See Taussig, *Mimesis and Alterity*, esp. 44–51.
65. See, e.g., the illustration for "Chain Hog Hoist no. 4." ANCO, *ANCO Machinery*, 47. Among the images in the equipment portion of the catalog, this one stands out not only because it is a photograph but because in the background, behind the machine, a group of workers stands, looking at both the machine and the camera. Unlike the workers in the front section of the catalog, these workers do not wear gleaming white uniforms but, rather their own dark street clothing and black caps. Their presence, though minimal, serves to belie many of the catalog's assertions about the clean, mechanistic nature of animal slaughter.
66. Donna Haraway, "Persistence of Vision," in Conboy, Medina, and Stanbury, *Writing on the Body*, 283.

67. Panofsky, *Perspective as Symbolic Form*, 29–30.
68. Panofsky, *Perspective as Symbolic Form*, 67–68.
69. Baudry, "Ideological Effects," 40.
70. Baudry, "Ideological Effects," 40.
71. Pachirat, *Every Twelve Seconds*, 33.
72. Pachirat, *Every Twelve Seconds*, 33.
73. *Libby's Atlas of the World.* A copy is stored in the Advertising Ephemera Collection, box 17, John W. Hartman Center for Sales, Advertising & Marketing History, David M. Rubenstein Rare Book and Manuscript Library, Duke University.
74. For more on advertising ephemera of the period, see "Emergence of Advertising."
75. See Kramer, "Empires, Exceptions, and Anglo-Saxons." For other parallels between human race and food animal breeding, see Rosenberg, "Race Suicide among the Hogs."
76. *Libby's Atlas of the World*, inside front cover.
77. Another page diagrams the comparative population sizes of nations, with the US rather modestly sized by comparison to China but placed at the top of the page. A different page shows "Notable High Buildings of the World," arranged so their comparative sizes are clear. The Statue of Liberty, though shorter than the dome of Moscow's St. Basil's Cathedral or the tallest monument, the Eiffel Tower, stands at the front, as though leading the pack.
78. By the 1920s, Libby was becoming increasingly identified with canned fruits, not meats. Even in this shift, however—particularly the company's embrace of canned pineapple as a signature product—the company continued to depend on metaphors of colonialist dominance and exploitation as it marketed its products' exoticism. For more on Libby's promotional strategies, including its embrace of purity as a marketing strategy, see Miller, "Pure Food Exhibits."
79. *Libby's Atlas of the World*, 38.
80. O'Brien, *Chicago Stock Yards*, 23, 37, 39, 53, 73.
81. O'Brien, *Chicago Stock Yards*, 37.
82. O'Brien, *Chicago Stock Yards*, 39.
83. O'Brien, *Chicago Stock Yards*, 73.
84. B. Heller, "Schwarzschild and Sulzberger Company." A copy is stored in the B. Heller and Company Collection, Hanna Holborn Gray Special Collections Research Center, University of Chicago Library.
85. For the current publication, see https://www.provisioneronline.com/.
86. "Hydro-Thermal Cold-Coil Refrigeration" advertisement, *National Provisioner* 121, no. 5 (July 30, 1949): 12. Special Collections and University Archives, Iowa State University Library.
87. "Marlo Coil Company" advertisement, *National Provisioner* 121, no. 15

(October 8, 1949): 5. Special Collections and University Archives, Iowa State University Library.

88. Barthes, *Mythologies*, 110–15.
89. "Marlo Coil Company" advertisement, 5 (ellipsis and parentheses in the original).
90. "Gebhardt Cold Air Circulators" advertisement, *National Provisioner* 121, no. 6 (August 6, 1949): 15. Special Collections and University Archives, Iowa State University Library.
91. Weems, "Scale, a Slaughterhouse View," 111.
92. Weems, "Scale, a Slaughterhouse View," 127–29.
93. Shukin, *Animal Capital*, 20–21.
94. Donna Haraway, "Persistence of Vision," in Conboy, Medina, and Stanbury, *Writing on the Body*, 285.

Chapter Two: Realizing Corporate Visions

1. For discussion of corporate archives, see, e.g., Nye, *Image Worlds*; Simpson, "Photographs"; Shipley, "Town That Photography Built."
2. The firm of Hedrich-Blessing Photographers, for instance, photographed on commission for businesses in Chicago and throughout the US for over seventy years. The Hedrich-Blessing archives are at the Chicago History Museum, accessed July 8, 2021, https://images.chicagohistory.org/groupitem/2531/.
3. See Durden, "Connections and Conflicts."
4. For his nature studies, see, e.g., Korling and Betz, *Prairie*; Korling and Voss, *Boreal Forest*; Korling and Petty, *Eastern Deciduous Forest*.
5. "New Frank Lloyd Wright Building."
6. For a more complete account of the Rath Packing Company's fiftieth-anniversary celebration, see Morgan, "*Waterloo Packer*: Selling Slaughter."
7. D. G. Schneider of Young & Rubicam to Byron Benson of Rath Packing Co., letter, 19 March 1941, 3, Rath Packing Company Records, collection MS 562, Series 1: Clippings, box 1, folder 1/53: "50th Year Anniversary Celebration, 1941, Young & Rubicam, Inc.," Special Collections and University Archives, Iowa State University Library (hereafter cited as Rath Records: Y&R).
8. See, e.g., Babcock, *First Fifty*. This commemorative book, a fiftieth-anniversary volume created for Standard Oil Company, incorporates photographs by Korling and other photographers.
9. On *Life* magazine and other photographically illustrated publications of this era, see Parr and Badger, *Photobook: A History*, 116–45.
10. *Waterloo Packer*, n.p. [first spread].
11. Advertising and public relations are two sides of the same coin, but advertising firms in the early to mid-twentieth century recognized that to be maximally effective they should not be perceived as such, either by the

press or by the public: a company's public activities and overall identity should not seem bound and defined entirely by selling. Advertising firms adjusted accordingly; when Young & Rubicam, for instance, decided to create a public relations division, it incorporated the new entity not as a department within the company but as a separate, wholly owned subsidiary, calling it the Bureau of Industrial Service. Employees of the Bureau could avoid use of the Young & Rubicam name when they wrote press releases or contacted magazines and newspapers on behalf of their clients in search of opportunities for free publicity. See Marchand, *Creating the Corporate Soul*, 246.

12. Torrey Stearns, "Report of Conference/Visit to the Rath Packing Company, Waterloo, Iowa," 10–20 March 1941, Rath Packing Company Records, collection MS 562, Series 5: Correspondence, box 2, folder 2/3: "50th Year Anniversary Celebration, Bureau of Industrial Service, Torrey Stearns," Special Collections and University Archives, Iowa State University Library (hereafter cited as Rath Records: Stearns).
13. Young & Rubicam's Bureau of Industrial Services, Inc., "A Publicity Program for the 50th Anniversary of the Rath Packing Company," 14 April 1941, 10–18, Rath Records: Y&R.
14. Young & Rubicam, "Publicity Program," 13, Rath Records: Y&R.
15. Young & Rubicam, "Publicity Program," 18, Rath Records: Y&R (capitalization in the original).
16. Young & Rubicam's Bureau of Industrial Services, Inc., "Publicity Suggestions for Fiftieth Anniversary of The Rath Packing Company," 17 December 1940, 3, Rath Records: Y&R.
17. See Davis, *Gospel of Kindness*, 179–208.
18. Torrey Stearns to Byron Benson, letter, 2 June 1941, Rath Records: Stearns.
19. R. A. Rath to J. W. Rath, letter, 21 July 1941, Rath Records: Stearns. As Davis notes, bloodless bullfights had relatively recently become a way for Americans, particularly in Texas, to enjoy the tradition of the bullfight while circumventing accusations of animal cruelty. *Gospel of Kindness*, 206–7.
20. R. A. Rath to Torrey Stearns, telegram, 24 July 1941, Rath Records: Stearns.
21. Two employees of the Mexican embassy in Washington, DC, appeared in his stead. See "Suggested Letter to Be Signed by the Mayor of Waterloo," Rath Packing Company Records, collection MS 562, Series 5: Correspondence, box 2, folder 2/7: "Invitations—and tickets—sent out for 50th Anniversary Celebration 1941," Special Collections and University Archives, Iowa State University Library.
22. See, e.g., "Mexicans Are Named 'Rath Week' Guests," *Des Moines Tribune*, August 1941. Clipping in Rath Packing Company Records, collection MS 562, Series 1: Clippings, box 2, folder 2/46: "Rath 50th Anniversary Celebration, 1941," Special Collections and University Archives, Iowa State University Library (hereafter cited as Rath Records: Newspapers).

23. Young & Rubicam had proposed that Rath should ally in some manner with the local Native American community, to connect its "Black Hawk" brand name and logo to the region's history; but the shape of the event as it unfolded was determined by Rath executives. See Stearns, "Report of Conference," 3–4, Rath Records: Stearns.
24. Byron G. Benson to J. W. Rath, "Indian Induction Ceremony," 11 June 1941, Rath Packing Company Records, collection MS 562, Series 5: Correspondence, box 2, folder 2/12: "Miscellaneous 50th Anniversary," Special Collections and University Archives, Iowa State University Library (hereafter cited as Rath Records: Miscellaneous).
25. Benson to Rath, "Indian Induction Ceremony," 3–4, Rath Records: Miscellaneous.
26. See, e.g., "Rath Packing Co. Launches Celebration of Half Century," *Des Moines Tribune*, August 12, 1941; "Rath Adopted into Sac and Fox Indian Tribe," *Waterloo Daily Courier*, August 13, 1941, 1. Clippings in Rath Records: Newspapers.
27. See Fehn, "'Only Hope We Had.'"
28. Horowitz, *Putting Meat on the American Table*, 67–68.
29. "Playing Indian is a persistent tradition in American culture. . . . It is, however, a tradition with limitations. Not surprisingly, these cling closely to the contours of power . . . largely the domain of white males." Deloria, *Playing Indian*, 7–8. Anthony, Clark, and Nagel argue that as a core aspect of American identity construction, from the late nineteenth century, "white American men expropriated cultural symbols from indigenous people whom they confined on reservations . . . to construct a masculinity into which all white men should . . . be blended to manufacture the illusion of a largely unified American national identity." "White Men, Red Masks," 111.
30. The Rath company maintained this dubious alignment with Native American identities and masculinities well past its anniversary celebration: archival films in the Rath company archives at Iowa State University show further induction ceremonies being performed by Meskwaki Nation elders, under Rath's sponsorship and with its encouragement—one depicts a member of a local baseball team, for which Rath was a key supporter, being inducted as a member of the Meskwaki Nation during a pause in a baseball game. In at least one television advertisement for Rath's "Black Hawk" products from the 1970s, an announcer gives a fake "translation" of a supposed Native American word for Rath's products and is then threatened by a fist, bearing a tomahawk, when he seems inclined to give away Rath's secret recipe. The announcer in the latter ad, an urbane white man, represents "civilized" masculinity, the tomahawk-bearing hand a threatening "savage" masculinity —both of which, the ad implies, are under Rath's command, operating to create products for the pleasure and convenience of the company's core consumer base, white women. See Rath Packing Co. Film Inventory, Rath

Packing Company Records, collection MS 562, Special Collections and University Archives, Iowa State University Library.

31. Conard provides a history of the Rath Packing Company in her pamphlet *Bringin' Home the Bacon.*
32. There were several meat industry labor unions in the early to mid-twentieth century, but by the 1940s, two dominated. Founded in 1897, the Amalgamated Meat Cutters and Butcher Workmen (AMCBW) was, by the mid-twentieth century, a veritable institution, affiliated with the American Federation of Labor (AFL) and regarded by workers and packers alike as relatively moderate. The United Packinghouse Workers of America (UPWA), first founded in 1939 as the Packinghouse Workers' Organizing Committee (PWOC) and affiliated with the Congress of Industrial Organizations (CIO), was younger and smaller than the AMCBW but more left-leaning and more aggressive. In the early 1940s there was some competition between the two, and thus pressure for each to generate results for its members.
33. A number of scholars have addressed the labor history of the meat industry, many focusing on issues of racial equality. See especially Fehn, " 'Only Hope We Had' "; Halpern, *Down on the Killing Floor*; Halpern and Horowitz, *Meatpackers*; Horowitz, *"Unite and Fight!"*; and Warren, *Struggling with "Iowa's Pride."*
34. PWOC newspaper the *Unionist* covered the process of unionization at Rath over several months. See, e.g., "Rath Workers"; "Rath Campaign"; "Local 46 at Raths"; "Rath Management." Copies of the *Unionist* are held in the Rath archives alongside materials related to the company's fiftieth-anniversary celebration in Rath Records: Newspapers.
35. Interview with Bert Hoffman, Rath Records: Miscellaneous.
36. Interview with R. A. Rath, 4, Rath Records: Miscellaneous.
37. "What is truly startling in this mass transition from life to death is the complete neutrality of the act. One does not experience, one does not feel; one merely observes. . . . Has this neutrality toward death had any further effect upon us? This broader influence does not have to appear in the land that evolved mechanized killing, or even at the time the methods came about. This neutrality toward death may be lodged deep in the roots of our time. It did not bare itself on a large scale until the War, when whole populations, as defenseless as the animals hooked head downwards on the traveling chain, were obliterated with trained neutrality." Giedion, *Mechanization Takes Command*, 246.
38. See Rath Packing Company Records, collection MS 562, Series 5: Correspondence, box 2, folder 2/18: "Congratulations and Acceptances to the Iowa Dinner 8/13," Special Collections and University Archives, Iowa State University Library.
39. "Proposed Photographs by Torkle [*sic*] Korling," Rath Packing Company

Records, collection MS 562, Series 5: Correspondence, box 2, folder 2/4, Special Collections and University Archives, Iowa State University Library. See also Morgan, "*Waterloo Packer*: Selling Slaughter," 140–41.

40. Swensen demonstrates how the US government deployed the photographs of Andreas Feininger to similar ends at almost exactly the same time: "Feininger's clear-eyed modernist style, in its apparent lack of politics, was particularly well suited for the distinctive strategy of persuasion adopted by the USA that emphasised the precision, scale, and human effort of US industry." "Strategy of Truth," 85.
41. "Nov. 1, 1941—List of People to Whom We Sent Deluxe Edition," Rath Packing Company Records, collection MS 562, Series 5: Correspondence, box 2, folder 2/5: "List for Mailing Out Books—Institutional," Special Collections and University Archives, Iowa State University Library (hereafter cited as Rath Records: List).
42. Oscar Mayer photograph album, Special Collections and Archives, Wisconsin Historical Society (accession number M2016-042). When I initially viewed the album, it had only recently been received, and there was no indication or documentation of who had made the photographs or for what purpose. Subsequently, archivists found loose images elsewhere in the Oscar Mayer archive, matching some of those in the album and bearing a credit to Korling on verso. Lee Grady, reference archivist, Wisconsin Historical Society, email message to author, May 29, 2019.
43. In exterior shots within the album, a banner above the door bears a star and the number 1241, indicating that the image was made during the Second World War and that to the date of creation, 1,241 employees of the plant had gone for active military service. A publication made in late 1943 to commemorate Oscar Mayer's sixtieth anniversary shows the same plant but with a lower number on its banner, 1,079, indicating that the unpublished album in the Wisconsin archives was made after 1943. See Oscar Mayer, *Sixty Years of Service*.
44. Celebration of the machine age and the "machine aesthetic" was fairly widespread in American literature and visual culture starting as early as the 1920s, and quite a number of studies have addressed the phenomenon. See, e.g., Sayre, "American Vernacular"; Lucic, "Charles Sheeler."
45. Sheeler's photo, titled *Criss-Crossed Conveyors*, is one of his best-known images and an icon of the machine aesthetic. See Charles Sheeler, *Criss-Crossed Conveyors, River Rouge Plant, Ford Motor Company*, 1927, gelatin silver print, Museum of Modern Art, New York (object number 224.1941), https://www.moma.org/collection/works/49016.
46. For more discussion of this phenomenon in Depression-era works made on industrial commission, see Fraser, "Hands Off the Machine."
47. Andreas Feininger, photographing American industry at the behest of the US government's Office of War Information (OWI) during the Second

World War in 1942, was similarly "encouraged to create close-up shots of hands. . . . Photographs of hands at work, the OWI knew, were synechdochic of the ways in which humans and machines were united in the war effort. A photograph of a worker's arm turning valves was more than a dramatic modernist composition, it connoted the beginning of a process." Swensen, "Strategy of Truth," 93–94.

48. Horowitz, *Putting Meat on the American Table*, 68–92.
49. This was not the case across industries. Photographing the aircraft industry during the Second World War, Andreas Feininger was encouraged to make images of women and men working side by side. See Swensen, "Strategy of Truth," 104–6.
50. See Morgan, "*Waterloo Packer*: Selling Slaughter," 143–44. Just like the photos Korling made for Rath, these, too, become hard to view with the hindsight brought by postwar revelations of Nazi concentration camp atrocities. Images intended to show the company's care of the pigs ultimately refer to something far different.
51. "Wieners are automatically banded in 'Cartridge Pack' machine, with yellow band bearing the imprint showing they have been US Government inspected." Oscar Mayer, *Sixty Years of Service*, n.p. [17], caption. By the time of the company's seventy-fifth anniversary in 1958, it was spelling the name "Kartridg-Pak," and touting the yellow band not as a sign of US government inspection but as a means of conveying "brand identification" through yellow bands "bearing the company trademark." See Oscar Mayer, *Tradition and a Company*, 4, caption.
52. On the role of scientific discovery in the development of corporate industrialism in the US, see Noble, *America by Design*, esp. 3–19, 110–66.
53. Carolyn Goldstein writes that "by 1925, many corporations . . . had established home economics or home service departments to help them understand consumers and control demand for their products." By 1940, the rosters of one professional organization of home economists indicated that over one-quarter of the organization's members worked for food companies, making them the largest professional employer of home economists. Carolyn Goldstein, "Part of the Package," in Stage and Vincenti, *Rethinking Home Economics*, 277. See also Goldstein, *Creating Consumers*, esp. 174–207.
54. See Mayer, *Oscar Mayer*, esp. 14–19.
55. See Marchand, *Creating the Corporate Soul*, esp. 26–34.
56. The company also released a smaller booklet in 1943, not long before the Korling album was created, to commemorate its sixtieth anniversary. Although it is illustrated with photographs, they are not the ones created by Korling. The names of the photographer(s) and author(s) of the text are not included within the publication. The book, comparatively brief, contains a number of historic photographs as well as contemporary ones. It shares some subject matter in kind with Korling's book—emphasis on female work-

ers in many images, for instance, including at the Cartridge-Pack machine—but offers a less comprehensive vision of plant operations overall. See Oscar Mayer, *Sixty Years of Service*.

57. A copy of *A Tradition and a Company* is retained in the J. Walter Thompson Company Archives, Corporation Vertical Files, box 92, John W. Hartman Center for Sales, Advertising & Marketing History, David M. Rubenstein Rare Book and Manuscript Library, Duke University (hereafter cited as J. Walter Archives).
58. An archive of Siegel's work, including negatives, prints, and manuscripts, resides in the Arthur Siegel Photograph Collection, call numbers 1980.0024 PH Skel; 1981.0033 PPL; 1982.0226 PPL, Chicago History Museum; online finding aid at http://chsmedia.org/media/fa/fa/1980/24.htm. Siegel's photographs, slides, and other materials are also archived at the Arthur Siegel (1913–1978) Papers, 1927–1985, Ryerson and Burnham Library, Art Institute of Chicago (accession number 1982.4); online finding aid at https://artic.contentdm.oclc.org/digital/collection/finding aids/id/27909/rec/.
59. For more about Siegel's life, see Viskochil, "Arthur Siegel."
60. Siegel was one of the last photographers hired by Stryker and did most of his work not for the Farm Security Administration (FSA) but for the Office of War Information (OWI), established at the start of the Second World War. See Rivard, *New Deal Photographs*, 210–11.
61. On Siegel's forty-year teaching career at the Chicago Institute of Design, see the essays in Travis and Siegel, *Taken by Design*. Siegel also taught at Black Mountain College, a famed incubator of modernism in Asheville, North Carolina, in the summer of 1951. See Thomson, *Begin to See*.
62. Oscar Mayer, *Tradition and a Company*, 16.
63. Oscar Mayer, *Tradition and a Company*, 16.
64. For an examination of masculinity and organizational culture, see Whyte, *Organization Man*. See also Collinson and Hearn, *Men as Managers*.
65. Meinhardt Raabe originated the role, and its longest occupant was George Molchan. See Oliver's obituary of Molchan: "George Molchan."
66. Oscar Mayer, *Tradition and a Company*, 48. Writing in 1956, Whyte notes that within American industry, such rhetoric of scientific curiosity and discovery was misleading and misrepresentative—the focus was on profit, not intellectual inquiry for its own sake: "By their own statements of policy the majority of corporations make it plain that they wish to keep their researchers' eyes focused closely on the cash register. . . . They discourage their scientists, sometimes forbid them, from publishing the results of their work in the learned journals or communicating them in any way to scientists outside the company preserve." *Organization Man*, 210.
67. Hine, *Men at Work*.
68. See Lazonick, "Organizational Capabilities."

Chapter Three: Purity, Harmony, Identity

1. Young & Rubicam, "Publicity Suggestions," 3, Rath Records: Y&R.
2. J. A. Muth, sales manager, "A Memento of Rath's 50th Anniversary," Rath Records: List.
3. Studies of consumer behavior and the effects of advertising by industry groups, advertising industry analysts, and other interested parties have been voluminous since the mid-twentieth century in most industries, meat production included. See, e.g., Lester B. Colby, "What Sound Cooperative Advertising Has Done for America's Meat Industry," *Sales Management*, January 1, 1942; see also Henry H. Haupt, "What Is the Real Cost of Spam Advertising?," *Advertising and Selling*, July 1946; clippings in McCann-Erickson Worldwide Vertical Files, box 22, John W. Hartman Center for Sales, Advertising & Marketing History, David M. Rubenstein Rare Book and Manuscript Library, Duke University (hereafter cited as McCann-Erickson Files). See also Horowitz, *Putting Meat on the American Table*, esp. 12–16.
4. Extensive documentation of market research conducted by advertising company J. Walter Thompson Co. on behalf of its clients, including Swift, Libby, and other meatpacking companies, is retained in the J. Walter Archives, Account Files 1885–2006.
5. Family size in advertising has likely been determined at least in part by available resources: it is more economical, in terms of both space and money, to create an image of a family of three than a family of five. See Brown, "Images of Family Life."
6. See Mehaffy, "Advertising Race/Raceing Advertising."
7. See, e.g., Goffman, *Gender Advertisements*; Zawisza-Riley, *Advertising, Gender and Society*; Lears, *Fables of Abundance*.
8. For more on the representation of scientists in meat industry publications, see chapter 2.
9. Brown, *Corporate Eye*, 163–68.
10. Parker, "Modern Style," 20.
11. Pollay, "Subsidizing Sizzle," 28.
12. Doss, *Looking at "Life" Magazine*, 1–21; Guimond, *American Photography*, 149–203.
13. Parker, "Modern Style," 21.
14. For an account of nineteenth-century food safety regulation in the US, see Booker, "Before *The Jungle*."
15. I discuss workers, worker photography, and labor organizations in the American meat industry in greater detail in chapter 4. For labor histories of the meat industry, see, e.g., Barrett, *Work and Community*; Fehn, "Striking Women"; Halpern, *Down on the Killing Floor*; Horowitz, *"Unite and Fight!"*; Warren, *Struggling with "Iowa's Pride."*

16. Due to price fixing, such noncompetition was virtually built into the structure of the meat industry at the turn of the century. "[In] 1903, six firms controlled 90 percent of the inspected cattle slaughter in the United States," until investigations during the First World War revealed the extent of the cooperation among companies. Horowitz, *Putting Meat on the American Table*, 31.
17. See Wyman, *SPAM: A Biography*; Armstrong and Black, *Book of Spam*.
18. The industry has long recognized the central importance of by-products to its business model. To give just one example, the 1944 Annual Report to stockholders of Swift & Company makes this quite clear: "Using 'everything but the squeal' . . . reflects one of the outstanding contributions of private enterprise operating under constant pressure of the competitive search for profits. . . . The greater values of livestock by-products that have been created during the past sixty years have been shared with the producer through higher prices paid for his livestock, and with the consumer through meat prices lower than what they otherwise would be." Swift & Company, *Year Book*, 15.
19. Halpern, *Down on the Killing Floor*, 14.
20. Halpern, *Down on the Killing Floor*, 14–15; see also Cronon, *Nature's Metropolis*, 234–44; Leslie, "Chicago's Other Skyscrapers," esp. 20–25.
21. Young, *Pure Food*, esp. 130–39.
22. McClintock, *Imperial Leather*, 210–11.
23. See Sinclair, *Autobiography*; Wade, "Problem with Classroom Use."
24. "Armour's 'Simon Pure' Leaf Lard" advertisement, 1911, Roy Lightner Collection of Antique Advertisements, 1871–2009, box 1, John W. Hartman Center for Sales, Advertising & Marketing History, David M. Rubenstein Rare Book and Manuscript Library, Duke University (hereafter cited as Roy Lightner Collection).
25. Bogle, *Toms*, 2.
26. On the construction of race through "sensory stereotypes about black people" beyond the visual, including stereotypes about the supposed sounds of Blackness, see Smith, *How Race Is Made*, 4. See also Craig, "Designing Ethnicity."
27. Bogle identifies the period in American films around the Great Depression as "the Age of the Negro Servant." Although the figure in Armour's ads was always intended to be a servant, the 1920s and 1930s saw the type gain in selling power. See Bogle, *Toms*, 28.
28. Armour was not the only meatpacking company to use plantations or plantation-derived imagery in its advertisements. The Morrell company used plantation imagery in its ham advertising into the mid-twentieth century. See, e.g., Pamphlet by Mrs. Marie Kimball, *Treasured Recipes of the Old South*, ca. 1941, Morrell Collection.
29. Discussing straightforward-seeming advertisements like these, Parker writes

in 1937 that even "a bank advertisement showing nothing but a financial statement is not quite devoid of everything save cold logic: the logical content is in itself a titilation [*sic*] to the acquisitive instinct. Advertising art or copy is built around the theorem that content may be either rationally or emotionally treated, but the effect of the content must be emotional." "Modern Style," 21–22. The emotional effects of the content in Armour's Star Ham ads that show the Black figure were clearly intended to carry through even into more rationally presented ads like this one that shows only a meal on a table.

30. Dougherty, *In Quest of Quality*, 35–37.
31. Dougherty, *In Quest of Quality*, 77.
32. "Hormel Dairy Hams and Bacon" advertisement, 1911, Roy Lightner Collection, box 5 (italics in the original).
33. See Morgan, "*Waterloo Packer*: Selling Slaughter," 137.
34. "Market for Canned Meats," Research Department, *McCall's Magazine*, McCann-Erickson Files.
35. A 1947 trade-publication advertisement for SPAM addressed this directly. Over a graph showing consistent growth in SPAM sales from 1940 to 1946, text of the ad reads: "Changes in Consumer Habits: Spam sales increase despite G.I. gripes. Proof that a good product can overcome adverse publicity is found in the sales records of spiced meat. G.I. Joe may have complained loudly about Spam coming out of his ears, BUT, store records show that the first thing he did on return—following a kiss to his wife or mother—was to rush out and buy a package of Spam!" Advertisement in *Sales Management*, March 15, 1947, McCann-Erickson Files. See also Wyman, *SPAM: A Biography*; Armstrong and Black, *Book of Spam*.
36. For records related to Oscar Mayer advertising, including its famed jingles, see J. Walter Archives, Account Files 1885–2006, Oscar Mayer.
37. "Oscar Mayer & Co.'s position in the [meat] industry . . . is somewhat atypical, primarily because we have concentrated our efforts on the production of sausages and other processed meats. The strong brand identification we have painstakingly established over the past eighty-seven years is predicated on the consistent, uniform quality and value that we have built into Oscar Mayer products." Mayer, *Oscar Mayer*, 9.
38. See Horowitz, *Putting Meat on the American Table*, 75–102.
39. See May, *Homeward Bound*.
40. Oscar Mayer & Co., already branded as a producer of hot dogs and lunch meat, chose the J. Walter Thompson Co. to handle its advertising in 1957. The collaboration proved wildly successful. J. Walter Thompson demonstrated its ability to build on the company's existing familiarity to consumers, recasting familiarity as nostalgia. The brand became associated with childhood, the past, and innocence. J. Walter Archives, Account Files 1885–2006, Oscar Mayer.
41. By the 1960s, Oscar Mayer had become a pioneer in television advertising.

Relatively early in the history of the medium, its iconic, catchy jingle "I Wish I Were an Oscar Mayer Wiener" became the archetype of the inescapable promotional song, so ubiquitous that within just years it became effectively a piece of Americana, elevating the brand itself beyond the status of food item to that of cultural icon. The jingle and its successors—notably the "My Bologna Has a First Name" advertisements of the 1970s—succeeded at embedding Oscar Mayer in the national psyche. The company made hot dogs and luncheon meat, but its catchy, hooky, adorable jingles—sung by children—associated it indelibly with American childhood experiences. The brand's print advertising, then, often had only to recall its TV advertising to jog the reader's memory. Some print ads included images in the shape of TV screens, bubbly, round-cornered rectangles. The print ad prompted the jingle to spring, unbidden, into mind nonetheless: print ad as earworm starter. The goal was ubiquity; print advertising reinforced the work the television was already doing. It is the repetitive sameness of its print ads that makes them so interesting—the most telling imagery, often, is the imagery that does not draw attention to itself, just as the most deeply held values are those one accepts without question or hesitation, even without thinking.

42. "Advertisements are never only about the product they are promoting. They are about how our culture is structured, what we believe about ourselves and others." Adams, *Pornography of Meat*, 14.
43. "Advertisements suggest that the proper way to do gender with the product is to *follow* traditional and transitional ideologies." Robinson and Hunter, "Is Mom Still Doing It All?," 482 (italics added). The reality of representation is, of course, more complicated. Representation, particularly of female bodies, does serve to reinforce, reproduce, and perpetuate power relationships; it is just as much proactive as reactive. "Representation . . . sets in play certain relations of power through which . . . discourses around sexual difference and subjects in and for those discourses are ongoingly produced. In this sense, representation may be regarded . . . as a strategy of normalization. Representation participates in the various relations of power with which we are surrounded and in which we are always in one way or another implicated. Representation can be understood, then, as a form of regulation." Annette Kuhn, "Body and Cinema," in Conboy, Medina, and Stanbury, *Writing on the Body*, 204.
44. A mockup of the ad in the J. Walter Thompson archive reads "Roto. Supplement" at the bottom. "Oscar Mayer Bacon" advertisement, 1961, J. Walter Archives, Domestic Advertisements, Oscar Mayer, box OM1.
45. See May, *Homeward Bound*.
46. A variant on this ad removed the smaller inset photograph and accompanied the main image with a different text: "How to be an extra-good mother at breakfast . . . serve 'em savory, protein-rich Oscar Mayer Bacon." J. Walter Archives, Domestic Advertisements, Oscar Mayer, box OM1.

47. In the 1950s in particular, Matthews argues, "barbecue worked to 'stabilize' the home . . . for barbecue's division of labor—'dad' is the chef whereas 'mom' is on prep and cleanup duty—clarified gender roles and reasserted the postwar male's position outside in the public sphere as a hunter while repositioning females inside the private sphere to tidy the detritus of masculine labor and success." "One Nation over Coals," 7.
48. Fathers often appear, even in contemporary advertising, in the backyard or engaged in typical activities associated with the backyard, not just grilling but also camping or playing with their children. See Robinson and Hunter, "Is Mom Still Doing It All?," 479. Other studies of mid-century advertising also note that fathers, if shown actively engaging with their children, often are depicted playing in outdoor environments. See Marshall et al., "Overt Provider," esp. 1665–66.
49. Matthews, "One Nation over Coals," 26.
50. "Often the father . . . stands a little outside the physical circle of the other members of the family, as if to express a relationship whose protectiveness is linked with, perhaps even requires, distance." Goffman, *Gender Advertisements*, 39.
51. Some other versions placed more emphasis on the packaging than the dogs themselves. J. Walter Archives, Domestic Advertisements, Oscar Mayer.
52. See J. Walter Archives, Domestic Advertisements, Oscar Mayer, box OM1.
53. "Our History," *Philadelphia Tribune*, accessed June 30, 2022, https://www.phillytrib.com/site/about.html.
54. "Idealized appeals to success and ethnicity are created by juxtaposing symbols of status with ethnic models and diacritica. . . . Advertisements tell minorities that they should believe and act upon the values of the dominant consumer culture when, in fact, many are discriminated against and their participation in that culture is thereby restricted. . . . African Americans, like everyone else, are represented as equal participants in the consumer culture." Craig, "Designing Ethnicity," 35.
55. Oscar Mayer's branding campaign, its self-identification with the hot dog, was resoundingly successful. Despite operating, for the whole of its history, a full-scale slaughtering operation at its headquarters in Madison, Wisconsin (the plant closed in 2017), up to and beyond the end of the twentieth century the company was widely perceived as a purveyor of hot dogs and lunch meat. In the 1990s, the company undertook a partial remediation of this image, in ads and marketing materials targeting large-scale food service customers. One pamphlet from 1973, designed for food service buyers, advertised Mayer's full range of portion-control steaks, chops, and other cuts by noting how limiting its public image had become: "If Oscar Mayer makes you think of hot dogs, think again." While it continued to deploy the iconic Wienermobile (most recently staffed not by "Little Oscar" but by teams of enthusiastic young "brand ambassadors") for popular public appearances,

within the institutional service branch of its business it endeavored to make buyers aware of the full range of its products. J. Walter Archives, Domestic Advertisements, Oscar Mayer.

56. For studies of home economics as a field and as a profession in the US, see, e.g., Dreilinger, *Secret History*; Goldstein, *Creating Consumers*; Shapiro, *Perfection Salad*; Shapiro, *Something from the Oven*; Stage and Vincenti, *Rethinking Home Economics*.
57. See Shapiro, "And Here She Is."
58. Other brand mascots had similar geneses, existing originally as brand identifiers or illustrations before being brought to life. "Originally, Elsie the Cow existed only on the pages of Borden's advertisements [as] the now well-known likeness of a friendly, contented cow with a wreath of daisies around her neck. . . . The decision to bring Elsie to life" occurred in 1939, to satisfy attendees at Borden's display for the 1939 New York World's Fair. Hajdik, "'Bovine Glamour Girl,'" 472.
59. Shapiro writes that Husted publicly expressed her frustration at General Mills' failure to pay her equally to male salespeople, given the importance of the Betty Crocker creation to the company's brand and success. "And Here She Is," 90–94.
60. Horowitz, *Putting Meat on the American Table*, 27–31.
61. Giedion, *Mechanization Takes Command*, 226.
62. Vernon D. Beatty, manuscript copy of talk given at Swift & Company, April 1940, J. Walter Archives, Account Files 1885–2006, box 41, Swift & Company.
63. "Swift's Veal" advertisement, 1945, J. Walter Archives, Domestic Advertisements, Swift & Co., box SF1. Indeed, to reinforce the connection between brand and quality, Swift also released comic strip–like advertisements showing young women being trained by older relatives to look for the Swift label. In one of these, "Auntie Sue escorts Judy the bride to the market," "Auntie Sue" and the male butcher teach the young woman how to buy and braise inexpensive cuts of meat. Swift's Veal advertisement, 1944. J. Walter Thompson Company Archives, Domestic Advertisements, Swift & Co., box SF1, John W. Hartman Center for Sales, Advertising & Marketing History, David M. Rubenstein Library, Duke University.
64. American consumers received competing messages about consumerism and frugality during the Second World War. Advertisements promoted the purchase of American goods as a patriotic duty, the consumer's contribution to economic health; but there were, simultaneously, robust campaigns encouraging frugality, thrift, and self-reliance. Advertisements for Swift and other meat purveyors sometimes encouraged buyers to pool their rations with friends and neighbors to buy larger cuts of meat, and many advertisements during the war years included an encouragement to buy war bonds. See advertisements ca. 1942–1944 in the J. Walter Archives, Domestic

Advertisements, Swift & Co. See also Witkowski, "World War II Poster Campaigns."

65. Scrapbook and photograph album, Martha Logan Photograph Collection, collection G1985:473, box 3/4, Chicago History Museum (hereafter cited as Martha Logan Collection).
66. "Martha Logan (Home Economics Division)," Swift and Company Records, collection MSS Lot Swift and Company, box 2, Chicago History Museum (hereafter cited as Swift Records).
67. Swift first used the name "Martha Logan" in 1933. In that year, it ran an advertisement in which a group of women "food specialists" sat around a table, learning about Swift's "ovenized" ham. Some of the women shown in the ad were real people—nutritionist Meta Given, for instance—but among them sat "Martha Logan" at the end of the table. Only a few years later did the company assign that name to its Home Economics Division and to all the women who worked within it. The Chicago History Museum retains a draft image of the ad: "7 of them came in a group to see how we 'ovenize' ham!" Martha Logan Collection, box 1/4.
68. "Martha Logan," script for presentation to Swift salesmen, ca. 1946, Swift Records.
69. "Martha Logan," 1958, Swift Records.
70. "Home Economics Division, Research Laboratories, Swift & Company," n.d., Swift Records.
71. "Meet Martha Logan . . . Meal Planner for Millions," pamphlet, n.d., Swift Records.
72. "Home Economics Division," Swift Records.
73. Scrapbook and photograph album, Martha Logan Collection.
74. "Meet Martha Logan," pamphlet, Swift Records; see also Scrapbook and photograph album, Martha Logan Collection.
75. Scrapbook and photograph album, Martha Logan Collection.
76. "Martha Logan," script, ca. 1946, Swift Records.
77. See images in the Martha Logan Collection.
78. Scrapbook and photograph album, Martha Logan Collection.
79. "Martha Logan," script, ca. 1946, Swift Records.
80. Shapiro, "And Here She Is," 88.
81. Frank Capra's film *It's a Wonderful Life*, starring Reed, came out in 1946, the same year this first image of Martha Logan went into circulation. Reed was younger, in the film, than Martha Logan appears to be; but the home economist's image seems designed to evoke a similar type of woman: the girl next door, grown into a competent and self-assured wife.
82. This image, too, echoes a headshot of McLean: in a headshot from the early 1950s, McLean appears wearing a smart jacket with sharp lapels, under which a triple strand of pearls is visible. But the resemblance is not so close as in the earlier images from the 1940s, a recognition, perhaps, of the

increasing power of media representation to shape ideals of motherhood and womanhood.

83. Quoted in "Cover Girl," *Swift News*, April 1970. Clipping in Swift Records. Similarly, Betty Crocker came, in later years, to be associated with convenience foods like cake mixes rather than with the expertise born of years' worth of kitchen experience. The persona, concomitantly, faded until it became little more than a brand. See Shapiro, "And Here She Is," 99.
84. "Meet Martha Logan," pamphlet, Swift Records.
85. An image in the Swift records depicting several of these employees includes an annotation on the back noting that "consumer comments and requests for information on Swift & Co. products are answered promptly by specially trained staff members under the supervision of Swift's professional home economist, Martha Logan." Martha Logan Collection, box 1/4.
86. "Meet 'Martha Logan,'" manuscript, n.d., Swift Records.
87. Martha Logan Collection, box 3/4.
88. Swift & Co. advertisement, 1946, J. Walter Archives, Domestic Advertisements, Swift & Co., box SF1.
89. "Martha Logan," manuscript, n.d. (after 1958), Swift Records.
90. "Home Economics Division," Swift Records.
91. McLean and Paris, *Young Woman in Business*, v. This text is unchanged from that of the earlier 1953 edition, authored by McLean.
92. The English translation of Simone de Beauvoir's book *The Second Sex* appeared in 1953, the same year as the first edition of McLean's *Young Woman in Business*. Germain contends that *The Second Sex,* in which de Beauvoir examines femininity and womanhood as constructs, would have received wide enough press coverage to enter into a general public awareness. See "Reading *The Second Sex*." It is impossible to know whether McLean read de Beauvoir's book or was even aware of it, but the caricature of the feminist evoked in this passage is sufficiently developed that McLean can evoke and reject it with just a few choice words, "militant" among them.
93. McLean and Paris, *Young Woman in Business*, 161–64.
94. In 1953, McLean instructed young women in how to enunciate for radio, with just a brief section addressing television; in 1962, she and Paris added a great deal more advice about television performance. See McLean, *Young Woman in Business*, 191–99; McLean and Paris, *Young Woman in Business*, 223–32.
95. "Home Economics Division," Swift Records.
96. Swift & Co. advertisement, 1947, J. Walter Archives, Domestic Advertisements, Swift & Co., box SF1.
97. "With pornography, fragmented body parts become sexualized so that someone can get pleasure from something. Yet that something—the woman used in pornography—was at some point someone, a very specific someone. Two mass terms are merged as one—individual animals into hamburgers, an individual woman into an object 'woman.' Through this doubling of

objectification, what we have before us is the butchering of women's subject status." Adams, *Pornography of Meat*, 25. See also Adams, *Sexual Politics of Meat*. For a critical overview of scholarly debates focused on gender in animal studies, see Fraiman, "Pussy Panic."

98. Both images are in the Martha Logan Collection, box 1/4.
99. Although public performance at workshops and demonstrations constituted one aspect of many Martha Logan home economists' work, the division also employed models and other professional performers. Stills from a series of films created in the 1950s and 1960s, for instance, show the division's second director, Paris, performing alongside and instructing women who, although they wear Martha Logan clothing, are models or actresses hired for the occasion. Martha Logan Collection, box 1/4.
100. No documentation has been preserved alongside this image to explain the event depicted. Martha Logan Collection, box 1/4. "Industrial" musicals would be commissioned by corporations for their company conventions, sales meetings, and other in-house events. The phenomenon was common in the US in the early to mid-twentieth century and was the subject of Dava Whisenant's 2018 documentary film *Bathtubs over Broadway*.

Chapter Four: Picturing Meatpacking Work

1. See Zuromskis, *Snapshot Photography*.
2. Brown, *Corporate Eye*, 16.
3. Brown, *Corporate Eye*, 80–81.
4. The Gilbreths' reports to their clients about worker participation often partook of racial, cultural, and gendered stereotypes, both to describe workers' attitudes toward the camera and to inject levity into their presentations. Lindstrom, "'Undiscovered Mary Pickford,'" 738–39. Brown further notes that the Gilbreths' photographic studies ultimately had more value as public relations material, helping to generate business for the Gilbreths' consultancy, than as a genuine means of increasing efficiency. *Corporate Eye*, 86.
5. Corwin, "Picturing Efficiency," 140.
6. Morrell Collection, box 30. Morrell was founded in England in the 1820s and began operating in the US in the 1870s. For a study of labor relations at Morrell, see Warren, *Struggling with "Iowa's Pride."*
7. See Commons, "Women in Unions"; Breckenridge and Abbott, "Women in Industry," 641–49.
8. Morrell was not alone in the creation of novel dress for women workers replacing men on the production lines during the First World War. A 1990 exhibition of women's work clothing at the Valentine Museum in Richmond, Virginia, included several uniforms designed specifically for women workers at munitions plants and the American Red Cross, among other organizations. See Pollard, "'Dressed for Work.'"

9. "Girls Help Solve Labor Shortage."
10. "Girls Help Solve Labor Shortage."
11. "What is called gender identity is a performative accomplishment compelled by social sanction and taboo. In its very character as performative resides the possibility of contesting its reified status." Butler, "Performative Acts," 520. See also Butler, *Gender Trouble.*
12. Men in industry-commissioned media rarely express such lightheartedness or gaiety. No one plays tug-of-war or sports a custom-designed coverall ensemble, certainly not one designed primarily to be photographed. Male packinghouse workers in meat industry photographs rarely even appear to relax or to shift focus away from their work; and male managers and executives labor at their white-collar thought-work with intense focus.
13. Pachirat, *Every Twelve Seconds*, 61.
14. Pachirat, *Every Twelve Seconds*, esp. 29–32.
15. Pachirat, *Every Twelve Seconds*, esp. 39.
16. Pachirat, *Every Twelve Seconds*, esp. 39.
17. Giedion, *Mechanization Takes Command*, 228. Machines and tools played a role at every stage; but, as Giedion points out, the work of slaughter proved difficult to mechanize fully, since animals varied and continue to vary so widely in size and shape within a single species, even within a breed. Some aspects of the work could be left to machines, but many jobs could not. Even through the intervening seven decades since Giedion's publication, even through increasing standardization in the sizes and shapes of animals' bodies, much of the work of slaughtering and breaking down larger meat animals such as cattle, pigs, and sheep still must be done by human beings with hand tools. Today, however, there are far more low-paying, unskilled and semiskilled jobs, and far fewer higher-paying jobs requiring higher levels of skill, than in past decades. See also Horowitz, " 'That Was a Dirty Job!,' " esp. 14–17.
18. In her memoir, Register notes that in the 1950s her father, a millwright and mechanic at Wilson & Co. in Albert Lea, Minnesota, received top pay, exceeded only by that of the skilled beef skinners. *Packinghouse Daughter*, 43.
19. Horowitz, *"Unite and Fight!,"* 18–19; *Putting Meat on the American Table*, 36–38.
20. See, e.g., Warren, *Struggling with "Iowa's Pride,"* esp. 42–44.
21. Horowitz, *Putting Meat on the American Table*, 83–89.
22. Sam Parks, quoted in Halpern and Horowitz, *Meatpackers*, 52.
23. Horowitz, *"Unite and Fight!,"* 23. See also Warren, *Great Packing Machine*, 102–6.
24. Sam Parks, quoted in Halpern and Horowitz, *Meatpackers*, 52. See also Virginia Houston's interview, in Halpern and Horowitz, *Meatpackers*, 77.
25. Warren, *Great Packing Machine*, 102.

26. Horowitz, *"Unite and Fight!,"* 24; Frank Wallace, quoted in Halpern and Horowitz, *Meatpackers*, 103.
27. Warren, *Great Packing Machine*, 76.
28. It was for this reason that the UPWA took as one of their early symbols an image of Black and white hands clasped; it was in the best interest of the union and of workers to oppose racism. See Horowitz, *"Unite and Fight!,"* esp. 4, 145–233.
29. Perry and Kegley, *Disintegration and Change*, 60.
30. Horowitz, *"Unite and Fight!,"* 22.
31. Warren offers a detailed analysis of these patterns as they occurred at the Morrell Meat Packing Company's facility in Ottumwa, Iowa, from the later nineteenth into the later twentieth century. See *Struggling with "Iowa's Pride."*
32. Horowitz, *"Unite and Fight!,"* 4.
33. Horowitz, *"Unite and Fight!,"* 12–19.
34. Horowitz, *"Unite and Fight!,"* esp. 145–233.
35. For a study of the challenges presented by archives containing objects, see Rudolph, "Separated at Appraisal."
36. Portelli, *Death of Luigi Trastulli*, 5–7.
37. Portelli, *Death of Luigi Trastulli*, 26.
38. Tagg, "Archiving Machine," 26.
39. "The archive, as printing, writing, prosthesis, or hypomnesic technique in general is not only the place for stocking and for conserving an archivable content *of the past* that would exist in any case, such as, without the archive, one still believes it was or will have been. No, the technical structure of the *archiving* archive also determines the structure of the *archivable* content even in its very coming into existence and in its relationship to the future. The archivization produces as much as it records the event." Derrida, *Archive Fever*, 16–17.
40. Many meat industry–related materials in the archives of the State Historical Society of Iowa, for instance, were donated by individuals, including former packinghouse workers and community members. The archives of the United Packinghouse, Food, and Allied Workers, though held as one collection at the Wisconsin Historical Society, similarly came via donations from multiple individuals. See the finding aid: http://digital.library.wisc.edu/1711.dl/wiarchives.uw-whs-mss00118.
41. See Peterson, "Visual Culture and Working-Class Community," 216.
42. Halpern, "Back of the Photograph," 192.
43. Focusing on the widely dispersed archival materials left by Dean Worcester, a US administrator in the Philippines at the turn of the twentieth century who used photography to document and catalog the peoples of the islands, Punzalan considers how diffusion of knowledge, information, and materials affects the ability to study or even frame a topic. It only complicates the situ-

ation when we recognize that a group's subject position or degree of power affects how, or even whether, its materials might have been archived. If a powerful individual such as Worcester has a widely dispersed archive, then the survival and maintenance of archives of less powerful individuals and organizations, not to mention entire subject or subjugated populations, may be even less consistent. "Archival Diasporas."

44. Archives of the Amalgamated Meat Cutters and Butcher Workmen (AMCBW), Wisconsin Historical Society.
45. Peterson, "Visual Culture and Working-Class Community," 216.
46. The AMCBW archives at the Wisconsin Historical Society include many examples of such images, often stamped with their source on verso. See, e.g., AMCBW Photographs, collection M80–118, Wisconsin Historical Society.
47. Ollmann, *Camera as Weapon*, 14.
48. Ollmann, *Camera as Weapon*, 19.
49. Wolf describes how, in the Soviet Union, clubs for workers' enrichment, including photography clubs, grew up often within workplaces, as amateur photographers were encouraged to document their own lives. See Erika Wolf, "Worker to Proletarian Photography," in Ribalta, *Worker Photography Movement*, 36–37.
50. Peterson, "Visual Culture and Working-Class Community," 227.
51. Quirke, *Eyes on Labor*, 2–12.
52. Quirke, *Eyes on Labor*, esp. ch. 5, 6.
53. Ollmann, *Camera as Weapon*, 72.
54. *Der Arbeiter-Fotograf*, in Ribalta, *Worker Photography Movement*, 99.
55. Rudolf Stumberger, "*AIZ*," in Ribalta, *Worker Photography Movement*, 87.
56. Erika Wolf, "Worker to Proletarian Photography," in Ribalta, *Worker Photography Movement*, 38.
57. See Anne Tucker, "Modernism and the Left," in Ribalta, *Worker Photography Movement*, 322–33.
58. Quirke, "Imagining Racial Equality," 98.
59. Quirke, "Imagining Racial Equality," 101–2.
60. Peterson, "Visual Culture and Working-Class Community," 220.
61. Peterson, "Visual Culture and Working-Class Community," 220.
62. Peterson, "Visual Culture and Working-Class Community," 216.
63. Peterson, "Visual Culture and Working-Class Community," 216.
64. Pallas photographed for Standard Oil Company, among others. See materials gathered and presented by his former assistant, Janet Ginsburg, at https://www.jaginsburg.com/mickey-pallas.html.
65. See "Biography," in "Finding Aid for the Mickey Pallas Archive," Center for Creative Photography, University of Arizona, Tucson, accessed July 20, 2021, https://aspace.ccp.arizona.edu/repositories/2/resources/261.
66. Photographs, AMCBW Archives, collection M79-517, Archives Division, Wisconsin Historical Society.

67. Although an enlarged copy of this image is labeled "Hormel, Austin, MN, 1950s" on the back, this label is probably incorrect—the image also appears on at least three separate contact sheets with Pallas's stamp on the back, all labeled "Tobin."
68. Advertisement, *Freedom* 2, no. 10 (October 1952), 8, accessed July 29, 2021, https://dlib.nyu.edu/freedom/books/tamwag_fdm000022/#8. A leftist newspaper serving the Black community from 1950 to 1955, *Freedom* was published in New York. Banks's advertisement in the paper suggests that he was soliciting commissions beyond Chicago, seeking to build a national reputation and client base.
69. One of the few sources to discuss Banks's work is an informal and incomplete online biography, written in the early 2000s by an acquaintance and admirer of Banks. It focuses more on his work as a musician and club owner than on his photography. Mendiola, "Jo Banks."
70. It is not clear whether the two photographers worked together to create the series or whether the separate bodies of work were brought together by a UPWA photo editor.
71. Halpern, *Down on the Killing Floor*, 199–200.
72. See Morgan, "Striking Images," esp. 158–66.
73. Morgan, "Striking Images," esp. 158–66.
74. For other instances of intimidation of strikebreakers by striking workers, see, e.g., Fehn, "Ruin or Renewal," 368; Warren, "Heyday of the CIO," 377–80.
75. Morgan, "Striking Images," 166.
76. Sam Parks, in Halpern and Horowitz, *Meatpackers*, 52.
77. Pachirat, *Every Twelve Seconds*, 27.
78. Sam Parks, in Halpern and Horowitz, *Meatpackers*, 52.
79. See Douglas, *Purity and Danger*.
80. Zuromskis, *Snapshot Photography*, 12.
81. Mary Bennett, special collections coordinator at the State Historical Society of Iowa, confirmed that these images came from the Tobin facility. Mary Bennett, email message to author, April 21, 2021.
82. See Strömberg and Karlsson, "Fun and Mischief." Strömberg and Karlsson's ethnographic study considers the use of humor among European meatpackers in the early twenty-first century, but their findings prove applicable to similar work, even across space and time.
83. Homer, *Odyssey*, bk. 10, lines 203–399.
84. Shakespeare, *Midsummer Night's Dream*, 251–83.
85. Multiple copies of the work exist; see, e.g., Marcel Duchamp, *L.H.O.O.Q. or La Joconde*, Norton Simon Museum, Pasadena, California (accession number P.1969.094), https://www.nortonsimon.org/art/detail/P.1969.094.
86. Stallybrass and White, *Politics and Poetics of Transgression*, 56–58 (italics in the original).

87. Thompson, *Customs in Common*, 509.
88. Richard Shusterman draws attention to the performative aspects of photography, distinguishing between photography as a process and the photograph as a kind of artifact. "The photograph and its aesthetic perception are only part of a larger complex of elements that constitutes photography as an activity and as an art." "Photography as Performative Process," 68. See also Baker, "For the Camera."
89. Wayne Brown, in *People of the Pack*, 89.
90. Jim Coleman, in *People of the Pack*, 94.
91. Pete Eisbach, in *People of the Pack*, 94.
92. "Flow" is often identified with a managerial standpoint. Flow "break[s] the connection between cause and effect, that is, between human actors and the aggregate products of their actions," facilitating a dehumanized, bureaucratic, organization-level vision of production. Rockefeller, "Flow," 566.
93. Thompson, *Customs in Common*, 399.
94. Strömberg and Karlsson, "Fun and Mischief," 645.
95. Strömberg and Karlsson, "Fun and Mischief," 632.
96. Stallybrass and White, *Politics and Poetics of Transgression*, 19 (italics in the original).
97. Stallybrass and White, *Politics and Poetics of Transgression*, 27.

Epilogue: Posing with Pigs

1. This occurs, for instance, in Korling's photographs for the Rath company's book *Waterloo Packer*, e.g. figure 30, or in advertisements like that shown in figure 69, in which cattle ranching occupies the first letter in the Swift company name.
2. See Bellet and Morgan, "Breed(ing) Narratives."
3. Paulus Potter's 1647 painting *The Bull* constitutes a notable exception; but although the painting shows an ordinary farmer and his ordinary bull (and his cow, ram, ewe, and lamb), the image itself is far from ordinary, measuring more than eleven feet in length and showing the bull nearly life-sized. Accessed December 1, 2021, https://www.mauritshuis.nl/en/our-collection/artworks/136-the-bull/#detail-data.
4. Nineteenth-century and early twentieth-century farming manuals and breeders' media frequently incorporated images of animals in which text and image work together to give a sense of the appropriate markings, sizes, behaviors, and other indices of various breeds. See, e.g., Ewart, *Meat Production*; Shaw, *Animal Breeding*.
5. Bellet and Morgan, "Breed(ing) Narratives," 206–20.
6. See Nordby, Beeson, and Fourt, *Livestock Judging Handbook*, esp. xiv–xv.
7. See Haraway, *When Species Meet*; Harrison, *Animal Machines*.

8. Side views of animals are found even in the most ancient of artworks. At the caves of Lascaux, in France, artists drew dozens of cows and horses in side view—potentially to evoke sympathetic magic that would render them harmless, and easier to kill, during an actual hunt. The side view turns the animal into an object, rendering it harmless and presenting it for human contemplation, delectation, and consumption. See "Prehistoric Sites and Decorated Caves of the Vézère Valley," UNESCO World Heritage Site List, accessed April 28, 2023, https://whc.unesco.org/en/list/85/.
9. See Franks and Grandin, "Leg Conformation."
10. Sheep would be shown unshorn, as the quality of the wool is a key aspect of the animal's value. Pigs might have the head raised or lowered, and a slight staggering of the hind legs might also allow for the display of the testicles, for a boar, or the teats, for a sow. While the side view is canonical, back views tend to be relatively common, as well. The back view permits better evaluation of the hind leg conformation and—for pigs—of the hams.
11. "When selecting breeding bulls from semen catalogs . . . visual appraisal is based on the picture supplied by the bull breeder." Franks and Grandin, "Leg Conformation." See also Cudworth, " 'Most Farmers Prefer Blondes,' " 37–39.
12. "If the herd is of some importance it is well to have a supply of good photographs on hand. . . . If no pictures are available the attendant or the breeder himself should be willing to make appointments with newspaper representatives for taking pictures. It is to the interest of the breeder that only good pictures be supplied. He should ascertain whether the representative is qualified to take pictures and in case he is not obtain the services of a livestock photographer . . . and have pictures made at his own expense. He should know how to pose his animals so that a good picture can be taken. He should practice posing his stock beforehand so that they will become accustomed to the meticulous handling necessary." Beeler, *Marketing Purebred Livestock*, 187–88.
13. Haraway, *When Species Meet*, 148.
14. "It is the capacity of animal life to be taken both literally and figuratively, as a material and symbolic resource of the nation, that constitutes its fetishistic potency. . . . The ambivalence of animal signs is for this reason a pivotal means of depoliticizing volatile contradictions between species and speculative currencies of capital and between capitalism's material and symbolic modes of production." Shukin, *Animal Capital*, 6.
15. "Dog people tend to see any 'defect' in their dogs as a defect in themselves." Haraway, *When Species Meet*, 114.
16. Some examples are also included in Arluke and Bogdan, *Beauty and the Beast*, esp. 83–108.
17. Education in agricultural practices often begins far earlier than other forms of professional training, and the educational experience is itself seen as

crucial to engendering support for industrial-scale agricultural practices. See, e.g., Adams et al., "Background Experience." For a history of American agricultural fairs and livestock exhibitions, see Kniffen, "American Agricultural Fair." Kniffen notes that agricultural fairs of the late 1940s often had the sponsorship of 4-H clubs, making clear that the educational aspects of agricultural fairs were still paramount. "American Agricultural Fair," 278. The 4-H program is federally owned and widespread across the American rural landscape, enrolling youths aged ten through twenty. Embodying complex relationships between rural communities, private industrial agricultural enterprise, and the federal government, 4-H educates American youth in both traditional and modern agricultural practices. Rosenberg writes that "4-H's deployment as a selectively embodied tactic of state power coincided with the near-total integration of national corporations into the production and consumption of foodstuffs." *4-H Harvest*, 10.

18. For a number of years at mid-century, the Rath Packing Company sponsored an annual meat animal show, and a series of these commemorative human-animal portraits is preserved within the company's archives. Rath Packing Company Records, collection MS 562, box 2, folder 4, Special Collections and University Archives, Iowa State University Library.
19. See Dickinson, *Photographing Livestock*; Nordby, Beeson, and Fourt, *Livestock Judging Handbook.*
20. See Bellet and Morgan, "Breed(ing) Narratives."
21. A gilt is a young female pig who has never borne a litter of piglets.
22. Numerous photographs from the 1940s and 1950s in the Rath Packing Company archives show aspects of these carcass contests: the same animal shown alive and as a carcass; candid images of entrants crowded around penned animals, paper and pencil in hand; images of expert judges and educators, explaining the results of the contests. See Rath Packing Company Records, collection MS 562, Series 13: Photographs, Special Collections and University Archives, Iowa State University Library.

Bibliography

Adams, A. L., G. A. Holub, W. S. Ramsey, and T. H. Friend. "Background Experience Affects Student Perceptions of the Livestock Industry." *NACTA Journal* 59, no. 1 (March 2015): 24–27.

Adams, Carol J. *The Pornography of Meat*. New York: Continuum, 2003.

Adams, Carol J. *The Sexual Politics of Meat: A Feminist-Vegetarian Critical Theory*. New York: Continuum, 1990.

Agamben, Giorgio. *Homo Sacer: Sovereign Power and Bare Life*. Stanford, CA: Stanford University Press, 1998.

Alexander, Jonathan. "*Labeur* and *Paresse*: Ideological Representations of Medieval Peasant Labor." *Art Bulletin* 72, no. 3 (September 1990): 436–52.

Allbright-Nell Company (ANCO). *ANCO Machinery and Equipment for the Meat Industry; Catalog No. 10*. Chicago: ANCO, 1923.

American Legislative Exchange Council (ALEC). "The Animal and Ecological Terrorism Act (AETA)." ALEC, 2004; 2013. Accessed April 11, 2023. https://alec.org/model-policy/the-animal-and-ecological-terrorism-act-aeta/.

American Meat Institute. *The Packing Industry: A Series of Lectures Given under the Joint Auspices of the School of Commerce and Administration of the University of Chicago and the Institute of American Meat Packers*. Chicago: University of Chicago Press, 1924.

Anderson, Joseph L. *Capitalist Pigs: Pigs, Pork, and Power in America*. Morgantown: West Virginia University Press, 2019.

Anderson, Virginia DeJohn. *Creatures of Empire: How Domestic Animals Transformed Early America*. New York: Oxford University Press, 2004.

Andersson Cederholm, Erika, Amelie Björck, Kristina Jennbert, and Ann-Sofie Lönngren, eds. *Exploring the Animal Turn: Human-Animal Relations in Science, Society and Culture*. Lund, Sweden: Pufendorf Institute for Advanced Studies, 2014.

Anthony, David, Tyeeme Clark, and Joane Nagel. "White Men, Red Masks: Appropriations of 'Indian' Manhood in Imagined Wests." In *Across the Great Divide: Cultures of Manhood in the American West*, edited by Matthew Basso, Laura McCall, and Dee Garceau, 109–30. New York: Routledge, 2001.

Arluke, Arnold, and Robert Bogdan. *Beauty and the Beast: Human-Animal Relations as Revealed in Real Photo Postcards, 1905–1935*. Syracuse, NY: Syracuse University Press, 2010.

Armstrong, Carol M. *Scenes in a Library: Reading the Photograph in the Book, 1843–1875*. Cambridge, MA: MIT Press, 1998.

Armstrong, Dan, and Dustin Black. *The Book of Spam: A Most Glorious and Definitive Compendium of the World's Favorite Canned Meat*. New York: Atria Books, 2007.

Artiga, Samantha, and Matthew Rae. "The COVID-19 Outbreak and Food Production Workers: Who Is at Risk?" *KFF*, June 3, 2020. Accessed June 2, 2023. https://www.kff.org/coronavirus-covid-19/issue-brief/the-covid-19-outbreak-and-food-production-workers-who-is-at-risk/.

Ayling, R. Stephen. *Public Abattoirs: Their Planning, Design, and Equipment*. London: E. and F. N. Spon; New York: Spon and Chamberlain, 1908.

Azoulay, Ariella. *The Civil Contract of Photography*. New York: Zone Books, 2008.

B. Heller and Company. "How to Produce a Fine Red Color in Bologna and Frankforts without the Use of Artificial Coloring." *Scientific Meat Industry* 1, no. 12 (December 1905): 13.

B. Heller and Company. "Killing on the Farm." *Scientific Meat Industry* 1, no. 12 (December 1905): 40–47.

B. Heller and Company. "Schwarzschild and Sulzberger Company." *Scientific Meat Industry* 1, no. 12 (December 1905): 7–12.

B. Heller and Company. *Secrets of Meat Curing and Sausage Making*. 4th ed. Chicago: B. Heller and Company, 1918.

Baatsen, Inneke, Bruno Blondé, and Julie De Groot. "The Kitchen Between Representation and Everyday Experience: The Case of Sixteenth-Century Antwerp." *Nederlands Kunsthistorisch Jaarboek (NKJ) / Netherlands Yearbook for History of Art* 64 (2014): 162–85.

Babcock, Franklin Lawrence. *The First Fifty, 1889–1939*. Chicago: Standard Oil Company (Indiana), 1939.

Baker, Simon. "For the Camera." *Aperture*, no. 221 (Winter 2015): 25–33.

Baker, Steve. *Picturing the Beast: Animals, Identity, and Representation*. Urbana: University of Illinois Press, 2001.

Baratay, Eric, and Elisabeth Hardouin-Fugier. *Zoo: A History of Zoological Gardens in the West*. London: Reaktion, 2002.

Barnard, John Levi. "The Bison and the Cow: Food, Empire, Extinction." *American Quarterly* 72, no. 2 (July 2020): 377–401.

Barrett, James R. *Work and Community in the Jungle: Chicago's Packinghouse Workers, 1894–1922*. Urbana: University of Illinois Press, 1987.

Barthes, Roland. *Camera Lucida: Reflections on Photography*. Translated by Richard Howard. New York: Hill and Wang, 1981.

Barthes, Roland. *Image, Music, Text*. Translated by Stephen Heath. New York: Hill and Wang, 1978.

Barthes, Roland. *Mythologies*. Translated by Annette Lavers. New York: Hill and Wang, 1972.

Batchen, Geoffrey, Mick Gidley, Nancy K. Miller, and Jay Prosser, eds. *Picturing Atrocity: Photography in Crisis*. London: Reaktion, 2012.
Baudry, Jean-Louis. "Ideological Effects of the Basic Cinematographic Apparatus." Translated by Alan Williams. *Film Quarterly* 28, no. 2 (Winter 1974–1975): 39–47.
Bazin, André. "The Ontology of the Photographic Image." 1945. Reprinted in *Classic Essays on Photography*, translated by Hugh Gray, edited by Alan Trachtenberg, 237–44. New Haven, CT: Leete's Island Books, 1980. Page references are to the 1980 edition.
Beeler, Maxwell N. *Marketing Purebred Livestock*. New York: Macmillan, 1929.
Beers, Diane L. *For the Prevention of Cruelty: The History and Legacy of Animal Rights Activism in the United States*. Athens: Ohio University Press, 2006.
Bellet, Camille. "Reconfiguring the Senses: Sensor Technologies and the Production of a New Sensorium in Cattle Farming." In *The Oxford Handbook of Animal Organization Studies*, edited by Linda Tallberg and Lindsay Hamilton, 101–14. Oxford: Oxford University Press, 2022.
Bellet, Camille, and Emily Kathryn Morgan. "Breed(ing) Narratives: Visualising Values in Industrial Farming." *Animal Studies Journal* 11, no. 1 (2022): 200–255.
Berger, John. *About Looking*. New York: Pantheon, 1980.
Berger, John. "Understanding a Photograph." In *The Look of Things*, 181–85. New York: Viking Press, 1974. Reprinted in *Classic Essays on Photography*, edited by Alan Trachtenberg, 291–94. New Haven, CT: Leete's Island Books, 1980. Page references are to the 1980 edition.
Berger, John. *Why Look at Animals?* New York: Penguin, 2009.
Blanchette, Alexander. "Herding Species: Biosecurity, Posthuman Labor, and the American Industrial Pig." *Cultural Anthropology* 30, no. 4 (November 2015): 640–69.
Blanchette, Alexander. *Porkopolis: American Animality, Standardized Life, and the Factory Farm*. Durham, NC: Duke University Press, 2020.
Boggs, Colleen Glenney. *Animalia Americana: Animal Representations and Biopolitical Subjectivity*. New York: Columbia University Press, 2013.
Bogle, Donald. *Toms, Coons, Mulattoes, Mammies, and Bucks: An Interpretive History of Blacks in American Films*. 5th ed. New York: Bloomsbury Academic, 2016.
Booker, Matthew. "Before *The Jungle*: The Atlantic Origins of US Food Safety Regulation." *Global Environment* 11, no. 1 (2018): 12–35.
Boyd, William. "Making Meat: Science, Technology, and the American Poultry Industry." *Technology and Culture* 42, no. 4 (October 2001): 631–64.
Brantz, Dorothee. "Slaughter in the City: The Establishment of Public Abattoirs in Paris and Berlin, 1780–1914." PhD diss., University of Chicago, 2003.
Breckenridge, Sophonisba, and Edith Abbott. "Women in Industry: The Chicago Stockyards." *Journal of Political Economy* 19, no. 8 (October 1911): 632–54.

Brothers, Caroline. *War and Photography: A Cultural History*. London: Routledge, 1997.
Brower, Matthew. *Developing Animals: Wildlife and Early American Photography*. Minneapolis: University of Minnesota Press, 2011.
Brown, Bruce W. "Images of Family Life in Magazine Advertising, 1920–1978." PhD diss., University of New Hampshire, 1979.
Brown, Elspeth H. *The Corporate Eye: Photography and the Rationalization of American Commercial Culture*. Baltimore: Johns Hopkins University Press, 2005.
Brunet, François, ed. *Circulation*. Terra Foundation Essays, vol. 3. Paris: Terra Foundation for American Art; Chicago: University of Chicago Press, 2017.
Bugos, Glenn E. "Intellectual Property Protection in the American Chicken-Breeding Industry." *Business History Review* 66, no. 1 (Spring 1992): 127–68.
Buller, Henry. "Individuation, the Mass and Farm Animals." *Theory, Culture, and Society* 30, no. 7–8 (December 2013): 155–75.
Bushnell, Charles Joseph. *The Social Problem at the Chicago Stock Yards*. Chicago: University of Chicago Press, 1902.
Butler, Judith. *Gender Trouble: Feminism and the Subversion of Identity*. 10th anniversary ed. New York: Routledge, 1999.
Butler, Judith. "Performative Acts and Gender Constitution: An Essay in Phenomenology and Feminist Theory." *Theatre Journal* 40, no. 4 (December 1988): 519–31.
Cartmill, Matt. *A View to a Death in the Morning: Hunting and Nature Through History*. Cambridge, MA: Harvard University Press, 1993.
Clark, Kenneth. *Animals and Men: Their Relationship as Reflected in Western Art from Prehistory to the Present Day*. New York: William Morrow, 1977.
Clemen, Rudolph A. *By-Products in the Packing Industry*. Chicago: University of Chicago Press, 1927.
Cockburn, Alexander. "A Short, Meat-Oriented History of the World: From Eden to the Mattole." *New Left Review* 1, no. 215 (January/February 1996): 16–42.
Coe, Sue. *Dead Meat*. New York: Four Walls Eight Windows, 1995.
Coleman, Kevin R., Daniel James, and Jayeeta Sharma. "Introduction: Photography and Work." *Radical History Review* 132 (October 2018): 1–22.
Collinson, David L., and Jeff Hearn, eds. *Men as Managers, Managers as Men: Critical Perspectives on Men, Masculinities and Managements*. London: Sage Publications, 1996.
Commons, John R. "Labor Conditions in Meat Packing and the Recent Strike." *Quarterly Journal of Economics* 19, no. 1 (November 1904): 1–32.
Commons, John R. "Women in Unions, Meat Packing Industry." *American Federationist* 13 (May 1906): 82.
Conard, Rebecca. *Bringin' Home the Bacon: The Rath Packing Company in Waterloo, 1891–1985*. Iowa City, Iowa: Tallgrass Historians, 2010.
Conboy, Katie, Nadia Medina, and Sarah Stanbury, eds. *Writing on the Body:*

Female Embodiment and Feminist Theory. New York: Columbia University Press, 1997.
Corwin, Sharon. "Picturing Efficiency: Precisionism, Scientific Management, and the Effacement of Labor." *Representations* 84, no. 1 (November 2003): 139–65.
Craig, Kenneth M. "Pieter Aertsen and *The Meat Stall*." *Oud Holland* 96, no. 1 (1982): 1–15.
Craig, Robert L. "Designing Ethnicity: The Ideology of Images." *Design Issues* 7, no. 2 (Spring 1991): 34–42.
Crary, Jonathan. *Techniques of the Observer: On Vision and Modernity in the Nineteenth Century*. Cambridge, MA: MIT Press, 1992.
Cronin, J. Keri. *Art for Animals: Visual Culture and Animal Advocacy, 1870–1914*. University Park, PA: Penn State University Press, 2018.
Cronon, William. *Nature's Metropolis: Chicago and the Great West*. New York: W. W. Norton, 1991.
Cudworth, Erika. "'Most Farmers Prefer Blondes': The Dynamics of Anthroparchy in Animals' Becoming Meat." *Journal for Critical Animal Studies* 6, no. 1 (2008): 32–45.
Cummins, John. *The Hound and the Hawk: The Art of Medieval Hunting*. London: Weidenfeld and Nicholson, 1988.
Damisch, Hubert. *The Origin of Perspective*. Translated by John Goodman. Cambridge, MA: MIT Press, 1994.
Daston, Lorraine, and Gregg Mitman, eds. *Thinking with Animals: New Perspectives on Anthropomorphism*. New York: Columbia University, 2005.
Data USA. "Butchers and Other Meat, Poultry, and Fish Processing Workers." Deloitte/Datawheel. Accessed June 2, 2023. https://datausa.io/profile/soc/butchers-other-meat-poultry-fish-processing-workers#.
Davenport, Arthur C. *The American Live Stock Market: How It Functions*. Chicago: Drovers Journal Print, 1922.
Davenport, F. Garvin. "The Sanitation Revolution in Illinois, 1870–1900." *Journal of the Illinois State Historical Society* 66, no. 3 (October 1973): 306–26.
Davis, Janet M. *The Gospel of Kindness: Animal Welfare and the Making of Modern America*. New York: Oxford University Press, 2016.
"Death Takes Three Employees." *Morrell Magazine* 1, no. 5 (November 1924): 21.
Deleuze, Gilles, and Félix Guattari. *A Thousand Plateaus: Capitalism and Schizophrenia*. Translated by Brian Massumi. Minneapolis: University of Minnesota Press, 1987.
Deloria, Philip J. *Playing Indian*. New Haven, CT: Yale University Press, 1998.
Derifield, Coreen. "'I Thought of the Money That We Could Use': Iowa Women and Industrial Wage Work, 1950–1970." *Annals of Iowa* 73, no. 1 (Winter 2014): 28–59.
Derrida, Jacques. "The Animal That Therefore I Am (More to Follow)." Translated by David Wills. *Critical Inquiry* 28, no. 2 (Winter 2002): 369–418.

Derrida, Jacques. *Archive Fever: A Freudian Impression*. Translated by Eric Prenowitz. Chicago: University of Chicago Press, 1996.

Deslippe, Dennis A. " 'We Had an Awful Time with Our Women': Iowa's United Packinghouse Workers of America, 1945–75." *Journal of Women's History* 5, no. 1 (1993): 10–32.

Dickinson, Darol. *Photographing Livestock: The Complete Guide*. Flagstaff, AZ: Northland Press, 1980.

Diderot, Denis, and Jean le Rond d'Alembert. *Encyclopédie, ou, Dictionnaire raisonné des sciences, des arts et des métiers*. Paris: Chez Briasson, 1751. Accessed April 1, 2022. https://archive.org/details/EncyclopeYdieouII Dide/page/350/mode/2up.

Donovan, Josephine, and Carol J. Adams, eds. *Beyond Animal Rights: A Feminist Ethic for the Treatment of Animals*. New York: Continuum, 1996.

Doss, Erika, ed. *Looking at "Life" Magazine*. Washington, DC: Smithsonian Institution Press, 2001.

Dougherty, Richard. *In Quest of Quality: Hormel's First 75 Years*. Austin, MN: Hormel, 1966.

Douglas, Mary. *Purity and Danger: An Analysis of Concepts of Pollution and Taboo*. London: Routledge and Kegan Paul, 1966. Reprinted as *Mary Douglas Collected Works*, vol. 2. London: Routledge, 2003. Page references are to the 2003 edition.

Dreilinger, Danielle. *The Secret History of Home Economics: How Trailblazing Women Harnessed the Power of Home and Changed the Way We Live*. New York: W. W. Norton, 2021.

Dudziak, Mary L. "The Numbers: Encountering Casualties in the Era of COVID-19." *Diplomatic History* 45, no. 3 (June 2021): 489–97.

Duffy, John. *The Sanitarians: A History of American Public Health*. Urbana: University of Illinois Press, 1990.

Durden, Mark. "Connections and Conflicts: Margaret Bourke-White's Corporate, Commercial, and Documentary Photography." In *Corporate Patronage of Art and Architecture in the United States, Late 19th Century to the Present*, edited by Monica E. Jovanovich and Melissa Renn, 63–76. New York: Bloomsbury Visual Arts, 2019.

Edwards, Elizabeth. "Tracing Photography." In *Made to Be Seen: Perspectives on the History of Visual Anthropology*, edited by Marcus Banks and Jay Ruby, 159–89. Chicago: University of Chicago Press, 2011.

Eisnitz, Gail A. *Slaughterhouse: The Shocking Story of Greed, Neglect, and Inhumane Treatment inside the U.S. Meat Industry*. Amherst, NY: Prometheus Books, 1997.

Eldridge, Mary Beth. *The Rath Packing Company Strike of 1948*. Waterloo, IA: printed by the author, 1990.

"Emergence of Advertising in America, 1850–1920." John W. Hartman Center

for Sales, Advertising & Marketing History, David M. Rubenstein Rare Book and Manuscript Library, Duke University. https://repository.duke.edu/dc/eaa.
Estes, Adam Clark. "America's Meat Shortage Is More Serious than Your Missing Hamburgers." *Vox*, May 8, 2020. Accessed May 4, 2023. https://www.vox.com/recode/2020/5/8/21248618/coronavirus-meat-shortage-food-supply-chain-grocery-stores.
Ewart, John. *Meat Production: A Manual for Producers, Distributors, and Consumers of Butchers' Meat* [. . .]. London: Crosby Lockwood, 1878.
Executive Order No. 13917. "Delegating Authority under the Defense Production Act with Respect to Food Supply Chain Resources during the National Emergency Caused by the Outbreak of COVID-19." 85 Fed. Reg. 26313 (May 1, 2020). Accessed May 3, 2023. https://www.govinfo.gov/content/pkg/FR-2020-05-01/pdf/2020-09536.pdf.
Fantasia, Rick. *Cultures of Solidarity: Consciousness, Action, and Contemporary American Workers*. Berkeley: University of California Press, 1988.
Federal Meat Inspection Act of 1906. U.S. Department of Agriculture: Food Safety and Inspection Service. https://www.fsis.usda.gov/policy/food-safety-acts/federal-meat-inspection-act.
Fehn, Bruce R. " 'Chickens Come Home to Roost': Industrial Reorganization, Seniority, and Gender Conflict in the United Packinghouse Workers of America, 1956–1966." *Labor History* 34, no. 2–3 (June 1993): 324–41.
Fehn, Bruce R. " 'The Only Hope We Had': United Packinghouse Workers Local 46 and the Struggle for Racial Equality in Waterloo, Iowa, 1948–1960." *Annals of Iowa* 54, no. 3 (Summer 1995): 185–216.
Fehn, Bruce R. "Ruin or Renewal: The United Packinghouse Workers of America and the 1948 Meatpacking Strike in Iowa." *Annals of Iowa* 56, no. 4 (Fall 1997): 349–78.
Fehn, Bruce R. "Striking Women: Gender, Race and Class in the United Packinghouse Workers of America (UPWA), 1938–1968." PhD diss., University of Wisconsin-Madison, 1991.
Fiddes, Nick. *Meat: A Natural Symbol*. London: Routledge, 1991.
Fink, Deborah. *Cutting into the Meatpacking Line: Workers and Change in the Rural Midwest*. Chapel Hill: University of North Carolina Press, 1998.
Foucault, Michel. *Discipline and Punish: The Birth of the Prison*. Translated by Alan Sheridan. New York: Vintage Books, 1995.
Foucault, Michel. *The History of Sexuality*. Vol. 1, *An Introduction*. Translated by Robert Hurley. New York: Vintage Books, 1980.
"Four Veterans Retired at Ottumwa." *Morrell Magazine* 1, no. 1 (July 1924): 4.
Fraiman, Susan. "Pussy Panic Versus Liking Animals: Tracking Gender in Animal Studies." *Critical Inquiry* 39, no. 1 (Autumn 2012): 89–115.
Franju, Georges, dir. *Blood of the Beasts*. Criterion Collection, 1949. Posted

October 20, 2021, under the title *Le sang des betes Georges Franju, 1949*, anyondra, YouTube, 22 min. Accessed June 5, 2022. https://www.youtube.com/watch?v=q0qE59rk3ic.

Franks, Marcy, and Temple Grandin. "Leg Conformation Is Too Important to Hide on Bull Photos." *Beef Magazine*, January 7, 2015. Accessed November 22, 2021. https://www.beefmagazine.com/cattle-genetics/leg-conformation-too-important-hide-bull-photos.

Fraser, Max. "Hands Off the Machine: Workers' Hands and Revolutionary Symbolism in the Visual Culture of 1930s America." *American Art* 27, no. 2 (Summer 2013): 94–117.

Fudge, Erica. *Brutal Reasoning: Animals, Rationality, and Humanity in Early Modern England*. Ithaca, NY: Cornell University Press, 2006.

Fudge, Erica. *Perceiving Animals: Humans and Beasts in Early Modern English Culture*. New York: St. Martin's, 2000.

Fulton, Gwynne. "*Phantasmatics*: Sovereignty and the Image of Death in Derrida's First *Death Penalty* Seminar." *Mosaic: An Interdisciplinary Critical Journal* 48, no. 3 (September 2015): 75–94.

Gallman, Robert E. "Pork Production and Nutrition during the Late Nineteenth Century: A Weighty Issue Visited Yet Again." *Agricultural History* 69, no. 4 (October 1995): 592–606.

Garval, Michael D. "Visions of Pork Production, Past and Future, on French Belle Époque Pig Postcards." *Nineteenth-Century Art Worldwide* 14, no. 1 (Spring 2015). Accessed May 18, 2015. http://www.19thc-artworldwide.org/index.php/spring15/garval-on-visions-of-pork-production-past-and-future-french-belle-epoque-postcards.

Geib, Paul E. "'Everything but the Squeal': The Milwaukee Stockyards and Meat-Packing Industry, 1840–1930." *Wisconsin Magazine of History* 78, no. 1 (October 1994): 2–23.

Genoways, Ted. *The Chain: Farm, Factory, and the Fate of Our Food*. New York: HarperCollins, 2014.

Germain, Rosie. "Reading *The Second Sex* in 1950s America." *Historical Journal* 56, no. 4 (December 2013): 1041–62.

Gerrard, Frank. *Meat Technology: A Practical Textbook for Student and Butcher*. London: Northwood, 1977.

Giedion, Siegfried. *Mechanization Takes Command: A Contribution to Anonymous History*. New York: Oxford University Press, 1948.

Gifford, Thomas. *The Dubuque Packing Company, 1965–1988*. Dubuque, IA: River Lights, 1997.

"Girls Help Solve Labor Shortage in Packing Industry." *Ottumwa Courier*, April 23, 1918, 7.

Godley, Andrew. "The Emergence of Agribusiness in Europe and the Development of the Western European Broiler Chicken Industry, 1945 to 1973." *Agricultural History Review* 62, no. 2 (January 2014): 315–36.

Goffman, Erving. *Gender Advertisements*. Cambridge, MA: Harvard University Press, 1979.

Goldberg, Eric J. "Louis the Pious and the Hunt." *Speculum* 88, no. 3 (July 2013): 613–43.

Goldberg, Vicki, ed. *Photography in Print: Writings from 1816 to the Present*. New York: Simon and Schuster, 1981.

Goldstein, Carolyn M. *Creating Consumers: Home Economists in Twentieth-Century America*. Chapel Hill: University of North Carolina Press, 2012.

Gombrich, Ernst H. *Art and Illusion: A Study in the Psychology of Pictorial Representation*. 4th ed. London: Phaidon Press, 1972.

Gordon, Steve C. "From Slaughterhouse to Soap-Boiler: Cincinnati's Meat Packing Industry, Changing Technologies, and the Rise of Mass Production, 1825–1870." *IA* 16, no. 1 (1990): 55–67.

Grabell, Michael, and Bernice Yeung. "Emails Show the Meatpacking Industry Drafted an Executive Order to Keep Plants Open." *ProPublica*, September 14, 2020. Accessed May 3, 2023. https://www.propublica.org/article/emails-show-the-meatpacking-industry-drafted-an-executive-order-to-keep-plants-open.

Grand, W. Joseph. *Illustrated History of the Union Stockyards*. Chicago: Thomas Knapp Printing and Binding, 1896.

Gray, David A. "Managing Motivation: The Seth Seiders Syndicate and the Motivational Publicity Business in the 1920s." *Winterthur Portfolio* 44, no. 1 (Spring 2010): 77–122.

Gregory, Neville, ed. *Animal Welfare and Meat Science*. New York: CABI, 1998.

Grey, Mark A. "Turning the Pork Industry Upside Down: Storm Lake's Hygrade Work Force and the Impact of the 1981 Plant Closure." *Annals of Iowa* 54, no. 3 (Summer 1995): 244–59.

Grodkowski, Grzegorz, Tomasz Sakowski, Kamila Puppel, and Ton Baars. "Comparison of Different Applications of Automatic Herd Control Systems on Dairy Farms—a Review." *Journal of the Science of Food and Agriculture* 98, no. 14 (November 2018): 5181–88.

Guimond, James. *American Photography and the American Dream*. Chapel Hill: University of North Carolina Press, 1991.

Hajdik, Anna Thompson. "A 'Bovine Glamour Girl': Borden Milk, Elsie the Cow, and the Convergence of Technology, Animals, and Gender at the 1939 New York World's Fair." *Agricultural History* 88, no. 4 (Fall 2014): 470–90.

Haladyn, Julian Jason. "Friedrich's 'Wanderer': Paradox of the Modern Subject." *RACAR: Revue d'Art Canadienne / Canadian Art Review* 41, no. 1 (January 2016): 47–61.

Halpern, Rick. "The Back of the Photograph: Making Meaning in the Archives." *Radical History Review* 2018, no. 132 (October 2018): 187–99.

Halpern, Rick. *Down on the Killing Floor: Black and White Workers in Chicago's Packinghouses, 1904–54*. Urbana: University of Illinois Press, 1997.

Halpern, Rick, and Roger Horowitz. *Meatpackers: An Oral History of Black Packinghouse Workers and Their Struggle for Racial and Economic Equality*. New York: Twayne Publishers, 1996.

Hamilton, Robert. "Image and Context: The Production and Reproduction of *The Execution of a VC Suspect* by Eddie Adams." In *Vietnam Images: War and Representation*, edited by Jeffrey Walsh and James Aulich, 171–83. London: Macmillan, 1989.

Haraway, Donna J. *Modest-Witness@Second-Millenium.FemaleMan-Meets-OncoMouse: Feminism and Technoscience*. London: Routledge, 1997.

Haraway, Donna J. *Primate Visions: Gender, Race, and Nature in the World of Modern Science*. New York: Routledge, 1989.

Haraway, Donna J. *Simians, Cyborgs, and Women: The Reinvention of Nature*. New York: Routledge, 1991.

Haraway, Donna J. *When Species Meet*. Minneapolis: University of Minnesota Press, 2008.

Harrison, Ruth. *Animal Machines*. London: CABI, 1964.

Hart, Stephen. "The Cultural Dimension of Social Movements: A Theoretical Reassessment and Literature Review." In "Sociology of Culture and Sociology of Religion." Special issue, *Sociology of Religion* 57, no. 1 (Spring 1996): 87–100.

Heinzerling, Lisa. "The Varieties and Limits of Transparency in U.S. Food Law." *Food and Drug Law Journal* 70, no. 1 (2015): 11–24.

Hemsworth, Paul H. "Human-Animal Interactions in Livestock Production." *Applied Animal Behaviour Science* 81, no. 3 (May 2003): 185–98.

Hewitt, William L. " 'So Few Undesirables': Race, Residence, and Occupation in Sioux City, 1890–1925." *Annals of Iowa* 50, no. 2 (Fall 1989): 158–79.

Hine, Lewis W. *Men at Work: Photographic Studies of Modern Men and Machines*. New York: Macmillan, 1932.

Hirsch, Robert. *Seizing the Light: A Social and Aesthetic History of Photography*. New York: Routledge, 2017.

Hoganson, Kristin. "Meat in the Middle: Converging Borderlands in the U.S. Midwest, 1865–1900." *Journal of American History* 98, no. 4 (March 2012): 1025–51.

Hoganson, Kristin. "SHGAPE Presidential Address: Mind the GAPE; Globality and the Rural Midwest." *Journal of the Gilded Age and Progressive Era* 19, no. 2 (2020): 176–90.

Holloway, Lewis, Carol Morris, Ben Gilna, and David Gibbs. "Biopower, Genetics and Livestock Breeding: (Re)Constituting Animal Populations and Heterogeneous Biosocial Collectivities." *Transactions of the Institute of British Geographers* 34, no. 3 (July 2009): 394–407.

Homer. *The Odyssey*. Translated by Peter Green. Berkeley: University of California Press, 2018.

Honig, Elizabeth Alice. "Making Sense of Things: On the Motives of Dutch Still Life." *RES: Anthropology and Aesthetics*, no. 34 (Autumn 1998): 166–83.

Horowitz, Roger. " 'It Wasn't a Time to Compromise': The Unionization of Sioux City's Packinghouses, 1937–1942." *Annals of Iowa* 50, no. 2–3 (Fall–Winter 1989): 241–68.

Horowitz, Roger. *"Negro and White, Unite and Fight!" A Social History of Industrial Unionism in Meatpacking, 1930–90*. Urbana: University of Illinois Press, 1998.

Horowitz, Roger. *Putting Meat on the American Table: Taste, Technology, Transformation*. Baltimore: Johns Hopkins University Press, 2006.

Horowitz, Roger. " 'That Was a Dirty Job!': Technology and Workplace Hazards in Meatpacking over the Long Twentieth Century." *Labor* 5, no. 2 (Summer 2008): 13–25.

Horowitz, Roger. " 'Where Men Will Not Work': Gender, Power, Space, and the Sexual Division of Labor in America's Meatpacking Industry, 1890–1990." In "Gender Analysis and the History of Technology." Special issue, *Technology and Culture* 38, no. 1 (January 1997): 187–213.

Horwitz, Richard P. *Hog Ties: Pigs, Manure, and Mortality in American Culture*. New York: St. Martin's, 1998.

Houghton, Charlotte. "This Was Tomorrow: Pieter Aertsen's *Meat Stall* as Contemporary Art." *Art Bulletin* 86, no. 2 (June 2004): 277–300.

Hribal, Jason. " 'Animals Are Part of the Working Class': A Challenge to Labor History." *Labor History* 44, no. 4 (November 2003): 435–53.

Hribal, Jason. *Fear of the Animal Planet: The Hidden History of Animal Resistance*. Oakland, CA: AK Press, 2010.

Hudson, John C. *Making the Corn Belt: A Geographical History of Middle-Western Agriculture*. Bloomington: Indiana University Press, 1994.

Humane Methods of Slaughter Act of 1958. 7 U.S.C. § 1901 et seq. (1958). https://www.govinfo.gov/content/pkg/USCODE-2015-title7/html/USCODE-2015-title7-chap48.htm.

Hurford, O. P. Hog hoist. US Patent 541,278, filed March 30, 1895, and issued June 18, 1895. United States Patent and Trademark Office. Accessed May 24, 2022. https://image-ppubs.uspto.gov/dirsearch-public/print/downloadPdf/0541278.

Igler, David. *Industrial Cowboys: Miller & Lux and the Transformation of the Far West, 1850–1920*. Berkeley: University of California Press, 2001.

Ingraham, Catherine. *Architecture, Animal, Human: The Asymmetrical Condition*. London: Routledge, 2006.

Jennings, Michael. "Agriculture, Industry, and the Birth of the Photo-Essay in the Late Weimar Republic." *October* 93 (Summer 2000): 23–56.

John Morrell and Company. *Your Trip Through Morrell's*. Ottumwa, IA: John Morrell and Company, n.d., ca. 1930.

Johnson, Keach. "The Corn Belt Meat Producers' Association of Iowa: Origins of a Progressive Pressure Group." *Annals of Iowa* 43, no. 4 (Spring 1976): 242–59.

Johnson, Robert. *The Art of Retouching Negatives and Finishing and Colouring Photographs*. London: Marion and Co., n.d., ca. 1900. 9th ed. Revised with additions by T. S. Bruce and Alfred Braithwaite. London: Marion and Co., 1918. Page references are to the 1918 edition.

Jones, Eleanor Lewis. "Deer Drinking: Reflections on a Watercolor by Winslow Homer." *Smithsonian Studies in American Art* 2, no. 3 (Autumn 1988): 55–65.

Kalof, Linda. *Looking at Animals in Human History*. London: Reaktion, 2007.

Kalof, Linda, and Amy Fitzgerald, eds. *The Animals Reader: The Essential Classic and Contemporary Writings*. New York: Berg, 2007.

Kalof, Linda, and Georgina M. Montgomery, eds. *Making Animal Meaning*. East Lansing: Michigan State University Press, 2011.

Keefe, Beatrice Radden. "Surveying Damage in the Walters 'Rose' (W.143)." *Journal of the Walters Art Museum* 68/69 (2010/2011): 97–106.

Kellogg, John Harvey. *Shall We Slay to Eat?* Battle Creek, MI: Good Health, 1899.

King, Caroline B. *Mr. Ham Goes to Town*. Ottumwa, IA: John Morrell and Company, 1939.

Klein, Naomi. *No Logo: Taking Aim at the Brand Bullies*. New York: Picador, 1999.

Klingender, Francis Donald. *Animals in Art and Thought to the End of the Middle Ages*. Cambridge, MA: MIT Press, 1971.

Kniffen, Fred. "The American Agricultural Fair: The Pattern." *Annals of the Association of American Geographers* 39, no. 4 (December 1949): 264–82.

Korling, Torkel, and Robert F. Betz. *The Prairie: Swell and Swale; From Nature*. Dundee, IL: printed by Torkel Korling, 1972.

Korling, Torkel, and Robert O. Petty. *Eastern Deciduous Forest: From Nature*. Evanston, IL: printed by Torkel Korling, 1977.

Korling, Torkel, and Edward G. Voss. *The Boreal Forest and Borders: From Nature*. Dundee, IL: printed by Torkel Korling, 1973.

Koslow, Susan. "Law and Order in Rubens's *Wolf and Fox Hunt*." *Art Bulletin* 78, no. 4 (December 1996): 680–706.

Kracauer, Siegfried. "Photography." 1927. Reprinted with translation by Thomas Y. Levin in *Critical Inquiry* 19, no. 3 (Spring 1993): 421–36. Page references are to the 1993 edition.

Kramer, Paul. "Empires, Exceptions, and Anglo-Saxons: Race and Rule between the British and United States Empires, 1880–1910." *Journal of American History* 88, no. 4 (March 2002): 1315–53.

Lambert, George. *A Trip through the Union Stock Yards and Slaughter Houses, Chicago, U.S.A.* Chicago, n.d., ca. 1890.

Landes, Joan B., Paula Young Lee, and Paul Youngquist, eds. *Gorgeous Beasts: Animal Bodies in Historical Perspective*. University Park: Pennsylvania State University Press, 2012.

Lauck, Jon K. "Competition in the Grain Belt Meatpacking Sector after World War II." *Annals of Iowa* 57, no. 2 (Spring 1998): 135–59.

Lazonick, William. "Organizational Capabilities in American Industry: The Rise

and Decline of Managerial Capitalism." *Business and Economic History* 19 (1990): 35–54.
Lears, T. J. Jackson. *Fables of Abundance: A Cultural History of Advertising in America*. New York: Basic Books, 1994.
Lee, Paula Young, ed. *Meat, Modernity, and the Rise of the Slaughterhouse*. Durham: University of New Hampshire Press, 2008.
Leslie, Thomas. "Chicago's Other Skyscrapers: Grain Elevators and the City, 1838–1957." *Journal of Urban History* 48, no. 1 (January 2022): 3–34.
Lewis, Sarah Elizabeth. "Where Are the Photos of People Dying of COVID?" *New York Times*, May 1, 2020. Accessed May 5, 2023. https://www.nytimes.com/2020/05/01/opinion/coronavirus-photography.html.
Leys, Tony. "Coronavirus Infects More Than 1,600 Workers at Four Iowa Meatpacking Plants." *Des Moines Register*, May 5, 2020. Accessed June 2, 2023. https://www.desmoinesregister.com/story/news/health/2020/05/05/coronavirus-infects-thousands-iowa-meatpacking-plant-workers-covid-19-waterloo-perry/5170796002/.
Libby's Atlas of the World. Chicago: Libby, McNeill, and Libby, n.d., ca. 1900.
Lindstrom, Richard. "'They All Believe They Are Undiscovered Mary Pickfords': Workers, Photography, and Scientific Management." *Technology and Culture* 41, no. 4 (October 2000): 725–51.
Lippit, Akira Mizuta. *Electric Animal: Toward a Rhetoric of Wildlife*. Minneapolis: University of Minnesota Press, 2000.
"Local 46 at Raths [*sic*] Elects Officers." *Unionist* (Austin, MN) 6, no. 19 (February 21, 1941): 1.
Lopez, Patricia J., and Kathryn A. Gillespie, eds. *Economies of Death: Economic Logics of Killable Life and Grievable Death*. London: Routledge, 2015.
Lucic, Karen. "Charles Sheeler and Henry Ford: A Craft Heritage for the Machine Age." *Bulletin of the Detroit Institute of Arts* 65, no. 1 (1989): 36–47.
Lucic, Karen. *Charles Sheeler and the Cult of the Machine*. London: Reaktion, 1991.
Lusty, Natalya. "Eli Lotar's Para-urban Visions." In *Photography and Ontology: Unsettling Images*, edited by Donna West Brett and Natalya Lusty, 87–103. New York: Routledge, 2019.
Malamud, Randy. *An Introduction to Animals and Visual Culture*. London: Palgrave Macmillan, 2012.
Marchand, Roland. *Advertising the American Dream: Making Way for Modernity, 1920–1940*. Berkeley: University of California Press, 1986.
Marchand, Roland. *Creating the Corporate Soul: The Rise of Public Relations and Corporate Imagery in American Big Business*. Berkeley: University of California Press, 1990.
Marling, Karal Ann. *As Seen on TV: The Visual Culture of Everyday Life in the 1950s*. Cambridge, MA: Harvard University Press, 1994.
Marshall, David, Teresa Davis, Margaret K. Hogg, Tanja Schneider, and Alan Petersen. "From Overt Provider to Invisible Presence: Discursive Shifts

in Advertising Portrayals of the Father in *Good Housekeeping*, 1950–2010." *Journal of Marketing Management* 30, no. 15–16 (November 2014): 1654–79.
Matthews, Kristin L. "One Nation over Coals: Cold War Nationalism and the Barbecue." *American Studies* 50, no. 3/4 (Fall/Winter 2009): 5–34.
May, Elaine Tyler. *Homeward Bound: American Families in the Cold War Era.* Rev. ed. New York: Basic Books, 2008.
Mayer, Oscar G., Jr. *Oscar Mayer & Company: From Corner Store to National Processor.* New York: Newcomen Society in North America, 1970.
McCarthy, Ryan, and Sam Danley. "Map: COVID-19 Meat Plant Closures." *Meat and Poultry*, June 23, 2020; last updated November 2020. Accessed May 4, 2023. https://www.meatpoultry.com/articles/22993-covid-19-meat-plant-map.
McClintock, Anne. *Imperial Leather: Race, Gender, and Sexuality in the Colonial Contest.* New York: Routledge, 1995.
McCoy, Liza. "Activating the Photographic Text." In *Knowledge, Experience, and Ruling Relations: Studies in the Social Organization of Knowledge*, edited by Marie L. Campbell and Ann Manicom, 181–92. Toronto: University of Toronto Press, 1995.
McGaw, Judith A. "Accounting for Innovation: Technological Change and Business Practice in the Berkshire County Paper Industry." *Technology and Culture* 26, no. 4 (October 1985): 703–25.
McHugh, Susan. *Animal Stories: Narrating across Species Lines.* Minneapolis: University of Minnesota Press, 2011.
McLean, Beth Bailey. *The Young Woman in Business.* Ames: Iowa State University Press, 1953.
McLean, Beth Bailey, and Jeanne Paris. *The Young Woman in Business.* 2nd ed. Ames: Iowa State University Press, 1962.
Mehaffy, Marilyn Maness. "Advertising Race/Raceing Advertising: The Feminine Consumer(-Nation), 1876–1900." *Signs* 23, no. 1 (Autumn 1997): 131–74.
Melosi, Martin V. *The Sanitary City: Urban Infrastructure in America from Colonial Times to the Present.* Baltimore: Johns Hopkins University Press, 2000.
Mendiola, Mojo. "Jo Banks: A Blues outside All Patterns." October 2006. Accessed August 24, 2021. http://www.steppeulvene.com/index.jo_banks_story_english.html.
Merchant, Carolyn. *The Death of Nature: Women, Ecology, and the Scientific Revolution.* New York: Harper and Row, 1980.
Midkiff, Ken. *The Meat You Eat: How Corporate Farming Has Endangered America's Food Supply.* New York: St. Martin's, 2004.
Miller, Bonnie M. "The Pure Food Exhibits in the 'Palace of Nibbling Arts': Culinary Pluralism at the Panama-Pacific International Exposition, San Francisco, 1915." In "Edible Matters: The Material and Visual Culture of Food in California, 1865–Present." Special issue, *Southern California Quarterly* 100, no. 2 (Summer 2018): 150–82.

Mitchell, W. J. T. *Cloning Terror: The War of Images, 9/11 to the Present*. Chicago: University of Chicago Press, 2011.

Mitchell, W. J. T. *What Do Pictures Want? The Lives and Loves of Images*. Chicago: University of Chicago Press, 2005.

Mizelle, Brett. "'I Have Brought My Pig to a Fine Market': Animals, Their Exhibitors, and Market Culture in the Early Republic." In *Cultural Change and the Market Revolution in America, 1789–1860*, edited by Scott C. Martin, 181–216. Lanham, MD: Rowman and Littlefield, 2005.

Mizelle, Brett. *Pig*. London: Reaktion, 2011.

Molloy, Claire. *Popular Media and Animals*. Houndmills, UK: Palgrave Macmillan, 2011.

Morgan, Emily Kathryn. "Striking Images: Photographs of Iowa Packinghouse Labor Conflict, 1948–1960." *Annals of Iowa* 77, no. 2 (Spring 2018): 151–89.

Morgan, Emily Kathryn. "'Those Truly Augean Stables': Photo-Textual Visions of Chicago's Packingtown." *Food and History* 16, no. 2 (2018): 107–34.

Morgan, Emily Kathryn. "*Waterloo Packer*: Selling Slaughter." *History of Photography* 42, no. 2 (May 2018): 128–45.

Moriarty, Catherine. "A Backroom Service? The Photographic Library of the Council of Industrial Design, 1945–1965." *Journal of Design History* 13, no. 1 (2000): 39–57.

Morin, Karen M. *Carceral Space, Prisoners and Animals*. Routledge Human-Animal Studies Series. London: Routledge, 2018.

Morse, Deborah Denenholz, and Martin A. Danahay, eds. *Victorian Animal Dreams: Representations of Animals in Victorian Literature and Culture*. Hampshire, UK: Ashgate, 2007.

Moxey, Keith. "Interpreting Pieter Aertsen: The Problem of 'Hidden Symbolism.'" *Nederlands Kunsthistorisch Jaarboek (NKJ) / Netherlands Yearbook for History of Art* 40 (1989): 29–39.

Muhm, Don. *Iowa Pork and People: A History of Iowa's Pork Producers*. Clive, IA: Iowa Pork Foundation, 1995.

Nadeau, Luis. *Encyclopedia of Printing, Photographic, and Photomechanical Processes*. Rev. ed. Fredericton, Canada: Atelier Luis Nadeau, 1997.

"New Beef Building, Sioux Falls Plant." *Morrell Magazine* 1, no. 8 (February 1925): 3–5.

"New Frank Lloyd Wright Building Shows Shape of Things to Come." *Life* 6, no. 19 (May 8, 1939): 15–17.

Newhall, Beaumont. *The History of Photography: From 1839 to the Present Day*. 4th ed. New York: Museum of Modern Art, 1978.

Nickel, Douglas R. *Francis Frith in Egypt and Palestine: A Victorian Photographer Abroad*. Princeton, NJ: Princeton University Press, 2004.

Nickel, Rod. "As Meat Plant Infections Rise, Canada Lets Packers Choose When to Close." Reuters, May 1, 2020. https://www.reuters.com/article/us-health

-coronavirus-canada-meatpacking/as-meat-plant-infections-rise-canada-lets -packers-choose-when-to-close-idUSKBN22D54F.

Noble, David F. *America by Design: Science, Technology, and the Rise of Corporate Capitalism*. New York: Alfred A. Knopf, 1977.

Nordby, Julius E., W. Malcolm Beeson, and David L. Fourt. *Livestock Judging Handbook*. 9th ed. Danville, IL: Interstate Publishers, 1962.

Nosowitz, Dan. "Meat Production Has Rebounded, but So Have Coronavirus Cases at Plants." *Modern Farmer*, June 12, 2020. https://modernfarmer .com/2020/06/meat-production-has-rebounded-but-so-have-coronavirus -cases-at-plants/.

Nye, David E. *Image Worlds: Corporate Identities at General Electric*. Cambridge, MA: MIT Press, 1985.

O'Brien, John. *Through the Chicago Stock Yards: A Handy Guide to the Great Packing Industry*. Chicago: Rand McNally, 1907.

Ogle, Maureen. *In Meat We Trust: An Unexpected History of Carnivore America*. Boston: Houghton Mifflin Harcourt, 2013.

Oliver, Myrna. "George Molchan, 82; Toured the Nation as 'Little Oscar.' " *Los Angeles Times*, May 2, 2005. https://www.latimes.com/archives/la-xpm -2005-may-02-me-molchan2-story.html.

Ollmann, Leah. *Camera as Weapon: Worker Photography between the Wars*. San Diego: Museum of Photographic Arts, 1991.

O'Rourke, John M. *The Cudahy Packing Company: A History, 1890–1924*. Omaha: Cudahy Packing, 1925.

Oscar Mayer & Co. *Sixty Years of Service, 1883–1943*. Chicago: Oscar Mayer & Co., 1943.

Oscar Mayer & Co. *A Tradition and a Company: Oscar Mayer & Co*. Chicago: Oscar Mayer & Co., 1958.

Owen, David. "Here's Looking at You." *New Yorker* 94, no. 41 (December 17, 2018): 28–33.

Pachirat, Timothy. *Every Twelve Seconds: Industrialized Slaughter and the Politics of Sight*. New Haven, CT: Yale University Press, 2011.

Pacyga, Dominic A. *Slaughterhouse: Chicago's Union Stock Yard and the World It Made*. Chicago: University of Chicago Press, 2015.

Panofsky, Erwin. *Perspective as Symbolic Form*. Translated by Christopher S. Wood. New York: Zone Books, 1997.

Parker, Paul. "The Modern Style in American Advertising Art." *Parnassus* 9, no. 4 (April 1937): 20–23.

Parr, Martin, and Gerry Badger. *The Photobook: A History*. London: Phaidon, 2004.

Peggs, Kay. *Animals and Sociology*. Houndmills, UK: Palgrave Macmillan, 2012.

People of the Pack: Stories of the Dubuque Packing Company. With interviews and preface by Will Hoyer. Dubuque, IA: TH Media, 2014.

Perren, Richard. *Taste, Trade and Technology: The Development of the International Meat Industry since 1840*. Aldershot, UK: Ashgate, 2006.

Perry, Charles R., and Delwyn H. Kegley. *Disintegration and Change: Labor Relations in the Meat Packing Industry*. Philadelphia: Industrial Research Unit, Wharton School, University of Pennsylvania, 1989.

PETA. "5 Videos That Will Change Your Life Forever." PETA, August 29, 2013. Accessed June 1, 2023. https://www.peta.org/living/food/factory-farm-video/.

Peterson, Larry. "Producing Visual Traditions among Workers: The Uses of Photography at Pullman." *International Labor and Working-Class History*, no. 42 (Fall 1992): 40–69.

Peterson, Larry. "Visual Culture and Working-Class Community: Photography and the Organizing of the Steelworkers' Union in Chicago." In *Looking for America: The Visual Production of Nation and People*, edited by Ardis Cameron, 212–37. Malden, MA: Blackwell, 2005.

Petrick, Gabriella M. "An Ambivalent Diet: The Industrialization of Canning." *OAH Magazine of History* 24, no. 3 (July 2010): 35–38.

Philo, Chris, and Chris Wilbert, eds. *Animal Spaces, Beastly Places: New Geographies of Human-Animal Relations*. New York: Routledge, 2000.

Pidgeon, Mary Elizabeth. *The Employment of Women in Slaughtering and Meat Packing*. Bulletin of the Women's Bureau, No. 88. Washington, DC: US Department of Labor, 1932.

Pollan, Michael. *The Omnivore's Dilemma: A Natural History of Four Meals*. New York: Penguin, 2006.

Pollard, Frances S. "Exhibit Notes: 'Dressed for Work: Women in the Work Force, 1900–1989.'" *History News* 45, no. 2 (March/April 1990): 22–23.

Pollay, Richard W. "The Subsidizing Sizzle: A Descriptive History of Print Advertising, 1900–1980." *Journal of Marketing* 49, no. 3 (Summer 1985): 24–37.

Portelli, Alessandro. *The Death of Luigi Trastulli and Other Stories: Form and Meaning in Oral History*. Albany: State University of New York Press, 1991.

Punzalan, Ricardo L. "Archival Diasporas: A Framework for Understanding the Complexities and Challenges of Dispersed Photographic Collections." *American Archivist* 77, no. 2 (Fall/Winter 2014): 326–49.

Quirke, Carol. *Eyes on Labor: News Photography and America's Working Class*. Oxford: Oxford University Press, 2012.

Quirke, Carol. "Imagining Racial Equality: Local 65's Union Photographers, Postwar Civil Rights, and the Power of the Real, 1940–1955." *Radical History Review* 2018, no. 132 (October 1, 2018): 96–125.

"Rath Campaign Has Workers Joining PWOC." *Unionist* (Austin, MN) 6, no. 17 (February 7, 1941): 1.

"Rath Management Meets with PWOC Representatives." *Unionist* (Austin, MN) 6, no. 23 (March 21, 1941): 1.

"Rath Workers at Waterloo Sign Up." *Unionist* (Austin, MN) 6, no. 15 (January 24, 1941): 1, 6.

Register, Cheri. *Packinghouse Daughter: A Memoir*. St. Paul: Minnesota Historical Society, 2000.

Rhoades, E. L. *Studies in the Packing Industry*. Chicago: University of Chicago, 1930.

Ribalta, Jorge, ed. *The Worker Photography Movement (1926–1939): Essays and Documents*. Madrid: Museo Nacional Centro de Arte Reina Sofia, 2011.

Ritvo, Harriet. *The Animal Estate: The English and Other Creatures in the Victorian Age*. Cambridge, MA: Harvard University Press, 1987.

Rivard, Betty. *New Deal Photographs of West Virginia, 1934–1943*. Morgantown: West Virginia University Press, 2012.

Robinson, Bryan K., and Erica Hunter. "Is Mom Still Doing It All? Reexamining Depictions of Family Work in Popular Advertising." *Journal of Family Issues* 29, no. 4 (April 2008): 465–86.

Robinson, Cyril. *Marching with Dr. King: Ralph Helstein and the United Packinghouse Workers of America*. Santa Barbara, CA: Praeger, 2011.

Rockefeller, Stuart Alexander. "Flow." *Current Anthropology* 52, no. 4 (August 2011): 557–78.

Romans, John R., William J. Costello, Wendell C. Carlson, Marion L. Greaser, and Kevin W. Jones. *The Meat We Eat*. 14th ed. Danville, IL: Interstate Publishers, 2001.

Ronan, Annie. "Capturing Cruelty: Camera Hunting, Water Killing, and Winslow Homer's Adirondack Deer." *American Art* 31, no. 3 (Fall 2017): 52–79.

Rosenberg, Gabriel N. *The 4-H Harvest: Sexuality and the State in Rural America*. Philadelphia: University of Pennsylvania Press, 2016.

Rosenberg, Gabriel N. "A Race Suicide among the Hogs: The Biopolitics of Pork in the United States, 1865–1930." *American Quarterly* 68, no. 1 (March 2016): 49–73.

Rosenberg, Nathan, and Manuel Trajtenberg. "A General-Purpose Technology at Work: The Corliss Steam Engine in the Late-Nineteenth-Century United States." *Journal of Economic History* 64, no. 1 (March 2004): 61–99.

Rosenblum, Naomi. *A World History of Photography*. 4th ed. New York: Abbeville Press, 2007.

Rothfels, Nigel, ed. *Representing Animals*. Bloomington: Indiana University Press, 2002.

Rudolph, Katie. "Separated at Appraisal: Maintaining the Archival Bond between Archives Collections and Museum Objects." *Archival Issues* 33, no. 1 (2011): 25–39.

Salter, David. *Holy and Noble Beasts: Encounters with Animals in Medieval Literature*. Cambridge: D. S. Brewer, 2001.

Sayre, Henry M. "American Vernacular: Objectivism, Precisionism, and the

Aesthetics of the Machine." *Twentieth Century Literature* 35, no. 3 (Autumn 1989): 310–42.
Scharnau, Ralph. "Workers, Unions, and Workplaces in Dubuque, 1830–1990." *Annals of Iowa* 52, no. 1 (Winter 1993): 50–78.
Scher, Isaac. "Almost 12,000 Meatpacking and Food Plant Workers Have Reportedly Contracted COVID-19: At Least 48 Have Died." *Business Insider*, May 8, 2020. Accessed May 4, 2023. https://www.businessinsider.com/meatpacking-plants-12000-coronavirus-cases-48-deaths-2020-5.
Schrepfer, Susan R., and Philip Scranton, eds. *Industrializing Organisms: Introducing Evolutionary History*. Hagley Perspectives on Business and Culture, vol. 5. New York: Routledge, 2003.
Sekula, Allan. "An Eternal Esthetics of Laborious Gestures." *Grey Room*, no. 55 (Spring 2014): 16–27.
Sekula, Allan. *Fish Story*. London: Mack, 2018.
Sekula, Allan. "The Traffic in Photographs." *Art Journal* 41, no. 1 (Spring 1981): 15–25.
Shakespeare, William. *A Midsummer Night's Dream*. In *The Riverside Shakespeare*, edited by Gwynne Blakemore Evans and John Joseph Michael Tobin, 251–83. 2nd ed. Boston: Houghton Mifflin, 1997.
Shapiro, Laura. "And Here She Is . . . Your Betty Crocker!" *American Scholar* 73, no. 2 (Spring 2004): 87–99.
Shapiro, Laura. *Perfection Salad: Women and Cooking at the Turn of the Century*. New York: Farrar, Straus and Giroux, 1986.
Shapiro, Laura. *Something from the Oven: Reinventing Dinner in 1950s America*. New York: Viking, 2004.
Sharp, Lesley A. *Animal Ethos: The Morality of Human-Animal Encounters in Experimental Lab Science*. Oakland: University of California Press, 2019.
Shaw, Thomas. *Animal Breeding*. New York: Orange Judd, 1907.
Shipley, Rosalind. "The Town That Photography Built: Images from the Consolidation Coal Company Photograph Collection, 1911–1946." *IA* 34, no. 1/2 (2008): 87–100.
Shukin, Nicole. *Animal Capital: Rendering Life in Biopolitical Times*. Minneapolis: University of Minnesota Press, 2009.
Shusterman, Richard. "Photography as Performative Process." In "The Media of Photography." Special issue, *Journal of Aesthetics and Art Criticism* 70, no. 1 (Winter 2012): 67–77.
Simmons, Laurence, and Philip Armstrong, eds. *Knowing Animals*. Leiden: Brill, 2007.
Simpson, Mildred. "Photographs in a Business Setting: Atlantic Richfield Company." *American Archivist* 45, no. 3 (Summer 1982): 315–21.
Sinclair, Upton. *The Autobiography of Upton Sinclair*. New York: Harcourt, Brace, and World, 1962.

Sinclair, Upton. *The Jungle*. New York: Doubleday, 1906.
Sinclair, Upton. "What Life Means to Me." *Cosmopolitan* 41 (October 1906): 541.
Singer, Peter. *Animal Liberation: A New Ethics for Our Treatment of Animals*. New York: New York Review, 1990.
Singer, Peter, and Lori Gruen. *Animal Liberation: A Graphic Guide*. London: Camden Press, 1987.
Skaggs, Jimmy M. *Prime Cut: Livestock Raising and Meatpacking in the United States, 1607–1983*. College Station: Texas A&M University Press, 1986.
[Smith, Adolphe]. "The Basle Model Municipal Abattoir." *Lancet* 169, no. 4368 (May 18, 1907): 1386–89.
[Smith, Adolphe]. "Chicago: The Dark and Insanitary Premises [. . .]." *Lancet* 165, no. 4246 (January 14, 1905): 120–23.
[Smith, Adolphe]. "Chicago: The Stockyards and Packing Town [. . .]." *Lancet* 165, no. 4245 (January 7, 1905): 49–52.
[Smith, Adolphe]. "Chicago: Tuberculosis among the Stockyard Workers [. . .]." *Lancet* 165, no. 4247 (January 21, 1905): 183–85.
[Smith, Adolphe]. "Chicago: Unhealthy Work in the Stockyards [. . .]." *Lancet* 165, no. 4248 (January 28, 1905): 258–60.
[Smith, Adolphe]. "The Municipal Slaughter-House of Brussels." *Lancet* 140, no. 3616 (December 17, 1892): 1409–11.
[Smith, Adolphe]. "The Paris Abattoir at La Villette." *Lancet* 169, no. 4350 (January 12, 1907): 123–26.
Smith, Mark M. *How Race Is Made: Slavery, Segregation, and the Senses*. Chapel Hill: University of North Carolina Press, 2006.
Sontag, Susan. *On Photography*. New York: Farrar, Straus and Giroux, 1977.
Sontag, Susan. *Regarding the Pain of Others*. Farrar, Straus and Giroux, 2002.
Specht, Joshua. *Red Meat Republic: A Hoof-to-Table History of How Beef Changed America*. Princeton, NJ: Princeton University Press, 2019.
Spiegel, Marjorie. *The Dreaded Comparison: Human and Animal Slavery*. New York: Mirror Books, 1996.
Stage, Sarah, and Virginia B. Vincenti, eds. *Rethinking Home Economics: Women and the History of a Profession*. Ithaca, NY: Cornell University Press, 1997.
Stallybrass, Peter, and Allon White. *The Politics and Poetics of Transgression*. London: Methuen, 1986.
Strömberg, Susanne, and Jan Ch. Karlsson. "Rituals of Fun and Mischief: The Case of the Swedish Meatpackers." *Employee Relations* 31, no. 6 (October 2, 2009): 632–47.
Stromquist, Shelton. *Solidarity and Survival: An Oral History of Iowa Labor in the Twentieth Century*. Iowa City: University of Iowa Press, 1993.
Stromquist, Shelton, and Marvin Bergman, eds. *Unionizing the Jungles: Labor and Community in the Twentieth-Century Meatpacking Industry*. Iowa City: University of Iowa Press, 1997.

Stull, Donald D., and Michael J. Broadway. *Slaughterhouse Blues: The Meat and Poultry Industry in North America*. Belmont, CA: Thomson/Wadsworth, 2004.
Stull, Donald D., Michael J. Broadway, and David Griffith. *Any Way You Cut It: Meat Processing and Small-Town America*. Lawrence: University Press of Kansas, 1994.
Swensen, James. "A Strategy of Truth: Andreas Feininger and the Creation of Propaganda for the Office of War Information, 1942." *History of Photography* 43, no. 1 (2019): 84–109.
Swift & Company. *Swift & Company Year Book*, 1944. Chicago: Swift & Company, 1944.
Tagg, John. "The Archiving Machine; or, The Camera and the Filing Cabinet." *Grey Room* 47 (Spring 2012): 24–37.
Tagg, John. *The Burden of Representation: Essays on Photographies and Histories*. Amherst: University of Massachusetts Press, 1988.
Tagg, John. *The Disciplinary Frame: Photographic Truths and the Capture of Meaning*. Minneapolis: University of Minnesota Press, 2009.
Taussig, Michael. *Mimesis and Alterity: A Particular History of the Senses*. New York: Routledge, 1993.
Taylor, Chloë. "Foucault and Critical Animal Studies: Genealogies of Agricultural Power." *Philosophy Compass* 8, no. 6 (June 2013): 539–51.
Taylor, Nik, and Richard Twine, eds. *The Rise of Critical Animal Studies: From the Margins to the Centre*. London: Routledge, 2014.
Thompson, E. P. *Customs in Common*. New York: New Press, 1993.
Thompson, William E. "Hanging Tongues: A Sociological Encounter with the Assembly Line." *Qualitative Sociology* 6, no. 3 (September 1983): 215–37.
Thomson, Julie J. *Begin to See: The Photographers of Black Mountain College*. Asheville, NC: Black Mountain College Museum and Arts Center, 2017.
Toffoli, Erica. "Capturing Capitalism's Work: Competing Photo-Narratives of the Bracero Program." *Radical History Review* 2018, no. 132 (October 1, 2018): 126–43.
Travis, David, and Elizabeth Siegel, eds. *Taken by Design: Photographs from the Institute of Design, 1937–1971*. Chicago: Art Institute of Chicago, 2002.
Trotter, Joe W. "The Continuing Transformation of Labor and Working-Class History: A Review Essay." *Annals of Iowa* 58, no. 1 (Winter 1999): 78–86.
Tuan, Yi-Fu. *Dominance and Affection: The Making of Pets*. New Haven, CT: Yale University Press, 1984.
Turner, James. *Reckoning with the Beast: Animals, Pain, and Humanity in the Victorian Mind*. Baltimore: Johns Hopkins University Press, 1980.
Turner, Katherine Leonard. *How the Other Half Ate: A History of Working-Class Meals at the Turn of the Century*. Berkeley: University of California Press, 2014.
Tyler, Tom, and Manuela Rossini, eds. *Animal Encounters*. Leiden: Brill, 2009.

Vialles, Noëlie. *Animal to Edible*. Translated by J. A. Underwood. Cambridge: Cambridge University Press, 1994.

Viskochil, Larry A. "Arthur Siegel: A Life in Photography, 1913–1978." *Chicago History* 10, no. 2 (Summer 1981): 66–73.

Wade, Louise Carroll. *Chicago's Pride: The Stockyards, Packingtown, and Environs in the Nineteenth Century*. Urbana: University of Illinois Press, 1987.

Wade, Louise Carroll. "The Problem with Classroom Use of Upton Sinclair's *The Jungle*." *American Studies* 32, no. 2 (Fall 1991): 79–101.

Wadiwel, Dinesh Joseph. *The War against Animals*. Leiden: Brill Rodopi, 2015.

Walsh, Margaret. "From Pork Merchant to Meat Packer: The Midwestern Meat Industry in the Mid Nineteenth Century." *Agricultural History* 56, no. 1 (January 1982): 127–37.

Walsh, Margaret. "Pork Packing as a Leading Edge of Midwestern Industry, 1835–1875." *Agricultural History* 51, no. 4 (October 1977): 702–17.

Walsh, Margaret. *The Rise of the Midwestern Meat Packing Industry*. Lexington: University of Kentucky Press, 1982.

Warren, Wilson J. "Evangelical Paternalism and Divided Workers: The Nonunion Era at John Morrell and Company in Ottumwa, 1877–1917." *Annals of Iowa* 56, no. 4 (Fall 1997): 321–48.

Warren, Wilson J. "The Heyday of the CIO in Iowa: Ottumwa's Meatpacking Workers, 1937–1954." *Annals of Iowa* 51, no. 4 (Spring 1992): 363–89.

Warren, Wilson J. *Struggling with "Iowa's Pride": Labor Relations, Unionism, and Politics in the Rural Midwest since 1877*. Iowa City: University of Iowa Press, 2000.

Warren, Wilson J. *Tied to the Great Packing Machine: The Midwest and Meatpacking*. Iowa City: University of Iowa Press, 2007.

Warren, Wilson J. "The Welfare Capitalism of John Morrell and Company, 1922–1937." *Annals of Iowa* 47, no. 6 (Fall 1984): 497–517.

Warren, Wilson J. "When 'Ottumwa Went to the Dogs': The Erosion of Morrell-Ottumwa's Militant Unionism, 1954–1973." *Annals of Iowa* 54, no. 3 (Summer 1995): 217–43.

Waterloo Packer: The Story of the Rath Packing Company. Waterloo, IA: Rath Packing Company, 1941.

Weems, Jason. *Barnstorming the Prairies: How Aerial Vision Shaped the Midwest*. Minneapolis: University of Minnesota Press, 2015.

Weems, Jason. "Scale, a Slaughterhouse View: Industry, Corporeality, and Being in Turn-of-the-Century Chicago." In *Scale*, edited by Jennifer L. Roberts, 106–43. Terra Foundation Essays, vol. 2. Paris: Terra Foundation for American Art; Chicago: University of Chicago Press, 2016.

Welty, Jeff. "Humane Slaughter Laws." *Law and Contemporary Problems* 70, no. 1 (January 2007): 175–206.

Whisenant, Dava, dir. *Bathtubs over Broadway*. Focus World, 2018. Film, 87 min. https://www.bathtubsoverbroadway.com/.

White, E. B. "Death of a Pig." *Atlantic Monthly* (January 1948): 28–33.
White, Sam. "From Globalized Pig Breeds to Capitalist Pigs: A Study in Animal Cultures and Evolutionary History." *Environmental History* 16, no. 1 (January 2011): 94–120.
Whyte, William. *The Organization Man*. New York: Simon and Schuster, 1956. Reprint, Philadelphia: University of Pennsylvania Press, 2002. Page references are to the 2002 edition.
Wilder, Fred William. *The Modern Packing House*. Chicago: Nickerson and Collins, 1905.
Williams, Anna Elizabeth. "Nothing but Bodies: Nineteenth-Century Representations of Animals in Georges Cuvier's Natural System and US Industrial Meat Production." PhD diss., University of Rochester, 2000.
Williams, Laura Anh. "Gender, Race, and an Epistemology of the Abattoir in *My Year of Meats*." In "Food and Ecology." Special issue, *Feminist Studies* 40, no. 2 (2014): 244–72.
Williams, Stephen J. "Georges Franju's 'Blood of the Beasts.'" *Stephen J. Williams: Writing and Art* (blog), February 17, 2015. Accessed June 2, 2023. https://stephenjwilliams.com/2015/02/17/blood-of-the-beasts/.
Wilson, Christopher. "The Making of a Best Seller, 1906." *New York Times*, December 22, 1985. https://www.nytimes.com/1985/12/22/books/the-making-of-a-best-selller-1906.html.
Witkowski, Terrence H. "World War II Poster Campaigns: Preaching Frugality to American Consumers." *Journal of Advertising* 32, no. 1 (Spring 2003): 69–82.
Wohl, Anthony S. *Endangered Lives: Public Health in Victorian Britain*. Cambridge, MA: Harvard University Press, 1983.
Wolfe, Cary. *Animal Rites: American Culture, the Discourse of Species, and Posthumanist Theory*. Chicago: University of Chicago Press, 2003.
Wyman, Carolyn. *SPAM: A Biography*. San Diego: Harcourt Brace, 1999.
Yeager, Mary. *Competition and Regulation: The Development of Oligopoly in the Meat Industry*. Greenwich, CT: Jai Press, 1981.
Young, James Harvey. "The Pig That Fell into the Privy: Upton Sinclair's *The Jungle* and the Meat Inspection Amendments of 1906." *Bulletin of the History of Medicine* 59, no. 4 (Winter 1985): 467–80.
Young, James Harvey. *Pure Food: Securing the Federal Food and Drugs Act of 1906*. Princeton, NJ: Princeton University Press, 1989.
Zawisza-Riley, Magdalena. *Advertising, Gender and Society: A Psychological Perspective*. London: Routledge, 2019.
Ziegler, P. Thomas. *The Meat We Eat*. Danville, IL: Interstate Publishers, 1948.
Zieren, Gregory. "'If You're Union, You Stick Together': Cedar Rapids Packinghouse Workers in the CIO." *Palimpsest* 76, no. 1 (Spring 1995): 30–49.
Zuromskis, Catherine. *Snapshot Photography: The Lives of Images*. Cambridge, MA: MIT Press, 2013.

Index

Iowa and the Midwest Experience

Angel De Cora, Karen Thronson, and the Art of Place: How Two Midwestern Women Used Art to Negotiate Migration and Dispossession
by Elizabeth Sutton

The Archaeological Guide to Iowa
by William E. Whittaker, Lynn M. Alex, and Mary C. De La Garza

Carnival in the Countryside: The History of the Iowa State Fair
by Chris Rasmussen

Dakota in Exile: The Untold Stories of Captives in the Aftermath of the U.S.-Dakota War
by Linda M. Clemmons

The Drake Relays: America's Athletic Classic
by David Peterson

Dubuque's Forgotten Cemetery: Excavating a Nineteenth-Century Burial Ground in a Twenty-First-Century City
by Robin M. Lillie and Jennifer E. Mack

Duffy's Iowa Caucus Cartoons: Watch 'Em Run
by Brian Duffy

Equal Before the Law: How Iowa Led Americans to Marriage Equality
by Tom Witosky and Marc Hansen

From Warm Center to Ragged Edge: The Erosion of Midwestern Literary and Historical Regionalism, 1920–1965
by Jon K. Lauck

Harvest of Hazards: Family Farming, Accidents, and Expertise in the Corn Belt, 1940–1975
by Derek S. Oden

Imaging Animal Industry: American Meatpacking in Photography and Visual Culture
by Emily Kathryn Morgan

Atlas of Iowa
by Robert C. Shepard, Patrick Bitterman, J. Clark Archer, and Fred M. Shelley

Iowa Past to Present: The People and the Prairie, Revised Third Edition
by Dorothy Schwieder, Thomas Morain, and Lynn Nielsen

The Iowa State Fair
by Kurt Ullrich

The Jefferson Highway: Blazing the Way from Winnipeg to New Orleans
by Lyell D. Henry Jr.

The Lost Region: Toward a Revival of Midwestern History
by Jon K. Lauck

Main Street Public Library: Community Places and Reading Spaces in the Rural Heartland, 1876–1956
by Wayne A. Wiegand

Necessary Courage: Iowa's Underground Railroad in the Struggle against Slavery
by Lowell J. Soike

On Behalf of the Family Farm: Iowa Farm Women's Activism since 1945
by Jenny Barker Devine

The Plea: A True Story of Murder and Redemption
by Patricia L. Bryan and Thomas Wolf

The Sacred Cause of Union: Iowa in the Civil War
by Thomas R. Baker

The Small-Town Midwest: Resilience and Hope in the Twenty-First Century
by Julianne Couch

A Store Almost in Sight: The Economic Transformation of Missouri from the Louisiana Purchase to the Civil War
by Jeff Bremer

Transcendental Meditation in America: How a New Age Movement Remade a Small Town in Iowa
by Joseph Weber

What Happens Next? Essays on Matters of Life and Death
by Douglas Bauer

Woman Suffrage and Citizenship in the Midwest, 1870–1920
by Sara Egge